ADVANCES IN CHEMICAL PHYSICS

VOLUME XXIII

Advances in
CHEMICAL PHYSICS

EDITED BY

I. PRIGOGINE

University of Brussels,
Brussels, Belgium

AND

STUART A. RICE

Department of Chemistry
and
The James Franck Institute
The University of Chicago
Chicago, Illinois

VOLUME XXIII

AN INTERSCIENCE® PUBLICATION

JOHN WILEY AND SONS

NEW YORK · LONDON · SYDNEY · TORONTO

An Interscience® Publication

Library of Congress Catalogue Card Number: 58-9935

ISBN 0-471-69927-6

Printed in the United States of America.

10 9 8 7 6 5 4 3 2 1

INTRODUCTION

In the last decades, chemical physics has attracted an ever-increasing amount of interest. The variety of problems, such as those of chemical kinetics, molecular physics, molecular spectroscopy, transport processes, thermodynamics, the study of the state of matter, and the variety of experimental methods used, makes the great development of this field understandable. But the consequence of this breadth of subject matter has been the scattering of the relevant literature in a great number of publications.

Despite this variety and the implicit difficulty of exactly defining the topic of chemical physics, there are a certain number of basic problems that concern the properties of individual molecules and atoms as well as the behavior of statistical ensembles of molecules and atoms. This new series is devoted to this group of problems which are characteristic of modern chemical physics.

As a consequence of the enormous growth in the amount of information to be transmitted, the original papers, as published in the leading scientific journals, have of necessity been made as short as is compatible with a minimum of scientific clarity. They have, therefore, become increasingly difficult to follow for anyone who is not an expert in this specific field. In order to alleviate this situation, numerous publications have recently appeared which are devoted to review articles and which contain a more or less critical survey of the literature in a specific field.

An alternative way to improve the situation, however, is to ask an expert to write a comprehensive article in which he explains his view on a subject freely and without limitation of space. The emphasis in this case would be on the personal ideas of the author. This is the approach that has been attempted in this new series. We hope that as a consequence of this approach, the series may become especially stimulating for new research.

Finally, we hope that the style of this series will develop into something more personal and less academic than what has become the standard scientific style. Such a hope, however, is not likely to be completely realized until a certain degree of maturity has been attained—a process which normally requires a few years.

At present, we intend to publish one volume a year, and occasionally several volumes, but this schedule may be revised in the future.

In order to proceed to a more effective coverage of the different aspects of chemical physics, it has seemed appropriate to form an editorial board. I want to express to them my thanks for their cooperation.

I. PRIGOGINE

CONTRIBUTORS TO VOLUME XXIII

J. C. BROWNE, Departments of Computer Science and Physics, The University of Texas, Austin, Texas

BRUCE H. MAHAN, Joint Institute for Laboratory Astrophysics, University ot Colorado, Boulder, Colorado

F. A. MATSEN, Departments of Chemistry and Physics, The University of Texas, Austin, Texas

C. BRADLEY MOORE, Department of Chemistry, University of California, Berkeley, California

MICHAEL R. PHILPOTT, Institute of Theoretical Science and Department of Chemistry, University of Oregon, Eugene, Oregon

P. M. RENTZEPIS, Bell Telephone Laboratories, Incorporated, Murray Hill, New Jersey

DAVID A. SHIRLEY, Department of Chemistry and Lawrence Berkeley Laboratory, University of California, Berkeley, California

CONTENTS

RECOMBINATION OF GASEOUS IONS

BRUCE H. MAHAN*

Joint Institute for Laboratory Astrophysics, University of Colorado, Boulder, Colorado

CONTENTS

I. INTRODUCTION

In the past fifteen years, there have been intensive interest and activity in the study of the reaction kinetics of partially ionized gases.[1,2] Much of the experimentation has dealt with elementary bimolecular ion–molecule processes such as charge transfer, electronic and vibrational excitation, and atom transfer phenomena. The importance of these bimolecular ion–molecule processes in electrical discharges, flames, the ionosphere, thermionic energy converters, and laboratory systems subject to ionizing radiation is well established. However, the quantitative influence of ion–molecule processes on the evolution of any specific system is dependent on ion concentration, and this in many circumstances is determined by the homogeneous "recombination" or mutual neutralization of charged particles. Thus if we are to be able to predict the behavior of ionized gases fully, we must understand charge recombination phenomena.

* Visiting Fellow, 1971–1972. Permanent address: Chemistry Department, University of California, Berkeley, California.

When we consider charge neutralization mechanisms, the processes by which positive ions and free electrons recombine[3] come immediately to mind. For only slightly ionized gases, at low pressures, the most important of these are radiative recombination

$$e + A^+ \rightarrow A + h\nu$$

and dissociative recombination

$$e + M^+ \rightarrow R_1 + R_2$$

where M^+ is a molecular ion, and R_1 and R_2 are its neutral fragments. The rate constants for radiative recombination of singly charged ions and electrons are of the order of 10^{-13} cm^3/sec at 300°K. Compared with other possible processes, ion–electron radiative recombination is rather slow, and although it may occur in highly dilute plasmas of monatomic gases, it is rarely important in systems occurring in the chemical laboratory. On the other hand, ion–electron dissociative recombination can be a very rapid process,[3] with typical second-order rate constants ranging from 10^{-7} to 10^{-6} cm^3/sec at 300°K. In gases of molecules that do not readily form negative ions, direct ion–electron dissociative recombination will be the dominant neutralization mechanism.

Despite the large rate constants associated with dissociative ion–electron recombination, this process often may not provide the most important path for charge neutralization. The bimolecular dissociative electron attachment reaction[4]

$$e + XY \rightarrow X + Y^-$$

also can be very rapid, with maximum rate constants approaching 10^{-7} cm^3/sec at 300°K. Moreover, three-body electron attachment reactions like

$$e + XY + M \rightarrow XY^- + M$$

are very rapid in many cases. In particular, if XY is a large polyatomic molecule with a substantial (~1 eV) electron affinity, the metastable negative ion formed as a result of a low-energy electron–molecule collision will live sufficiently long before electron ejection that the probability of stabilization by collision with another molecule will approach unity even at gas pressures of the order of 1–10 torr. Under these circumstances, the effective second-order electron attachment coefficient may lie in the range from 10^{-11} to 10^{-6} cm^3/sec. Thus in many electron attaching gases at pressures above a few torr and under conditions of low charge concentration ($n \leq 10^{12}$ ions/cm^3), low-energy electrons will tend to be converted to negative ions more rapidly than they are removed by direct combination with positive ions. When this occurs, the homogeneous rate of disappearance of ions will be governed by ion–ion neutralization processes.

The importance of ion recombination was first recognized by J. J. Thomson and E. R. Rutherford[5] in 1896, and 1897, Rutherford[6] performed the first experiments in which an attempt was made to measure ion recombination rates. In 1903 Langevin[7] also measured recombination rates, and developed an important theory which pertains to recombination in dense, highly ionized gases. Other very important theoretical contributions were made by Thomson[8] in 1924, and by Harper[9] in 1932. Experiments performed in the years prior to 1935 are of very dubious validity, since vacuum and gas purification techniques were not well developed. References to and discussion of this early experimental work can be found in the books by the Thomsons[10] and Loeb.[11]

Experiments performed in the late 1930s were more reliable and led to the recognition that there are three regimes of gas density in which the mechanism of ion recombination differs appreciably. The experiments measured the ion recombination coefficient α, defined by

$$-\frac{dn_+}{dt} = -\frac{dn_-}{dt} = \alpha(n_+)(n_-)$$

where $n_\pm$ is the concentration of charged particles. Experiments carried out by Mächler[12] in 1936 showed that at room temperature, α is inversely proportional to pressure in the range from 5 to 15 atm. This finding suggests that at these elevated gas densities, the rate of ion recombination is limited by the speed at which oppositely charged ions can diffuse together. Since Langevin[7] and Harper[9] had separately put forth theories which related α to the ion mobility and diffusion coefficients, respectively, it is appropriate to refer to these high-pressure recombination phenomena as occurring in the Langevin-Harper regime.

In 1938 Sayers[13] and Gardener[14] separately performed the first quantitatively reliable experiments on ion recombination in the pressure range from 0.1 to 2 atm. They showed that at the lower end of this pressure range, the ion recombination coefficient α at first increased approximately linearly with increasing pressure, and then reached what appeared to be a limiting value of 2×10^6 cm^3/sec at approximately 1 atm. This type of behavior had been predicted in 1924 by Thomson,[8] and this range of intermediate pressures is now known as the Thomson regime.

It is not clear when it was first generally realized that in the limit of vanishing gas density, a direct bimolecular mutual neutralization reaction should occur. Apparently the first serious attempt to calculate the rate of such a process was by Bates and Massey[15] in 1943. In 1958 Yeung[16] reported attempts to measure the rate of bimolecular ion neutralization in several systems. However, the conditions of his experiments make it seem likely that an important, if not dominant, ion loss mechanism was diffusion to the walls of the containing vessel, and therefore that the derived rate constants do not faithfully represent the bimolecular neutralization process. Mahan and Person,[17] Greaves,[18] and Carlton and Mahan[19] measured bimolecular recombination coefficients for several systems in 1964, and found values of the order of 2×10^{-7} cm^3/sec at 300°K. Subsequent measurements using the merged ion beam method and a combined radio-frequency conductivity and mass spectrometric technique have given similar results. Thus the existence and general magnitude of the rate of the bimolecular mode of ion recombination are well established.

As reliable experimental results have appeared, there have been a number of papers that have dealt with the theory of ion recombination. In what follows, we shall discuss these experiments and the theory of ion recombination for each of the three density regimes. A brief review of ion–ion recombination appeared[20] in 1962, and more recently, Flannery[21] has provided an extensive review of the theory of termolecular ion recombination.

II. EXPERIMENTAL METHODS

Many ion recombination rate data have been obtained by the use of charge collection techniques. A gas between plane-parallel electrodes is uniformly ionized by a pulse of either X-rays or vacuum ultraviolet radiation. After a preselected delay time, a voltage pulse is applied to one electrode, and the ions remaining in the intraelectrode volume are collected at the other electrode and measured by a current integrating device. In the early work of Sayers[13] and Gardener,[14] and in the more recent work of Ebert et al.[22] and McGowan,[23] X-rays were used to produce ions in initial concentrations up to approximately 10^6 cm^{-3}. Under these conditions, the number of ions collected after any single pulse was small, so repetitive ionizing and collecting pulses had to be used to accumulate a readily measurable charge. In the experiments of Mahan and coworkers,[17,19,24] ions were produced by vacuum ultraviolet photolysis, and the concentrations (10^7–10^8 cm^{-3}) were sufficient so that a single pulse yielded one point on the ion concentration–time curve.

The ion collection technique is straightforward and can be used over a very wide range of ambient gas pressures. However, care must be taken to define the ion collection volume through the use of guard rings on the collecting electrode. In addition, compensating circuitry must be employed, or a correction made, for false ion collection current, which arises from the finite capacitance of the collecting electrodes and leads. These precautions were taken by all the experimenters mentioned above.

When X-rays or γ-rays are used to ionize a gas, most of the ionization is created by the impact of secondary electrons produced along the trajectory of the very energetic primary photoelectron. Thus the initial ionization tends to be spatially inhomogeneous. If the delay time between the ionizing and collecting pulses is short compared to the time required for diffusional processes to make the ion concentration uniform, the apparent value of the recombination coefficient α will be quite large and will decrease with time. Reliable values of α can be obtained only by using delay times that are long compared to the time it takes ions to diffuse a distance that is approximately equal to their macroscopic average separation.

The difficulties with initially inhomogeneous ion concentrations are avoided if the experiments are done by selectively photoionizing a gas that is in minor concentration in an inert buffer gas. The ion–electron pairs are then produced randomly throughout the gas, and the electrons diffuse rapidly through the buffer gas (deliberately chosen to be non-electron attaching) until they become attached to a molecule of a minor component. Thus in order to study the recombination of NO^+ with NO_2^-, Mahan and Person[17] used the krypton 1236 Å resonance line (10 eV) to photoionize mixtures of approximately 0.1 torr NO and 0.01 torr NO_2 in 20–700 torr of an inert gas.

The concentration of electron attaching gas must be chosen so that electron attachment is rapid compared with ion–electron recombination. If the effect bimolecular attachment rate constant is k_a, and the concentrations of attaching gas and positive ions are n_a and n_+, the condition for sufficiently rapid attachment is

$$n_a \gg \frac{\alpha}{k_a} n_+$$

Since αn_+ is usually of the order 1–100 sec^{-1}, and k_a is usually larger than 10^{-10} cm^3/sec, the condition is easily achieved.

If homogeneous mutual neutralization is the only ion loss process in a macroscopically neutral plasma, the rate of change of the ion concentration is given by

$$\frac{dn}{dt} = -\alpha n^2$$

The integral rate law is then

$$\frac{1}{n} - \frac{1}{n_i} = \alpha t$$

where n_i is the initial ion concentration. The ion recombination coefficient, α can be derived from the slope of a plot of $1/n$ vs. t. In practice, however, complications due to ion loss by diffusion almost always arise. Under these circumstances, the appropriate differential equation for ion loss is

$$\frac{\partial n}{\partial t} = -\alpha n^2 + D\nabla^2 n \tag{2.1}$$

where D is the mean diffusion coefficient for positive and negative ions, if $D_+ \cong D_-$. Diffusional loss becomes important when the ion concentration, the recombination coefficient, or the vessel dimension is small, and the diffusion coefficient large (low pressures). More precisely, McGowan[25] has shown that for parallel plate collection electrodes large compared with their separation S, the parameter

$$\beta = \alpha n_i \frac{S^2}{D} \tag{2.2}$$

must be greater than 10^3 if the fractional error in α derived from a $1/n$ vs. t plot is to be no greater than 10%.

Some indication of the importance of diffusional loss can be gleaned from the linearity of the plots of $1/n$ vs. t. Diffusional loss causes the expected straight line to be concave upward, increasingly so in the latter stages of the decay. If the plots are not linear over concentration changes of at least a factor of ten, diffusional effects are important and the α derived from the apparently linear portion of the curve will be too large. In this situation, one can get a better approximation to the recombination coefficient by applying the correction factors computed by McGowan,[25] particularly if the diffusional correction is small. If the diffusion correction is large, it is preferable to fit the experimental data directly to computer solutions of the loss equation. This is the procedure followed by Fisk et al.[24]

The concentration of ions may be followed as a function of time by measuring the dielectric constant or electrical conductivity of the gas. The complex dielectric constant κ of an ionized gas is given by [26]

$$\kappa = 1 - \frac{e^2}{\epsilon_0} \sum_i \frac{N_i}{m_i(\omega^2 + \nu_i^2)} \left(1 + j\frac{\nu_i}{\omega}\right)$$

where e is the fundamental charge, ϵ_0 is the permittivity of free space, N_i, m_i, and ν_i are the concentration, mass, and collision frequency, respectively, of the ith charged species, and j is the imaginary quantity. The quantity ω is the angular frequency of the applied electric field, and when this greatly exceeds the collision frequencies, we get

$$\kappa = 1 - \frac{e^2}{\epsilon_0 \omega^2} \sum_i \frac{N_i}{m_i}$$

Thus if the dielectric constant can be measured as a function of time, and if the masses and relative concentrations of all charged species are known, the absolute value of the recombination coefficient can be obtained. In addition to the problems associated with diffusional loss of ions, one must take into account the spatial distribution of the probing electric field, which is in general not uniform. The measured dielectric constant determines an average ion concentration (as does the charge collection method) but with the concentrations in regions of high electric field weighted most heavily. This problem has been treated by a number of authors in connection with the measurement of electron concentrations at microwave frequencies.

The only reported determinations of ion recombination coefficients by dielectric constant measurements are those by Yeung[16] and Greaves.[18] Greaves measured the decay of ions in iodine vapor at pressures between 0.03 and 1 torr. The probing electric field had a frequency of 8 MHz, which satisfied the condition $\omega \gg \nu$ for the range of pressures employed. A small mass spectrometer was used to show that the principal ions were I_2^+ and I^-. In the earlier work of Yeung,[16] a mass spectrometer was not used, so that proper interpretation of the dielectric measurements was not possible.

Dielectric constant measurements must be carried out under conditions where $\omega \gg \nu_i$. This implies that either the neutral gas pressure is low, which leads to diffusional losses, or the frequency is high, in which case sensitivity is lost. If the condition $\omega \gg \nu$ is not met, or indeed is reversed, it is preferable to speak of the electrical conductivity of the gas. The complex conductivity is given by[26]

$$\sigma = e^2 \sum_i \frac{N_i}{m_i} \left(\frac{\nu_i - j\omega}{\omega^2 + \nu_i^2} \right)$$

and in the case that $\omega \ll \nu$, this reduces to

$$\sigma = e^2 \sum_i \frac{N_i}{m_i \nu_i}$$

Thus the measurement of the conductivity can give the ion concentration, if the masses, relative concentrations, and the collision frequencies of the individual ionic species are known. The one reported application of this technique is by Eisner and Hirsch,[27] who measured the conductivity of an NO^+–NO_3^- plasma with a 1 kHz probing field at total presssures of air in the range of 15–20 torr ($\nu \cong 10^8$).

The most elegant technique for studying mutual neutralization is the ion–ion beam method. The experiments of Aberth, Peterson, Moseley, and others at the Stanford Research Institute (SRI) involve the merged beam arrangement,[28] which has been reviewed in detail by Neynaber.[29] In the SRI apparatus, positive and negative ion beams of approximately 3 keV energy are separately mass-analyzed and then merged. After traveling together for 33 cm, the positive and negative beams are electrostatically separated and measured. The neutral particles formed in the interaction region continue on a straight path to strike a stainless steel surface and are detected by the resulting secondary electron emission.

The great advantage of the merged beam technique is that very low relative energies with little dispersion can be obtained even though the laboratory energy of the beams is in the keV range. The relative energy of two ions is given by

$$E_r = \frac{1}{2}\mu\,(v_1^2 + v_2^2 - 2v_1v_2\cos\theta) \tag{2.3}$$

where μ is their reduced mass, and θ is the angle between their velocity vectors $\mathbf{v}_1$ and $\mathbf{v}_2$. In the merged beam experiment, $\cos\theta$ is unity, so

$$E_r = \frac{1}{2}\mu\,(v_1 - v_2)^2 = \mu\left[\left(\frac{E_1}{M_1}\right)^{1/2} - \left(\frac{E_2}{M_2}\right)^{1/2}\right]^2 \tag{2.4}$$

where E_i and M_i are the laboratory energies and masses of the ions, respectively. If for simplicity the ion masses are taken to be equal, and if the difference in energy between the beams ΔE is small compared to E_i, then (2.4) reduces to the approximation

$$E_r \cong \frac{(\Delta E)^2}{8\bar{E}}$$

where $\bar{E}$ is the mean laboratory energy of the beams. Thus beams of 2.9 and 3.1 keV equal-mass ions produce a relative energy of 1.67 eV. Further analysis shows, as might be expected intuitively, that the spread in relative energy due to the finite spreads in the beam energies is reduced to a very small value.

The fact that both ion beams operate in the kilovolt region allows one to avoid some of the problems associated with space charge, and makes detection of the neutral products relatively easy. The difficulties associated with the merged beam method stem principally from beam aligment problems, definition of the interaction region, and elimination of false neutral signal from charge exchange or collisional detachment interaction with the background gas. In addition, the vibrational and sometimes even the electronic states of the ions which come from duoplasmatron and other electron impact sources are not well defined.

The merged beam technique has also been used by Berry and co-workers[30,31] to study the radiation emitted upon neutralization of Na^+ by O^-. In the initial phase[30] of this work an underestimate of the ion interaction region viewed by the photon detector led to a considerable overestimate of the recombination cross sections. In the more recent and complete report,[31] this error was corrected, and cross sections of more reasonable size were obtained.

Harrison and co-workers[32–34] have measured mutual neutralization cross sections by the inclined beam method. In this work, the positive and negative ion beams intersect at an angle of 20°, so some energy deamplification is obtained. The inclined beam method has the advantages that the collision region is fairly well defined, and the neutral products from the positive and negative beams for the most part appear in different places (along the respective trajectories of their parent beams). Therefore, in the H^+–H^- experiment, measurement of the neutral signal in the direction of the proton beam gave (along with background noise) the number of atoms formed by electron transfer

$$H^+ + H^- \rightarrow H + H$$

whereas the neutral signal in the direction of the H^- beam contained H atoms formed by this reaction and by electron detachment

$$H^+ + H^- \rightarrow H^+ + H + e$$

Thus a better estimate of the charge transfer cross section could be obtained from measurements of the neutral particle flux in the direction of the positive beam.

Because the energy deamplification in the inclined beam configuration is less than in the merged beam experiments, it is better suited for the higher energy experiments, whereas the merged beam method is best for lower relative energy experiments. The operating ranges overlap, and in this region the agreement between the data of the SRI and Harrison groups is spectacularly good.[35,36]

III. BIMOLECULAR ION RECOMBINATION

The experiments of Sayers[13] and of Gardener[14] in 1938 established the fact that the overall recombination coefficient α increases with pressure above approximately 200 torr, and consequently demonstrated the existence of the three-body or Thomson recombination process. However, neither investigator was able to extend his measurements successfully to lower pressures and thereby determine whether α went to a finite or zero value as the pressure approached zero. Thus the occurrence of bimolecular recombination could not be demonstrated by these early experiments. Theoretical work in the 1940–1955 period indicated, however, that the bimolecular ion recombination coefficient k_0 should be of the order of 10^{-7} cm^3/sec.

A. Experimental Results

In 1958, Yeung[16] reported attempts to measure α at pressures of the order of 1 torr. Rate constants were obtained by measuring $n(t)$ over a change of slightly more than a factor of two, and consequently the influence of diffusional loss of ions was not discernible, but was probably important. Also this pioneering work was marred by a miscalibration, which led to values of α that are too small by approximately one order of magnitude. The corrected values have been reported by Sayers[20] and Greaves.[18]

Early in 1964, Mahan and Person,[17] Greaves,[18] and Carlton and Mahan[19] reported measurements of the bimolecular ion recombination rate constant for a number of ion pairs. The experiments in Mahan's laboratory were performed using the photoionization–charge collection technique. Values of α were measured at a number of pressures of inert gas, and a plot of α vs. p was linearly extrapolated to zero pressure to obtain k_0. In Greaves's experiments, I_2 vapor was ionized by a strong radio-frequency pulse, and the charge concentration decay followed by the dielectric constant method. Variation of the total pressure from 0.23 to 1 torr did not change the rate constants, although there was evidence that diffusional loss was important at the lower pressures. Rate constants determined at the higher pressures were taken to be equal to k_0.

Table I is a collection of bimolecular recombination rate constants for a number of ion pairs, measured by the charge collection, dielectric constant, and mass spectrometer probe techniques in bulk gases. Very few systems have been investigated in more than one laboratory, but in the cases where this has been done, the agreement is quite good. In particular, the values of k_0 for I_2^+–I^- obtained by Greaves[18] and Carlton and Mahan[19] are virtually identical, and within 25% of Yeung's[16] corrected value. Although this agreement is encouraging, it should be pointed out that in these three experiments, the condition given by (2.4) for diffusional contributions to be negligible is only marginally satisfied. Thus all three results may be high by as much as 20–30%.

The values of k_0 measured at 300°K for the systems NO^+–NO_2^-, NO^+–SF_6^-, $C_6H_6^+$–SF_6^-, and I_2^+–I^- all fall in the range of $1–2 \times 10^{-7}$ cm^3/sec, as Table I shows. In the first two of these systems, it is likely (see Section IV.E) that the positive ions were in fact $(NO)_2^+$, and in the

TABLE I

Bimolecular Recombination Coefficients Measured in Homogeneous Systems

System	Probable ions	T (°K)	$k_0 \times 10^7$ (cm³/sec)
Photoionized NO, (NO_2)	$(NO)_2^+$, NO_2^-, NO_3^-	300	2.1 ± 0.6[a]
Photoionized NO, (SF_6)	$(NO)_2^+$, SF_6^-	300	1.28 ± 0.5[b]
Photoionized C_6H_6, (SF_6)	$C_6H_6^+$, SF_6^-	300	2.2 ± 0.8[b]
Photoionized I_2	I_2^+, I^-	300	1.45 ± 0.5[b]
Discharged I_2	I_2^+, I^-	296	1.22 ± 0.03[c]
Discharged I_2	I_2^+, I^-	293	1.47 ± 0.07[c,d]
Irradiated air	NO^+, NO_2^-	300	1.75 ± 0.6[e]
Irradiated air	NO^+, NO_3^-	300	0.34 ± 0.12
Photoionized TlI	Tl_2I^+, TlI_2^-	530	0.40 ± 0.1[f]
Photoionized TlBr	Tl_2Br^+, $TlBr_2^-$	570	0.51 ± 0.1[f]
Photoionized TlCl	Tl_2Cl^+, $TlCl_2^-$	610	0.56 ± 0.1[f]
Photoionized TlI, (NO_2)	$Tl_2NO_3^+$, $Tl(NO_3)_2^-$	550	0.53 ± 0.1[f]

[a] Ref. 17.
[b] Ref. 19.
[c] Ref. 18.
[d] Ref. 16.
[e] Ref. 27.
[f] Ref. 24.

first system, it is possible that NO_3^- as well as NO_2^- was present. The systems are similar in that the ionization energies of the parent of the positive ion are all near 9.2 eV, but the electron affinities of the negative ions range from 1.5 to 3.0 eV. Since the rate constant is largely determined by the positions of crossings of the ionic and neutral potential surfaces, and these are in turn determined by the ionization energy, electron affinity, and excited state energy level scheme of the neutrals, the order of magnitude agreement among these rate constants is not too surprising. The recombination rate constants measured in the temperature range 550–600°K for the system Tl_2X^+, TlX_2^-, where X is I, Br, Cl, and NO_2, are noticeably smaller, lying in the range 0.4–0.6 × 10^{-7} cm³/sec. This is undoubtedly a reflection of the high electron affinity of the halide and sparse lower energy level spectrum of thallium, which are factors that tend to make the ionic–neutral crossing occur at smaller internuclear separations, and also of the elevated temperature, which diminishes the effect of the attractive Coulomb forces on the rate constant.

Figure 1 shows the mutual neutralization cross section for the H^+–H^- system as a function of relative velocity measured by the merged and inclined beam methods. The experiments of Moseley, Aberth, and Peterson[35]

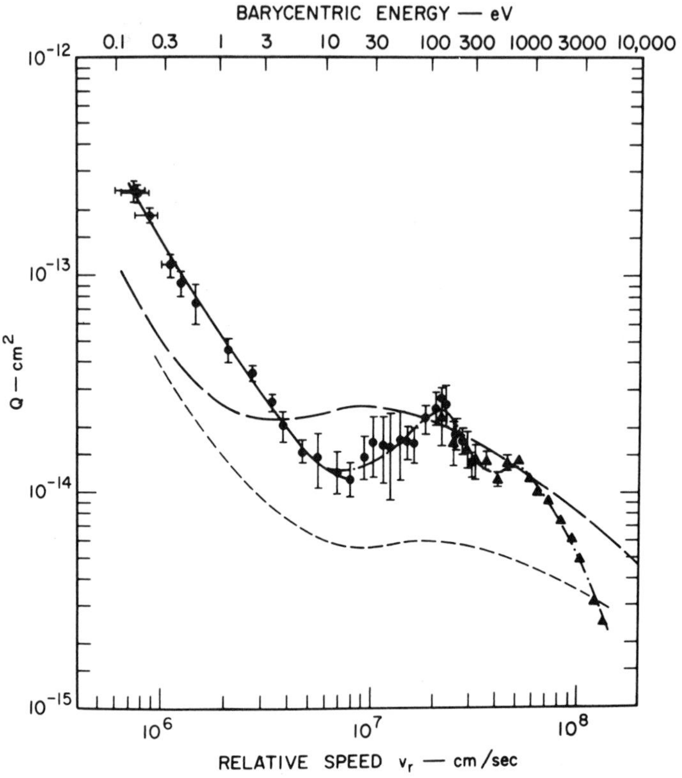

Fig. 1. Ion–ion mutual neutralization cross sections vs. velocity for $H^+ + H^-$. Circles, Ref. 35; triangles, Ref. 32; short-dashed curve, Ref. 41; long-dashed curve, Ref. 42.

covered the relative energy range from 0.2 to 250 eV, and Rundel, Aitken, and Harrison[32] extended the measurements from 200 eV to 10 keV. It is clear that the data of the two groups are in excellent agreement in the range of overlapping energies. Similar good agreement has been obtained by the two groups[34,36] in measurements of the He^+–H^- and He^+–D^- cross sections.

The SRI researchers have determined neutralization rates as a function of energy for a number of other atomic[36] and molecular[37] ion pairs, which are listed in Table II. By fitting the cross sections to a four-term polynomial expression, the data were extrapolated to 300°K, and a thermal rate constant was computed. The resulting rate constants are of the same general magnitude as those found in bulk gas experiments. However, for the NO^+–NO_2^- system the merged beam results are over twice as large as

TABLE II

Bimolecular Recombination Coefficients Measured with Merged Beams[a]

System	$k_0 \times 10^7$ (cm³/sec)	System	$k_0 \times 10^7$ (cm³/sec)
H^+, H^-	3.9 ± 2.1	O_2^+, O_2^-	4.2 ± 1.3
H_2^+, D^-	4.7 ± 1.5	N_2^+, O_2^-	1.6 ± 0.5
Na^+, O^-	2.1 ± 1.0	O_2^+, NO_2^-	4.3 ± 1.3
O^+, O^-	2.7 ± 1.3	NO^+, NO_2^-	5.1 ± 1.5
N^+, O^-	2.6 ± 0.8	O_2^+, NO_3^-	1.3 ± 0.4
O_2^+, O^-	1.0 ± 0.4	NO^+, NO_3^-	8.1 ± 2.3
NO^+, O^-	4.9 ± 2.0		

[a] See Ref. 45 for reference to the original work.

the constants obtained from bulk gas measurements,[17,27] and for the NO^+–NO_3^- system, the merged beam measurements are ten times the bulk gas results.[27] It is not yet clear which technique gives the more reliable thermal average rate constant. It is possible that the ions involved in the merged beam experiments are slightly excited vibrationally, and that this leads to rate constants that are too large. On the other hand, the bulk measurements[27] on the NO^+–NO_3^- recombination were obtained with a rather complicated chemical system (air irradiated with 1 MeV electrons) using a mass spectrometer sampling technique which may be rather difficult to calibrate absolutely. More complete experiments on the bulk gas systems and work with different ion sources in the beam experiments should resolve the discrepancy.

Weiner, Peatman, and Berry[31] have reported merged beam measurements of the energy dependents of the total cross sections for Na^+–O^- neutralization followed by light emission. The specific processes detected are

$$\begin{aligned} Na^+ + O^- &\rightarrow O(^3P) + Na(4p) \rightarrow Na(3s) + h\nu \\ &\rightarrow O(^3P) + Na\ (3d) \rightarrow Na(3p) + h\nu' \\ &\rightarrow O(^3P) + Na(3p) \rightarrow Na(3s) + h\nu'' \end{aligned}$$

The imprecision of the data at low energies precludes computation of thermal rate constants, but in the 0.1–7 eV range, cross sections of the order of 10^{-14} cm² for the endoergic production of Na(4p), and of the order of 10^{-13} for formation of Na(3d) and Na(3p) were obtained. Most of the formation of Na(3p) may be by radiative cascade from Na(3d). The energy dependence of the cross sections shows features which suggest that

$$\begin{aligned} Na^+ + O^- &\rightarrow Na(3p) + O(^1D) \\ &\rightarrow Na(3p) + O(^1S) \end{aligned}$$

may also occur. This interesting investigation is the first which the electronic states of the final products of ion recombination have been determined.

B. Interpretation

The usual theoretical approach to bimolecular mutual neutralization involves consideration of pseudocrossings[38–39] of the initial and final adiabatic potential surfaces for the process. These pseudocrossings arise when two approximate states of the Hamiltonian of simple molecular orbital structure, such as the pure ionic and covalent states, have the same energy at a particular internuclear separation. These two diabatic states can cross each other smoothly. However, the adiabatic states derived from them, and constructed by diagonalizing the electronic Hamiltonian, can not cross if they are of the same symmetry. The adiabatic surfaces approach each other closely, with a minimum separation determination to first order by the matrix element of the Hamiltonian between the two diabatic states. Motion of the nuclei with finite velocity in the vicinity of the pseudocrossing can cause the system to undergo a transition from one surface to another, and thereby lead to a mutual neutralization.

Figure 2 shows that the approximate condition for the occurrence of a pseudocrossing for the process

$$A^+ + B^- \rightarrow A + B$$

where A or B may be electronically excited, it is that the energy change ΔE of the reaction equal the Coulomb energy at the crossing point, e^2/R_x. When the electron affinity of B is small, most of these crossings will occur at large R and will involve molecular states which dissociate to highly excited Rydberg states of A. If the electron affinity of B is large, the crossings tend to occur at smaller R and lead to the lower excited states of A and B.

If the initial conditions of a collision are such that the ions pass through one or more of the pseudocrossing regions, they may, because of their finite velocity and closeness of the two curves, transfer from the ionic to to covalent surface on the incoming or outgoing leg of the trajectory. If only one or, in general, an odd number of such transitions occurs before the ions separate, a bimolecular neutralization has occurred.

Accurate values of the ionization energies, electron affinities, and the excitation energies of relevant electronic states of the neutrals are needed to locate the positions of the pseudocrossings of the ionic surface with the

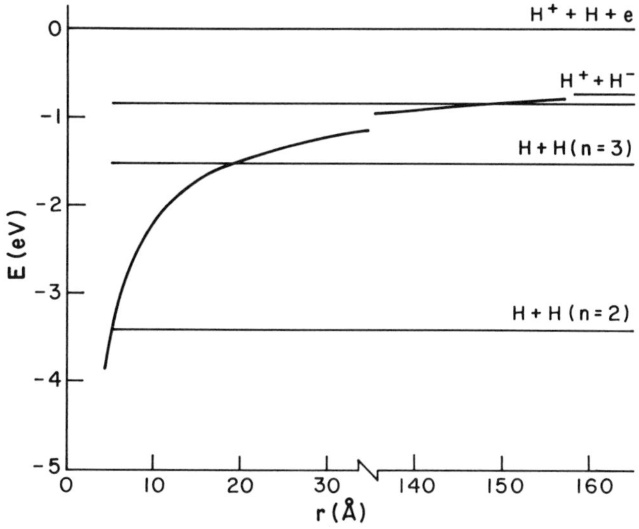

Fig. 2. Curve crossings for the H^+–H^- system.

higher excited states of the neutral products. The Landau-Zener theory[38,39] is most frequently used to compute the probability of transitions between surfaces. This theory requires knowledge of the matrix elements between the diabatic states at their crossing points. For many ions pairs, particularly molecular ions, most of these quantities will be unknown or uncertain.

In such circumstances, it still may be valuable to analyze the rate data in terms of a primitive collision model. Assume that there is a distance R_c outside of which several diabatic curve crossings have occurred, so that systems that reach R_c will have essentially unit probability of undergoing neutralization. Then, elementary collisional mechanics gives for the greatest impact parameter b_m which allows oppositely charged ions with initial relative energy E_r to reach R_c the expression

$$b_m{}^2 = R_c{}^2\left(1 + \frac{e^2}{R_c E_r}\right)$$

The quantity $\pi b_m{}^2$ is the reaction cross section, and the thermal rate constant is

$$k_0 = \pi \int g b_m{}^2 f(g)\, dg$$

where g is the initial relative ion speed, and $f(g)$ is the Boltzmann distribution function. The result of the integration is

$$k_0 = R_c^2 \left(\frac{8\pi kT}{\mu}\right)^{1/2} \left(1 + \frac{e^2}{R_c kT}\right) \tag{3.1}$$

where μ is the reduced mass of the ions, and T is the temperature.

The first term of this expression represents a "hard sphere" contribution to the rate constant resulting from the geometric cross-sectional area πR_c^2. The second term represents the additional contribution to the rate constant that results from the Coulomb attractive force between the ions. For typical crossing distances ($R_c \cong 10$ Å) and at 300°K, this second term is of the order of 25, and dominates the rate constant. Thus the very large values ($\sim 10^{-7}$ cm^3/sec) which are typical of k_0 at room temperature are due principally to the strong attractive Coulomb forces between ions, and the large distances at which diabatic curve crossings occur. If k_0 from (3.1) is equated to measure experimental data and solved for R_c, the result is a lower limit for the region in which transitions from ionic to covalent surfaces actually occur, since the probability of transfer in (3.1) has been taken equal to unity.

Several attempts have been made to evaluate a priori mutual neutralization cross sections. These involve the expression for the total cross section

$$\sigma = 2\pi \int_0^\infty P(b,g) b \, db$$

where $P(b,g)$ is the probability of neutralization as a function of impact parameter and initial relative speed. If only one pseudocrossing occurs, and the system passes twice through this region, once each on the incoming and outgoing legs of the trajectory, one has

$$P(b,g) = 2p_{if}(1 - p_{if})$$

where p_{if} is the probability of a transition from the initial to the final adiabatic curve. Most frequently, the Landau-Zener expression[38,39]

$$p_{if} = \exp\left[-\frac{2\pi H_n^2}{h v_n S_n}\right]$$

has been used to find p_{if}. Here H_n is the matrix element of the Hamiltonian between the initial and final diabatic states at the nth crossing, v_n is the radial speed, and S_n is the absolute value of the difference of the slopes of the diabatic states at the crossing point.

In an early application of this procedure, Magee[40] treated the O^+–O^- and O^+–O_2^- systems, estimating H_n by use of approximate wave functions, and taking account of several ion–neutral crossings. Only approximate values of the electron affinities were available to Magee, so his results have at best qualitative significance. Nevertheless, cross sections that yield rate constants of approximately 10^{-7} cm^3/sec were obtained.

Bates and Lewis[41] calculated the cross sections and rate constants for the H^+–H^- neutralization reactions. The principal reaction at low energies is production of H(1*s*) and H(3*s*,*p*,*d*), whereas at relative energies above 50 eV, the most important process is production of H(1*s*) and H(2*s*,2*p*). Although the ionic curve crosses the neutral surfaces that lead to H(1*s*) and *H*(4*s*,*p*,*d*,*f*), the crossing is at such a large distance that the matrix elements are vanishing small. Thus according to the Landau-Zener theory, the transition probability between adiabatic states is near unity on *both* the incoming and outgoing legs of the trajectory, so the production of H(n = 4) is negligible. More recently, Dalgarno et al.[42] have computed the same cross sections but with a more refined estimate of the potential curves and splittings. This calculation and that of Bates and Lewis are compared with the experimental data of Aberth et al., and Harrison and co-workers, in Fig. 1. The two calculations effectively bracket the experimental data at high energies, reproduced the main undulation at approximately 100 eV relative energy, and are too small in the low energy regime.

The quantitative failure of the theoretical calculation at low energies may be a consequence of the neglect of transitions to the n = 4 levels of hydrogen. Although reasoning based on the Landau-Zener theory indicates that these transitions should be negligible,[41] there is experimental evidence that this conclusion may not be sound. Berry and co-workers[31] observed large ($\sim 10^{-13}$ cm^2) cross sections for population of the Na(3*d*) state from Na^+–O^- collisions, even though the crossing is at 288 Å. They also observed population of Na(4*p*), even though there is no pseudo-crossing possible with this state, since it lies above the asymptotic energy of the separated ions. More theoretical work is needed to elucidate the role of transitions between surfaces that lie close to each other over extended internuclear distances.

To deal with more complex systems, approximate semiempirical versions of the Landau-Zener theory are valuable. Bates and Boyd[43] computed the splittings at the crossing points for a number of neutralization processes in the H^-–alkali ion systems, and noted that when the logarithm of the splitting was plotted as a function of the exoergicity of the process, all values fell on a common smooth curve. This correlation of H_n with ΔE thus allows one to estimate H_n for other processes for which ΔE is known from spectroscopic, electron affinity, and ionization energy data. Once ΔE and H_n are known, the neutralization rate constant depends only on the ion reduced mass and the temperature, and Bates and Boyd give a graphical method for finding k_0. This method of using the Landau-Zener theory has recently been restated by Bates,[44] with a comparison to some of the more recent experimental data. In general, this semiempirical application of Landau-Zener theory underestimates the values of the measured recombination rate constants by factors somewhat greater than two.

Olson et al.[36] have recently taken a very similar approach to the application of Landau-Zener theory to atomic ion neutralization. From 68 experimental and theoretical values of matrix elements at crossing points found in the literature, they parameterized H_n in terms of the crossing point distance and the electron binding energy of the negative ion. The transition probabilities were then determined for the many crossings that occur in the systems He^+–H^-, N^+–O^-, O^+–O^-, and H^+–H^-, and the neutralization cross sections calculated for a range of energies. The agreement with the experimental results is quite good, with most of the theoretical values being low by factors of two or three over most of the collision energy range.

More recently, Olson[45] has proposed an absorbing sphere model to deal with the neutralization of molecular ions, where many psuedocrossings can occur. By using the dependence of the matrix elements on intermolecular separation found empirically, and the Landau-Zener transition probability, a critical distance R_c is found at which the total probability of neutralization is a maximum. The total neutralization probability for ions reaching this distance is taken equal to unity, and the corresponding cross section, after an upward adjustment of 10%, is taken as the neutralization cross section. The calculated thermal rate constants agree rather well with the available experimental values, usually with the experimental and calculational uncertainty, with the most notable discrepancies attributable to possible participation of vibrationally excited states of negative molecular ions, or to failure of the assumption that many final states of the products are available.

A model similar in spirit but rather different in execution has recently been outlined briefly by Smirnov.[46]

In summary, the general magnitude of the bimolecular ion recombination coefficient seems well established experimentally. The dependence of the coefficient on the internal excitation of the ions is still an open question, as is the nature of the final electronic states or chemical identity of the final neutralization products. From the theoretical point of view, the numerical accuracy of ab initio calculations of the recombination coefficient is better than an order of magnitude, but can not be counted on to be more accurate than a factor of three. Semiempirical methods are more successful. The question of the effectiveness of pseudocrossings at large distances (>100 Å) needs attention.

IV. TERMOLECULAR ION RECOMBINATION

Experiments show that as the neutral gas pressure is increased from 0.1 to 1 atm at room temperature, the effective ion–ion recombination coefficient α at first increases linearly with pressure, then rises more slowly, and eventually reaches a maximum or saturation value of approximately 2×10^{-6} cm^3/sec. This type of neutral gas pressure dependence is indicative of a termolecular recombination process which saturates to a second-order rate law at high pressure. Upon consideration of the very large values of the typical bimolecular recombination coefficients, one may wonder how a termolecular process can compete with and exceed the bimolecular rate at such modest pressures.

We note that the measured rate constants for bimolecular ion neutralization show that ions must generally approach to within 10–50 Å before electron transfer is highly probable. The mutual Coulomb potential energy of the ions becomes comparable to their average thermal energy at 300°K when they are closer than 400 Å. Thus there are many ion–ion collisions in which there is a strong Coulomb interaction, but which, in the absence of other effects, do not lead to neutralization.

J. J. Thomson[8] first realized the consequence of the considerable kinetic energy colliding ions acquire as a result of their Coulomb attraction. He reasoned that if an ion with such excess kinetic energy were to collide with a neutral, there would be a great probability that the ion would lose a considerable fraction of its kinetic energy, and might therefore become unable to separate from its oppositely charged partner. The resulting closed orbit of the ion pair might lead to the final neutralization process either directly or after further ion–neutral collisions. Thus the role of the third body is to convert ion–ion collisions in which the Coulomb attraction is strong, but in which an effective pseudocrossing is not reached, into processes which eventually lead to neutralization. The fact that the Coulomb potential is strong at distances at which the electron transfer rate is negligible makes the termolecular contribution large compared to the bimolecular contribution, even at modest neutral gas pressures.

A. The Thomson Ion Recombination Theory

Thomson[8] was able to formulate an expression for the termolecular recombination rate which is in rather good agreement with experimental findings, both qualitatively and quantitatively. A little reflection, reinforced by subsequent theoretical work, shows that the quantitative success of the Thomson theory is largely fortuitous and the qualitative basis of the theory is unsound. Nevertheless, it is worth recounting the theory; for it was much in common with some current models for the recombination of neutral atoms. Moreover, the need for and strength of more recent theories of the recombination process become clearer in the light of the deficiencies of the Thomson model.

In discussing Thomson's formulation of the recombination coefficient, it is useful to consider the recombination rate as the product of two factors:

1. The rate $R(\mathbf{v}_1,\mathbf{v}_2,b)\ d^3v_1\ d^3v_2\ db$ of collisions of ions with initial velocities $\mathbf{v}_1$, $\mathbf{v}_2$, and impact parameter b.
2. The probability $P(\mathbf{v}_1,\mathbf{v}_2,b)$ that either of the ions will undergo a collision with a neutral which leads to stable binding of the ion pair.

The rate of ion recombination will be the product of these two factors, integrated over all velocities and impact parameters, so

$$\alpha_T = \frac{1}{n_+n_-}\int P(\mathbf{v}_1,\mathbf{v}_2,b)\ R(\mathbf{v}_1,\mathbf{v}_2 b)\ d^3v_1\ d^3v_2\ db$$

It should be noted that even in the definition of P we have made the important assumption that *single* ion–neutral collisions are responsible for production of stable ion pairs, and thus recombination.

The collision rate R can be written down directly as

$$\frac{R}{n_+n_-}\ d^3v_1\ d^3v_2\ db = 2\pi b\,|\mathbf{v}_1 - \mathbf{v}_2|\,f(\mathbf{v}_1)f(\mathbf{v}_2)d^3v_1\ d^3v_2\ db$$

where $f(\mathbf{v})$ is the Maxwell-Boltzmann velocity distribution function. The correct choice of $P(\mathbf{v}_1,\mathbf{v}_2,b)$ is not so obvious. Thomson[8,10] felt that the probability of an ion undergoing such a deactivating collision with a neutral would be near unity if the ion and its charged partner were at a distance such that their mutual Coulomb potential energy (and hence their relative kinetic energy) was equal or greater than the mean thermal energy of the neutrals, $3/2kT$. Thus P is set equal to ϵ, the probability that either ion undergoes a collision while the ion–ion distance is less than or equal to a distance d, defined by

$$\frac{e^2}{d} = \frac{3}{2}kT$$

This intuitive and rather arbitrary definition of the Thomson critical radius d removes much of the explicit dependence of the deactivation probability on velocity, and with this very considerable simplification, it is possible to write

$$\alpha_T n_+ n_- = 2\pi \int_0^n \epsilon(b) b \, db \int |v_1 - v_2| f(\mathbf{v}_1) f(\mathbf{v}_2) d^3v_1 \, d^3v_2$$

and so

$$\alpha_T \cong \pi \, d^2 \bar{g} \bar{\epsilon}$$

where $\bar{g}$ is the mean relative ion speed, and $\bar{\epsilon}$ is the probability that either of the ions will undergo a collision with a neutral while the ion–ion distance is less than d, averaged over the ion–ion impact parameter. The evaluation of $\bar{\epsilon}$ has been discussed by Loeb,[11] who shows that if one makes the approximation that the ions pass each other by following straight line paths, then

$$\bar{\epsilon} = \omega_+ + \omega_- - \omega_+ \omega_-$$

where

$$\omega_\pm = 1 - \frac{\lambda_\pm^2}{2d^2}\left[1 - e^{-2d/\lambda_\pm}\left(\frac{2d}{\lambda_\pm} + 1\right)\right]$$

and $\lambda_\pm$ is the mean free path for the positive or negative ion. The function $\bar{\epsilon}$ has been tabulated,[11] and since

$$\lambda = \frac{1}{N\sigma}$$

where N is the neutral gas density and σ is the ion–neutral diffusion or mobility cross section, it is possible to calculate α_T as a function of neutral gas density.

It is useful to note that at low pressures we find that

$$\lim_{d/\lambda \to 0} \bar{\epsilon} = \frac{2}{3} dN(\sigma_+ + \sigma_-)$$

and consequently

$$\alpha_T \cong \frac{4}{3} \pi \, d^3 \bar{g} (\sigma_+ + \sigma_-) N \qquad \lambda \gg d \tag{4.1}$$

Therefore, the recombination coefficient increases linearly with pressure in the low-pressure regime. At high pressures, the inequality $d \gg \lambda$ holds, and the high-pressure limit of $\bar{\epsilon}$ is unity, so

$$\alpha_T \cong \pi \, d^2 \bar{g} = \frac{16}{9} \sqrt{\frac{2\pi}{\mu}} \frac{e^4}{(kT)^{3/2}} \qquad \lambda \ll d \tag{4.2}$$

This high-pressure second-order rate law limit corresponds to the situation in which every ion pair that reaches the separation d undergoes a deactivating collision.

Equations (4.1) and (4.2) make it clear that the magnitude of the recombination coefficient is very sensitive to the value of the critical radius d. Thomson's method of selecting this critical radius was more intuitive than rigorous. In subsequent papers by Natanson,[47] Brueckner,[48] and Parks,[49] other more firmly based estimates of the critical radius have been proposed. The values

$$d = \frac{2}{3}\frac{e^2}{kT} \qquad \text{Thomson}$$

$$= \frac{5}{12}\frac{e^2}{kT} \qquad \text{Natanson}$$

$$= \frac{1}{4}\frac{e^2}{kT} \qquad \text{Brueckner}$$

proposed for ions and neutrals of equal mass lead to third-order rate constants which vary by up to a factor of 19, and span the available experimental results, with Thomson's value giving the best agreement. Thus as the estimation of the critical radius has become more rigorous, the agreement between theory and experiment has diminished. This suggests that the idea of a single critical capture radius is not an appropriate basis for the theory of ion recombination.

Despite its rather obvious conceptual deficiencies, the Thomson theory is a lucid, easy to use model which is reasonably accurate numerically, particularly when the ions and neutrals have comparable masses. The idea that ions which are to recombine must enter a region of low potential energy so as to increase their kinetic energy and thereby increase the probability that one of them will lose energy upon collision with a neutral is an important concept. Even though the Thomson theory is based on this solid idea, it neglects such important aspects of the recombination process as the instability of weakly bound ion pairs to dissociation by subsequent collisions, the possible contribution to the rate of collisions which occur when the ion–ion distance is greater than the critical value, and the less than unit probability that ion–neutral collisions which occur within the critical distance will in fact produce bound ion pairs. The more detailed theories to be discussed show that all these factors are important, and thus the numerical success of the Thomson theory and its close relatives is largely fortuitous.

B. Numerical Calculations

Mahan and Person[50] attempted a numerical calculation of the ion recombination coefficient in 1964. Colliding ions were assumed to follow hyperbolic trajectories, and the rate constant for ion–neutral collisions which produced bound ion pairs was calculated as a function of the ion–ion distance at which the ion–neutral collision occurred. As expected, the deactivation rate constant was largest when the ions were close and moving rapidly, for such ions are apt to lose large amounts of energy in ion–neutral collisions. The calculation showed, however, that the deactivation rate was appreciable for ion pairs whose separation was considerably greater than the Thomson critical radius. In fact, the deactivation rate for widely separated ion pairs was great enough that the calculation of the recombination coefficient failed to converge to a finite value. This behavior was evidently a consequence of the neglect of possible dissociation of weakly bound ion pairs by subsequent ion–neutral collisions. Even though a rather limited set of initial values of the ion–ion impact parameter and relative speed were explored in the calculation, it was clear that the concept of a single critical radius for deactivation and the neglect of subsequent dissociative collisions were serious deficiencies of the Thomson theory.

Feibelman[51] performed a more complete numerical calculation of the three-body ion recombination rate using Monte Carlo techniques. His results show very clearly the importance of ion–neutral collisions which tend to dissociate weakly bound ion pairs. Ion–ion impact parameters as large as 1800 Å made large contributions to the rate constant if the criterion was merely that the ions become bound as a result of one ion–neutral collision. Feibelman then imposed a requirement that an ion pair contributed to the recombination rate only if it became bound after an initial collision, and if its binding energy had increased after 15 subsequent ion–neutral collisions. This stability criterion essentially eliminated the contribution to the recombination rate coefficient of ion pairs which had undergone their first deactivating collision at separations greater than 700 Å, and thereby led to a finite value of 1.84×10^{-25} cm^6/sec for the third-order recombination coefficient for O_2^+–O_2^- in oxygen at room temperature. This compares rather favorably with the experimental value of 1.86×10^{-25} deduced from the experiments of McGowan.[23] Moreover, Feibelman's calculation shows that even with the stability criterion, ion pairs that undergo deactivating collisions at separations greater than the Thomson radius to make a significant contribution to the recombination rate, and some pairs first deactivated within the Thomson radius eventually dissociate.

C. Multicollision Theories of Recombination

In contrast to the single "strong collision" models of recombination, of which the Thomson theory is an example, one can regard recombination as a multicollision process. In this picture, recombining pairs randomly gain or lose energy by collisions with neutrals, and the finite recombination rate is a consequence of a net flux downward in the energy of ion pairs. The recombination rate remains proportional to the first power of the neutral density, despite the multicollision aspect of the process, because the recombination rate depends linearly on the transition rate between energy levels, which is in turn proportional to the neutral gas density. This multicollision picture is the basis for the theories of ion recombination advanced by Pitayevski,[52] Landon and Keck,[53] Mahan, [54] Bates and Moffett,[55] and Bates and Flannery, [56,57] These theories have been recently reviewed in detail by Flannery,[21] so we present here only an outline of the major points of derivation and results.

To fix the ideas, it is useful to refer to Fig. 3, an energy level diagram for ion pairs. The (negative) energies of the bound pairs are regarded as quasicontinuous in the eventual classical calculation of the rate. The energy $|E_s|$ is the binding energy of the ion pair at which the recombination is made permanent by charge transfer, and the energy $|E_c|$ is the greatest binding energy an ion pair may have. The energies E_i, E_f, and E are arbitrary, and $n(E)\,dE$ is the number density of ion pairs with energies between E and $E + dE$. The collisional rate co-efficient K_{if} is defined such that $n(E_i)\,dE_i\,NK_{if}\,dE_f$ is the total number of ion pairs per unit volume per unit time making transitions from the energy interval dE_i about E_i to the interval dE_f about E_f. Here again N is the number density of neutral molecules.

The recombination rate is simply the net flow down past some arbitrary negative energy E. Writing this net flow as the difference of the flow down

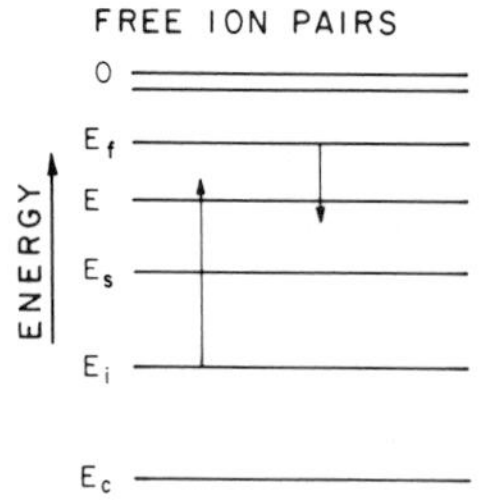

Fig. 3. Energy level diagram for ion recombination.

from E_f to all energies below E, integrated over all E_f, and the flow up from E_i to all energies above E, integrated over all E_i, we get

$$\alpha(n_+)(N_-) = N\int_{E_i=E_c}^{E} \int_{E_f=E}^{\infty} (n_f\, K_{fi} - n_i\, K_{if})\, dE_i\, dE_f \qquad (4.3)$$

as the fundamental equation for the recombination coefficient. Letting the ratio of the actual number density at energy E_i to the number density which would exist at thermal equilibrium be

$$\begin{aligned} \rho_i &= \frac{n_i}{n_{iT}} \leq 1 \\ &= 1 \qquad E > 0 \\ &= 0 \qquad E < E_s \end{aligned}$$

and using detailed balance in the form

$$n_{iT}\, K_{if}\, dE_i\, dE_f = n_{fT} K_{fi}\, dE_f\, dE_i \qquad (4.4)$$

we can reduce (4.3) to

$$\alpha(n_+)(n_-) = N\int_{E}^{\infty} n_{iT}\, dE_i \int_{E_c}^{E} (\rho_f - \rho_i)\, K_{if}\, dE_f \qquad (4.5)$$

Thus the calculation of the recombination coefficient must involve determination of the nonequilibrium distribution function for combining ion pairs $\rho(E)$ and the rate coefficient for collision energy change K_{if}. The calculation of K_{if} is a formidable problem in classical mechanics which has been treated in detail by Flannery,[21] and Bates and Flannery,[56] and will not be discussed here. Various theories differ in the manner in which the distribution function $\rho(E)$ is determined, and we turn to this problem now.

The most convenient point of departure is a classical master equation for the time rate of change of the population n_f of a given bound level as a result of collisions:

$$\frac{\partial n_f}{\partial t} = N\left[\int_{E_s}^{\infty} n_i K_{if}\, dE_i - n_f \int_{E_c}^{\infty} K_{fi}\, dE_i\right] \qquad (4.6)$$

Introducing the distribution function ρ, and using detailed balance (Eq. (4.4)) gives

$$n_{fT}\frac{\partial \rho_f}{\partial t} = N\int_{-\infty}^{\infty} (\rho_i - \rho_f) n_{iT} K_{if}\, dE_i \qquad (4.7)$$

where the lower limits of the integrals have both been set equal to $-\infty$.

In the theory of Keck and Carrier[58] as applied by Landon and Keck,[53] it is assumed that K_{if} is appreciable only when $E_i - E_f$ is small, and consequently ρ_i is appreciable only when $E_i - E_f$ is small, and consequently ρ_i can be expressed as an expansion in E about E_f

$$\rho_i = \rho_f + \frac{\partial \rho_f}{\partial E}\Delta E + \frac{\partial^2 \rho_f}{\partial E^2}\frac{\Delta E^2}{2}, \quad \Delta E = E_i - E_f$$

with higher terms neglected. Introduction of this expansion into (4.7) gives

$$n_T \frac{\partial \rho}{\partial t} = \Delta_1 \frac{\partial \rho}{\partial E} + \frac{\Delta_2}{2}\frac{\partial^2 \rho}{\partial E^2} \tag{4.8}$$

with

$$\Delta_n = N \int n_{iT}\, K_{if}\, (E_i - E_f)^k \, dE_i$$

Equation (4.8) is a Fokker-Planck equation for the distribution function in terms of the moments of the energy change per unit time. Keck and Carrier[58] show that if the energy change per collision is small enough, the relation

$$\Delta_1 \cong \frac{1}{2}\frac{\partial \Delta_2}{\partial E}$$

holds, and so (4.8) reduces in this approximation to

$$n_T \frac{\partial \rho}{\partial t} = \frac{\partial}{\partial E}\left(\frac{\Delta_2}{2}\frac{\partial \rho}{\partial E}\right) \tag{4.9}$$

which is a diffusion equation to be solved for ρ and for the flux of recombining particles. The theory of Pitayevski[52] as applied to ion recombination by Mahan[54] is also based on a solution of the diffusion equation.

The steady-state distribution function can be found by setting (4.9) equal to zero, and integrating, using the boundary conditions

$$\rho(0) = 1, \qquad \rho(E_s) = 0$$

The result is

$$\rho(E) = 1 - \frac{\int_E^0 dE_0/\Delta_2}{\int_{-E_s}^0 dE_0/\Delta_2} \tag{4.10}$$

The recombination coefficient can be found by integrating (4.9) between $-E_s$ and 0, using the additional boundary condition $(\partial\rho/\partial E)_{-E_s} = 0$:

$$\frac{\partial}{\partial t}\int_{-E_s}^0 n(E)\, dE = -\alpha n_+ n_- = -\frac{1}{2}\left[\int_{-E_s}^0 \frac{dE_0}{n_T(E_0)\Delta_2(E_0)}\right]^{-1}$$

or

$$\alpha = N\left[\int_{-E_S}^{0} \frac{dE_0}{\frac{1}{2}f(T,E_0)\Delta_2(E_0)}\right]^{-1} \tag{4.11}$$

where $f(T,E_0)$ is the equilibrium ratio of bound pairs at E_0 to the product of free ion concentrations, or the recombination equilibrium constant.

Thus, very much as the spatial diffusional flux is dependent on the mean square displacement of a particle per unit time, the rate of recombination is related to the mean square energy change per unit time. The condition for the validity of this diffusional formulation of the recombination is that the fractional change in the energy of a bound pair upon collision may be small. This will be true for ion neutral collisions of the grazing type, regardless of the masses. However, much of the total ion–neutral collision cross section comes from more nearly head-on collisions. Here the criterion of small energy change will be met only if the mass of the neutral is much less than that of either ion, or if one of the ions has a much smaller mass than the other ion and the neutral.

The accuracy of the diffusion theory and closely related approximations has been discussed by Bates and Jundi.[59] They find that the Landon-Keck procedure yields rate constants that are 40% larger than those computed by the more accurate theory of Bates and Flannery for mass ratios, which should best satisfy the small energy change criterion of the diffusion theory. Bates and Jundi propose an "effective gradient" and "modified effective gradient" theories, which are very much in spirit of the diffusion theory but are more accurate. The relation between these various approximate theories has been discussed by Flannery[60] from the point of view of Markov processes.

D. The Quasiequilibrium Theory of Recombination

Bates and Moffet[55] and Bates and Flannery[56] have developed what is apparently a highly accurate theory of three-body ion recombination. They first determined accurate binary rate coefficients K_{if} for the change in the internal energy of the ion pairs through either resonant charge transfer collisions[55] with neutrals, or through ion–neutral collisions which occur on the Sutherland-Langevin type of potential[56] (hard sphere repulsion, polarization attraction). Then the exact steady-state distribution function was determined numerically by setting (4.6) equal to zero to obtain the integral equation

$$\rho_f \int_{E_C}^{\infty} K_{fi}\, dE_i = \int_{E_S}^{\infty} \rho_i K_{if}\, dE_i \tag{4.12}$$

This integral equation was converted by a numerical quadrature formula to a set of linear equations, which were in turn solved in terms of the rate coefficients. The recombination coefficient was then computed using (4.5).

The results of this very extensive numerical work can be expressed in the form[61]

$$\alpha = ep^{1/2}\, C[M_{13}^{-1/2}\, \gamma(a,A) + M_{23}^{-1/2}\, \gamma(b,B)] \tag{4.13}$$

where e is the fundamental charge, p is the polarizability of the neutral, and M_{13}, M_{23} are the reduced masses of the neutral with ions 1 and 2, respectively. The dimensionless parameter C is given by

$$C = \frac{\pi^{3/2}e^6}{2(kT)^3} N = 1.30 \times 10^{-8}\, \mathrm{NT}^{-3}$$

where T is the Kelvin temperature, k is Boltzmann's constant, and N is the neutral gas number density. Knowledge of the mass ratios

$$a = \frac{M_2 M_3}{M_1\,(M_1 + M_2 + M_3)}, \qquad b = \left[\frac{M_1}{M_2}\right]^2 a$$

and of the ion–neutral interaction potential parameters

$$A = \left(\frac{2kT}{pe^2}\right)^{1/2} S_{13}^2 \qquad B = \left(\frac{S_{23}}{S_{13}}\right)^2 A$$

where the S_{i3} are the ion–neutral hard sphere radii, allows one to select values of γ, a dimensionless coefficient determined from the calculations. Table III reproduces the values of γ found by Bates and Flannery.[56] Use

TABLE III

The Dimensionless $\gamma(c,C)$ Coefficients of Bates and Flannery[a]

			c		
C	0.2	0.3	0.4	0.5	0.9
0.3	0.947	1.271	1.447	1.575	1.765
0.9	0.897	1.219	1.399	1.519	1.72
1.5	1.092	1.540	1.784	1.947	2.206
2.0	1.352	1.927	2.231	2.435	2.754

c	0.1	0.6	1.2	1.6	2.0	5.0	10.0
$\gamma(c,0.9)$	0.428	1.598	1.739	1.714	1.633	1.151	0.746

C	0.01	0.1	0.2	10
$\gamma(\frac{1}{3},C)$	1.418	1.364	1.343	9.667

[a] Ref. 56.

of this table allows α to be determined for a large range of mass ratios and ion–neutral radii. The authors[57] suggest that departures of the ion–neutral interaction from the Langevin form can be largely accounted for by replacing $\gamma(a,A)$ and $\gamma(b,B)$ by

$$\gamma'(a,A) = \frac{\kappa_{L1}(A)}{\kappa_1}\,\gamma(a,A)$$

$$\gamma'(b,B) = \frac{\kappa_{L2}(B)}{\kappa_2}\,\gamma(b,B)$$

where κ_1 and κ_2 are the experimental mobilities of the positive and negative ions, $\kappa_{L1}(A)$ and $\kappa_{L2}(B)$ are the corresponding mobilities calculated from the Langevin mobility theory

$$\kappa_{Li}(A) = \frac{g(A)}{4\pi p M_{i3}^{1/2}\,N}$$

where $g(A)$ is a tabulated function.[62]

Several points concerning the comparison of other theories to the Bates-Flannery work should be noted. First, depending on the interaction parameters, the value of α for the equal ion–neutral mass case ranges from 0.56 to 0.6 of the value predicted by the Thomson theory. Bates and Flannery[56] show, however, that this agreement of the Thomson theory with their more accurate treatment is entirely fortuitous. As noted earlier by Mahan and Person[50] and Feibelman,[51] the downward flux of free to weakly bound ions is extremely large. The Thomson theory succeeds by rather abitrarily neglecting the proper large fraction of these free to bound transitions, and totally ignoring the large number of bound to free transitions. The results of Feibelman's Monte Carlo calculation of $\alpha/N = 1.8 \times 10^{-25}$ for O_2^+–O_2^- recombining in O_2 through hard sphere ion–neutral collisions is reasonably close to the value of 1.4×10^{-25} deduced from the Bates-Flannery work. Finally, the results of the diffusion theory of Landon and Keck[53] are higher by approximately 40% than the results of Bates and Flannery for those cases in which the diffusion theory is most apt to be valid (light neutral, heavy ions) and are even higher relatively as the mass of the neutral increases.

E. Experimental Results

The experimental data with which the various theories can be compared are those of Mahan and Person,[17] McGowan,[23] and Fisk, Mahan, and Parks.[24] The situation is somewhat unsatisfactory in that in no experiments in which termolecular recombination dominates have the identities of the ions been monitored *simultaneously* with measurements of the recombination rate. Fisk et al.[24] did auxiliary mass spectrometric measurements to show that in the thallium iodide system, where the primary ions are Tl^+ and I^-, the principal recombining ions are Tl_2I^+ and TlI_2^-.

In the experiments of Mahan and Person,[17] the primary ions were certainly NO^+ and NO_2^-, but mobility measurements indicated that the recombining ions had a greater molecular weight, possibly in the range 60–76. Subsequent observations of photoionized NO by Lineberger and Puckett[63] and Weller and Biondi[64] have clarified the situation somewhat. Weller and Biondi noted that for pressures of NO of 0.2 torr in 15 torr of neon (roughly the conditions of Mahan and Person), the primary NO^+ ion was replaced by $(NO)_2^+$ after 10^{-3} sec. This is consistent with the rate constant deduced for the NO^+–NO association reaction found by Lineberger and Puckett.[63] Thus it appears highly likely that in the experiments of Mahan and Person, $(NO)_2^+$ was the dominant ion.

The evidence for the nature of the negative ion is even less direct. The equilibrium constant[65] for the hydration of NO_2^- is small enough so that if the pressure of water vapor were as high as 10^{-4} torr, which is unlikely, most of the NO_2^- would not be hydrated. No evidence of a cluster $NO_2 \cdot NO_2^-$ has been reported, but Fehsenfeld et al.[66] have found that the reaction

$$NO_2^- + NO_2 \rightarrow NO_3^- + NO$$

has a rate constant of approximately 4×10^{-12} cm^3/sec. With the concentrations of NO_2 employed by Mahan and Person,[17] it is likely that NO_2^- was converted to NO_3^- at a rate comparable to the ion recombination rate. Thus the negative ions in these experiments were probably NO_2^- and NO_3^- in comparable amounts.

Similar considerations apply to the recombinations studies of McGowan[23] in pure oxygen. The equilibrium constant of Yang and Conway[67] for

$$O_2^+ + O_2 = O_2^+$$

is large enough so that O_4^+ should be the dominant positive ion even at the lowest pressures investigated by McGowan. On the other hand, the investigations of Voshall, Pack, and Phelps[68] for

$$O_2^- + O_2 = O_4^-$$

indicate that even up to atm, O_2^- is the principal negative ion. However, the more recent results of Conway and Nesbit[69] show that the equilibrium constant for this reaction is 1.58 torr^{-1} at 293°K, and consequently it is more likely that in McGowans experiments O_4^- was the dominant negative ion.

TABLE IV
Experimental and Theoretical Termolecular Recombination Coefficients

System	Third body	Theory[a] $k \times 10^{25}$ (cm^6/sec	Experiment $k \times 10^{25}$ (cm^6/sec)
NO^+, NO_2^-	H_2	0.28	0.58 ± 0.06[b]
	D_2	0.398	0.61 ± 0.06
	He	0.256	0.41 ± 0.06
	Ne	0.628	1.04 ± 0.05
	N_2	1.36	2.10 ± 0.1
	Ar	1.37	1.45 ± 0.1
	Kr	1.69	1.75 ± 0.1
	Xe	2.18	2.78 ± 0.2
O_4^+, O_2^-	O_2	1.50	1.86 ± 0.3[c]

[a] Ref. 61.
[b] Ref. 17.
[c] Ref. 23.

Table IV lists the available experimental termolecular recombination rate constants. Also shown are the theoretical values of Flannery[61] calculated with the Bates-Flannery[56] theory. For the oxygen system, the ions assumed in the calculation were in fact O_4^+ and O_2^- rather than O_4^+ and O_4^-. Moreover, for the nitric oxide–nitrogen dioxide system, Flannery's calculations are based on NO^+ and NO_2^- as the recombining ions. The arguments given above indicate that $N_2O_2^+$ and a composite of NO_2^- and NO_3^- would be more realistic. This choice would tend to lower the calculated recombination coefficients by approximately 25%, thereby decreasing the agreement between theory and experiment. Even with this qualification, the agreement between experiment and theory is fairly satisfactory, considering the several sources of experimental error, the crude nature of the Langevin potential, and the neglect of inelastic collisions.

The need for further experimental work on termolecular ion recombination is clear. In particular, systems in which the identity of the (preferably monatomic) ions is unequivocal should be investigated, and the effect of ion complexing agents like water and other polar molecules should be explored systematically.

V. ION RECOMBINATION AT HIGH PRESSURE

As noted earlier, the experiments of Mächler[12] show that in the pressure range of 5–20 atm at room temperature, the effective recombination coefficient α varies inversely with pressure. There are two early theories that are consistent with this inverse pressure dependence of the recombination coefficient.

In 1903, Langevin[7] proposed that recombination occurs as a result of ions drifting together under the influence of their mutual Coulomb attraction, with the drift inhibited by collisions of the ions with the neutral gas molecules. If we consider a single positive ion, the flux of negative ions through a sphere of radius r is

$$\Phi = 4\pi r^2 v_d n_-(r)$$

where v_d is the relative drift velocity, and $n_-(r)$ is the number density of negative ions at a distance r. As an approximation for $n_-(r)$ we take the bulk concentration of negative ions. The relative drift velocity of the ions can be expressed as the electric field e/r^2 at the negative ion times the sum of the ion mobilities, or

$$v_d = \frac{e}{r^2}(\kappa_+ + \kappa_-)$$

The recombination coefficient is then

$$\alpha_L = \frac{\Phi}{n_-} = 4\pi e(\kappa_- + \kappa_+) \tag{5.1}$$

This is Langevin's[7] expression for the recombination coefficient. The approximation of setting the ion concentration in the vicinity of an ion equal to the bulk concentration is somewhat dubious, and will be explored in more detail subsequently.

An expression closely related to Langevin's was developed by Harper.[9] He reasoned that in an homogeneous weakly ionized gas, most ions will be separated by such large distances that their mutual Coulomb interactions will be negligible compared to their average thermal kinetic energy. Thus the rate at which ions recombine might well be limited by the rate at which they can diffuse to distances at which their mutual Coulomb attraction assures recombination.

If recombination is assumed to be certain if an ion pair diffuses to a separation r_H, then the recombination coefficient should be expressed as

$$\alpha_H = 4\pi r_H{}^2 u$$

where u is the mutual diffusional drift velocity. To find u, we use the result that if two particles are initially at r_0, after a time t their mean square displacement will be given by[69]

$$\overline{r^2} = r_0{}^2 + 6\,Dt$$

where D is the sum of the individual diffusion coefficients. Differentiation with respect to time gives

$$\frac{1}{2r}\frac{d\overline{r^2}}{dr} = u = \frac{3D}{r}$$

Thus the recombination coefficient becomes

$$\alpha_H = 12\pi r_H(D_+ + D_-)$$

To proceed further, a value for r_H must be selected. Harper[9] reasoned that this critical radius would be the distance at which the drift velocity induced by mutual Coulomb attraction became equal to the diffusional drift velocity. Thus r_H is defined by

$$\frac{\kappa e}{r_H{}^2} = \frac{3D}{r_H}$$

$$r_H = \frac{e\kappa}{3D} = \frac{e^2}{3kT}$$

where we have used the Einstein relation between the diffusion and mobility coefficients. The final results for the recombination coefficient can be expressed as either

$$\alpha_H = \frac{4\pi e^2}{kT}(D_+ + D_-) \tag{5.2}$$

or

$$\alpha_H = \alpha_L = 4\pi e(\kappa_+ + \kappa_-)$$

Thus Harper's result is identical with Langevin's. The introduction of an arbitrary trapping radius is an unsatisfactory feature of Harper's derivation, and discussions of the manner in which the best choice of this radius is made do not seem profitable. In what follows we present a unified approach to the problem based on the general theory of diffusion controlled reactions[70] which clarifies both the problems of the critical radius and of the effective concentration of ions.

Again consider oppositely charged ions diffusing together under the influence of the mutual potential U which vanishes at infinity. Fick's first law of diffusion for this case becomes

$$\frac{dN}{dt} = 4\pi r^2 D'\left[\frac{dn}{dr} + \frac{n}{kT}\frac{dU}{dr}\right] \tag{5.3}$$

where the left side is the number of negative ions that approach a radius r of one positive ion per unit time, D' is the coefficient of relative diffusion, and n is the negative ion number density. Replacing dN/dt by Φ and solving the linear differential equation we get

$$n_r = e^{-U/kT}\left[n_\infty - \frac{\Phi}{4\pi D'}\int_r^\infty e^{U/kT}\frac{dr}{r^2}\right] \tag{5.4}$$

where n_∞ is the bulk average negative ion concentration. Equation (5.4) shows that the concentration n_r of negative ions at a distance r from a positive ion will be less than local equilibrium value $n_\infty \exp(-U/kT)$ if a net recombination is occurring, and will in general be different from the bulk concentration. This depletion of the local ion concentration is enhanced if D' is small.

An effective rate constant k' can be defined by the phenomenological equation

$$k' = \frac{\Phi}{n_\infty} \tag{5.5}$$

Also, let k be the second-order rate constant which would apply if diffusion were fast enough or reaction slow enough so that the ion concentration at a separation r would be given by the equilibrium expression $n_\infty \exp[-U(r)/kT]$. Then

$$k' = k\,\frac{n_r\,(\text{actual})}{n_r\,(\text{equil})} = k\,\frac{n_r}{n_\infty}\,e^{U/kT}$$

Combining this with (5.4) and (5.5) gives

$$k' = \frac{k}{1 + \dfrac{k}{4\pi r^* D'}} = \frac{4\pi r^* D'}{1 + \dfrac{4\pi r^* D'}{k}} \tag{5.6}$$

where

$$\frac{1}{r^*} = \int_{r_c}^\infty e^{U/kT}\frac{dr}{r^2} \tag{5.7}$$

Equation (5.6) makes it clear that if the condition

$$4\pi r^* D' \ll k$$

holds, diffusional effects will determine the phenomenological rate constant k', whereas if the inequality is reversed, diffusional effects will be negligible. For fast reactions (no activation energy) between *neutral* particles, the rate constant k can be expressed approximately as $\pi\sigma^2\bar{v}$ where σ is of the order of a molecular diameter, and $\bar{v}$ is the mean relative speed. Also, the diffusion coefficient D' can be approximated by $\frac{1}{3}\lambda\bar{v}$ where λ is the mean free path. It is reasonable to take r^* equal to λ for this case, and so disregarding factors of order unity, we have

$$\frac{k}{4\pi r^* D'} \cong \left(\frac{\sigma}{\lambda}\right)^2$$

which is of the order of 10^{-5} at atmospheric pressure. Thus diffusional effects are not noticeable for homogeneous gas reactions between neutrals at ordinary pressures. For ion–ion recombination, however, it is reasonable to replace σ and r^* with d, the Thomson radius, so that we have

$$\frac{k}{4\pi r^* D'} \cong \frac{d}{\lambda}$$

for this case, if conditions are near STP. Also, d and λ are of the same magnitude at 1 atm, and thus ion recombination will be diffusion limited at pressures much above 1 atm. Again it is the strength of the Coulomb interaction at great distances that distinguishes the ion–ion systems qualitatively from all others.

We return now to (5.6) and consider the high density limit in which the recombination reaction is diffusion controlled:

$$\lim_{N\to\infty} k' = 4\pi r^* D' = \alpha_{LH}$$

To evaluate α_{LH}, we can set D' equal to the sum of the individual ion diffusion coefficients and use the Coulomb potential in (5.7) to find that

$$r^* = \frac{e^2}{kT}\left[1 - \exp\left(\frac{-e^2}{r_c kT}\right)\right]^{-1} \tag{5.8}$$

To proceed further, we must specify the distance r_c, which physically is the distance to which ions must approach if neutralization is to occur free from further diffusional effects. The lower limit for r_c must be the distance at which electron transfer becomes a virtual certainty, which in the high-pressure regime will be somewhat smaller than the Thomson radius d. An upper limit for r_c would be approximately the ionic mean free path. Thus for r_c we take

$$r_c \cong \gamma d + \lambda$$

where γ is a parameter less than unity. At pressures of the order of 10 atm or greater and at 300°K, we find

$$\frac{e^2}{r_c kT} = \frac{e^2}{\gamma\, dkT} > 1$$

Therefore,

$$r^* \simeq \frac{e^2}{kT}$$

and

$$\alpha_{LH} = \frac{4\pi e^2}{kT}(D_+ + D_-) = 4\pi e(\kappa_+ + \kappa_-)$$

Thus when proper account is taken of diffusion and drift, the Langevin-Harper expression for α is recovered without the necessity for specifying an exact critical radius.

The physical situation for pressures near 1 atm is more complicated, since the conditions for neither free termolecular recombination nor fully diffusion limited recombination are met. Natanson[47] has treated this transition region between the Thomson and Langevin-Harper regimes by using a critical radius theory of the Thomson type with the diffusion-limited reaction rate theory just outlined. Natanson's work has been discussed extensively by McDaniel,[62] and we present here only a brief outline and somewhat simplified derivation of this result.

The rate coefficient k for termolecular deactivation can be expressed approximately as

$$k = \pi d_N^2 \bar{g} \bar{\epsilon} \left(1 + \frac{e^2}{d_N kT}\right) \tag{5.9}$$

Here $\bar{g}$ is the relative speed of the ions, $\bar{\epsilon}$ is the probability that they will undergo a collision with a neutral while their separation is less than d_N, which is the Natanson[47] critical radius, given by

$$d_N = \frac{\lambda}{2}\left[\left(1 + \frac{5e^2}{3\lambda kT}\right)^{1/2} - 1\right] \tag{5.10}$$

where λ is the ionic mean free path. The Natanson radius depends on the ion mean free path, and in the limit of low pressures becomes

$$d_N = \frac{5}{12}\frac{e^2}{kT} \qquad \lambda \gg d \tag{5.11}$$

which is of the same magnitude but somewhat smaller than the Thomson radius. The last factor in (5.9) accounts for the mutual attraction of the ions, which increases the effective cross section of the critical sphere over its geometric value of πd_N^2.

By using (5.11) we can write the termolecular rate constant as

$$k = \frac{17}{5}\pi d_N^2 \bar{g} \bar{\epsilon}$$

This expression is to be used in (5.6) to find the phenomenological constant k'. Also, in (5.7) the radius r_c at which diffusional effects cease can be set equal to $d_N + \lambda$. Rather simpler expressions result if we use (5.10) to show that

$$\frac{e^2}{(\lambda + d_N)kT} = \frac{12}{5}\frac{d_N}{\lambda}$$

Making these substitutions in (5.6) we get

$$\alpha_N = \frac{17}{5}\pi d_N{}^2 \bar{g}\bar{\epsilon}\, C\left[1 + \frac{17 d_N{}^2 \bar{g}\bar{\epsilon} kT(C - 1)}{20e^2(D^+ + D^-)}\right]^{-1} \tag{5.12}$$

where

$$C = \exp\left(\frac{12}{5}\frac{d_N}{\lambda}\right)$$

The value of $\bar{\epsilon}$ is found just as in Thomson's theory, but with d_N replacing the Thomson critical radius.

The Natanson expression for the recombination coefficient goes over to the Langevin-Harper form in the high-pressure limit, whereas in the low-pressure regime it has the same form but is 0.83 times as large as the Thomson expression. Its behavior in the intermediate pressure region around 1 atm is consistent with the rather meager data that are available.[62] Thus, aside from the fact that it involves a somewhat arbitrary trapping radius d_N, the Natanson expression satisfactorily describes the behavior of the recombination coefficient over a wide range of pressures.

Bates and Flannery[57] have pointed out that it would be exceedingly difficult to extend their theory of the termolecular recombination coefficient rigorously into the pressure range near and above 1 atm. They make explicit use of an assumption concerning the equality of the populations of accessible elements of the phase space of bound ion pairs, which requires that the mean time between collisions be long compared to the orbiting times of the ion pairs. Removing this assumption, which fails at moderate and high densities, would increase the computational time involved in the theory beyond what is practical.

Bates and Flannery[57] have proposed a modification scheme for the Natanson expression which removes some of the arbitrariness of the choice of d_N, and allows results computed by their quasiequilibrium theory to be extended to higher pressures. The essential procedure involves replacing the Natanson trapping radius with

$$d' = \left(\frac{\alpha_{BF}}{\alpha_N}\right)^{1/3} d_N$$

which is in general different for the two ions. Here α_{BF} is the computed Bates-Flannery partial termolecular recombination coefficient, multiplied by the ratio of Langevin to experimental ion mobility. The Natanson expression is also modified to take account of the fact that there are two different trapping radii, and two different iron–neutral mean free paths.

This method of estimating α over the entire pressure range has been applied only to the recombination of ions in pure oxygen.[57,21] Comparison with McGowan's experiment data[23] showed agreement to within approximately 10% over the pressure range from 0.1 to 1 atm, if the recombining ions were taken to be O_4^+ and O_2^-. This latter choice is not fully consistent with mass spectrometric investigations[67,68] of the nature of the ions in pure oxygen gas. No meaningful quantitative test of the theory in the high-pressure regime is possible, since the only data are those of Mächler[12] for unknown ionic species in air. Nevertheless, the qualitative behavior of the calculated value of α is consistent with Mächler's results.

The experimental and theoretical situations in this intermediate- to high-pressure regime are perhaps the least satisfactory of all. There are no data from a single laboratory which span this pressure range, and the only data available at high pressures were collected with the techniques available 35 years ago. No positive chemical or mass identification of the ions in any of the higher-pressure experiments has been made. The grafting of the Bates-Flannery low-pressure theory to the framework of the Natanson theory shows numerical promise, but is aesthetically unsatifying. Also, it is not clear that low field mobilities suffice to represent the ion–neutral collision properties in situations where the ions diffuse to within small distances before they recombine. The dissociation equilibria of clustered ions may in addition be affected by the acceleration produced by their mutual Coulomb attraction.

References

1. M. Venugopalan, *Reactions Under Plasma Conditions*, Vol. 2, Wiley, New York, 1971.
2. L. Friedman and B. Reuben, *Advances in Chemical Physics*, I. Prigogine and S. Rice, Eds., Vol. 19, Wiley, New York, 1971, p. 141.
3. J. Bardsley and M. Biondi, *Advances in Atomic and Molecular Physics*, D. Bates and I. Esterman, Eds., Vol. 6, Academic Press, New York, 1970.
4. L. G. Christophoru, *Atomic and Molecular Radiation Physics*, Wiley, New York, 1971.
5. J. J. Thomson and E. Rutherford, *Phil. Mag.*, **42**, 392 (1896).
6. E. Rutherford, *Phil. Mag.*, **47**, 422 (1897).
7. P. Langevin, *Ann. Chim. Phys.*, **28**, 289, 443 (1903).
8. J. J. Thomson, *Phil. Mag.*, **47**, 337 (1924).
9. W. R. Harper, *Proc. Cambridge Phil. Soc.*, **28**, 219 (1932).

10. J. J. Thomson and G. P. Thomson, *Conduction of Electricity through Gases*, 3rd ed., Vol. 1, Cambridge University Press, Cambridge, 1928.
11. L. B. Loeb, *Basic Processes of Gaseous Electronics*, University of California Press, Berkeley, 1961.
12. W. Mächler, *Z. Physik*, **104**, 1 (1936).
13. J. Sayers, *Proc. Roy. Soc. (London) Ser. A*, **169**, 83 (1938).
14. M. E. Gardener, *Phys. Rev.*, **53**, 75 (1938).
15. D. R. Bates and H. S. W. Massey, *Phil. Trans. Roy. Soc. (London) Ser. A*, **239**, 269 (1943).
16. T. H. Y. Yeung, *Proc. Phys. Soc. (London)*, **71**, 341 (1958).
17. B. H. Mahan and J. C. Person, *J. Chem. Phys.*, **40**, 392 (1964).
18. C. Greaves, *J. Electron. Control*, **17**, 171 (1964).
19. T. S. Carlton and B. H. Mahan, *J. Chem. Phys.*, **40**, 3683 (1964).
20. J. Sayers, in *Atomic and Molecular Processes*, D. R. Bates, Ed., Academic Press, New York, 1962.
21. M. R. Flannery, in *Case Studies in Atomic Collision Physics*, M. C. R. McDowell and E. A. McDaniel, Eds., Vol. 2, Wiley, New York, 1972.
22. H. G. Ebert, J. Booz, and R. Koepp, *Z. Physik*, **181**, 187 (1964).
23. S. McGowan, *Can. J. Phys.*, **45**, 439 (1967).
24. G. A. Fisk, B. H. Mahan, and E. K. Parks, *J. Chem. Phys.*, **47**, 2649 (1967).
25. S. McGowan, *Can. J. Phys.*, **45**, 429 (1967).
26. C. B. Wharton, *Plasma Diagnostic Techniques*, R. Huddletone and S. Leonard, Eds., Academic Press, New York, 1965.
27. P. M. Eisner and M. N. Hirsh, *Phys. Rev. Letters*, **26**, 874 (1971).
28. W. H. Aberth and J. R. Peterson, *Phys. Rev.*, **1**, 158 (1970).
29. R. Neynaber, *Advances in Atomic and Molecular Physics*, D. Bates and I. Esterman, Eds., Vol. 5, Academic Press, New York, 1969.
30. J. Weiner, W. Peatman, and R. S. Berry, *Phys. Rev. Letters*, **25**, 79 (1970).
31. J. Weiner, W. Peatman, and R. S. Berry, *Phys. Rev. A*, **4**, 1824 (1971).
32. R. D. Rundel, K. L. Aitken, and M. F. A. Harrison, *J. Phys. B*, **2**, 954 (1969).
33. T. D. Gaily and M. F. A. Harrison, *J. Phys. B*, **3**, L25 (1970).
34. T. D. Gaily and M. F. A. Harrison, *J. Phys. B*, **3**, 1098 (1970).
35. J. Moseley, W. Aberth, and J. Peterson, *Phys. Rev. Letters*, **24**, 435 (1970).
36. R. E. Olson, J. Peterson, and J. Moseley, *J. Chem. Phys.*, **53**, 3391 (1970).
37. J. Peterson, W. Aberth, and J. Moseley, *Phys. Rev. A*, **3**, 1651 (1971).
38. L. Landau, *Phys. Ztg. Sow. Union*, **2**, 46 (1932).
39. C. Zener, *Proc. Roy. Soc. (London) Ser. A*, **137**, 696 (1932).
40. J. L. Magee, *Discussions Faraday Soc.*, **12**, 33 (1952).
41. D. R. Bates and J. T. Lewis, *Proc. Phys. Soc. (London)*, **A68**, 173 (1955).
42. A. Dalgarno, G. Victor, J. Browne, and T. Webb, unpublished data quoted in Refs. 35 and 36.
43. D. R. Bates and T. J. M. Boyd, *Proc. Phys. Soc. (London)*, **A69**, 910 (1956).
44. D. R. Bates, *Comments Atom. Molec. Phys.*, **2**, 107 (1970).
45. R. E. Olson, *J. Chem. Phys.*, **56,** 2979 (1972).
46. A. A. Radtsig and B. M. Smirnov, *Abstr. VIIth Int. Conf. Phys. Electr. At. Collisions*, North-Holland, Amsterdam, 1971, p. 481.
47. G. L. Natanson, *Soviet Phys.-Tech. Phys. (English Transl.)*, **4**, 1263 (1959).
48. K. A. Brueckner, *J. Chem. Phys.*, **42**, 439 (1964).
49. E. K. Parks, *J. Chem. Phys.*, **48**, 1483 (1968).
50. B. H. Mahan and J. C. Person, *J. Chem. Phys.*, **40**, 2851 (1964).

51. P. J. Feibelman, *J. Chem. Phys.*, **42**, 2462 (1965).
52. L. P. Pitayevski, *Soviet Phys. JETP* (*English Transl.*), **15**, 919 (1962).
53. S. A. Landon and J. C. Keck, *J. Chem. Phys.*, **48**, 374 (1968).
54. B. H. Mahan, *J. Chem. Phys.*, **48**, 2629 (1968).
55. D. R. Bates and R. J. Moffett, *Proc. Roy. Soc.* (*London*) *Ser. A*, **291**, 1 (1966).
56. D. R. Bates and M. R. Flannery, *Proc.. Roy. Soc.* (*London*) Ser. *A*, **302**, 367 (1968).
57. D. R. Bates and M. R. Flannery, *J. Phys. B*, **2**, 184 (1969).
58. J. C. Keck and G. Carrier, *J. Chem. Phys.*, **43**, 2284 (1968).
59. D. R. Bates and Z. Jundi, *J. Phys. B*, **1**, 1145 (1968).
60. M. R. Flannery, *Ann. Phys.* (*N.Y.*), **37B**, 67 (1971).
61. M. R. Flannery, *Phys. Rev. Letters*, **21**, 1729 (1968).
62. E. W. McDaniel, *Collision Phenomena in Ionized Gases*, Wiley, New York, 1964.
63. W. C. Lineberger and L. J. Puckett, *Phys. Rev.*, **186**, 116 (1969).
64. C. S. Weller and M. A. Biondi, *Phys. Rev.*, **172**, 198 (1968).
65. J. D. Payzant, R. Yamdagni, and P. Kebarle, *Can. J. Chem.*, **49**, 3308 (1971).
66. F. C. Fehsenfeld, E. E. Ferguson, and D. K. Bohme, *Planetary Space Sci.*, **17**, 1759 (1969).
67. Y. Yang and D. Conway, *J. Chem. Phys.*, **40**, 1729 (1964).
68. R. Voshall, J. Pack, and A. Phelps, *J. Chem. Phys.*, **43**, 1990 (1965).
69. D. C. Conway and L. E. Nesbit, *J. Chem. Phys.*, **48**, 509 (1968).
70. R. M. Noyes, *Progr. Reaction Kinetics*, **1**, 129 (1961).

VIBRATION→VIBRATION ENERGY TRANSFER

C. BRADLEY MOORE

Department of Chemistry, University of California, Berkeley, California

CONTENTS

I. INTRODUCTION

The transfer of energy among the vibrational, rotational, and translational degrees of freedom of a gas is a problem of great practical importance in the kinetics of chemically reacting systems, in shock wave and gas dynamic phenomena, in electrical discharges, and in numerous other non-equilibrium systems. Vibrational relaxation is also a challenging and interesting problem in molecular dynamics. Vibrational energy transfer fits into the spectrum of molecular collision phenomena between elastic and chemically reactive scattering. We are concerned with collisions in which changes in the internal quantum states of the molecules take place but in which no rearrangement of chemical bonds occurs. These changes are caused by the simultaneous dependence of the intermolecular potential upon the vibrational, rotational, and translational coordinates. The variety of intermolecular potentials and of molecular structure parameters which occurs in nature produces a wealth of observable phenomena. The problem of inelastic scattering presents interesting theoretical challenges beyond the elastic scattering problem but is not so difficult as reactive scattering. For the most part exact calculations are precluded by the number of quantum states available and by the paucity of intermolecular potential data. Thus much of our qualitative understanding of energy transfer mechanisms has been gained from empirical correlations of data combined with approximate calculations on simple model systems. During the past few years the combined effect of new experimental techniques, of theoretical advances, and of the need for data in the development of gas lasers has led to much new work on vibrational energy transfer.

The transfer of energy between vibration and translation ($V \rightarrow T$) has been studied extensively by ultrasonic, shock tube, and other gas dynamic techniques. The vast literature on this subject has been reviewed several times.[1–6] In these studies it became clear that transfer of energy among different vibrational degrees of freedom of a gas, $V \rightarrow V$ transfer, usually proceeded much more rapidly than did $V \rightarrow T$ energy transfer, but no rates for $V \rightarrow V$ processes had been measured before 1962. Since that time a number of spectroscopic experiments, often with lasers, and some ultrasonic and shock tube experiments have produced a great wealth of data. Early $V \rightarrow V$ work has been reviewed by two of the first investigators, Callear and Lambert.[1] Data on a great variety of $V \rightarrow V$ energy transfer processes are now available and a start has been made on a systematic qualitative understanding of many types of $V \rightarrow V$ transfers. The importance of long-range forces in near-resonant transfers has been demonstrated both theoretically and experimentally.

The study of vibrational relaxation has also been broadened beyond the original studies of $V \rightarrow T$ energy transfer by investigations of the role of the rotational degrees of freedom. In 1962 Cottrell and his co-workers[7] first suggested that vibrational energy might be transferred into rotational degrees of freedom. Since that time there have been two direct observations of the transfer of vibrational energy into rotation.[8,9] In addition many series of vibrational relaxation rates may be explained reasonably only by the coupling of vibrational energy into rotation.[10–19] Recent theoretical approaches[20,21] combined with an increasing body of experimental data may soon yield a convincing description of the distribution of transferred vibrational energy among the translational and rotational degrees of freedom.

This chapter is devoted primarily to a discussion of $V \rightarrow V$ energy transfers. Some newly developed experimental methods are first described. The deduction of microscopic rate constants from macroscopic relaxation rates is then discussed. The final section describes the types of V ⟶ V energy transfer processes, the data now available, and its interpretation.

II. EXPERIMENTAL METHODS

During the past decade new experimental tools have made the study of V ⟶ V energy transfer possible. Spectroscopic techniques for exciting vibrational states and for monitoring concentrations as a function of time have improved greatly. Experiments utilizing laser sources have opened many new areas.[22] Flash spectroscopic techniques developed in many directions at Cambridge[1,19,23] have produced much interesting V ⟶ V data. Sensitive, fast-response, infrared detectors have made it possible to monitor vibrational excitation in shock tubes.[24] Recently, Smith [25]has developed an infrared emission technique that allows the relaxation of the highly excited species produced in chemical reactions to be studied.

Some of the experimental methods that have long been used in V ⟶ T studies have been applied with imagination and care to yield interesting V ⟶ V energy transfer data. The spectrophone, after decades of nonsense, has finally produced reliable information.[26] Multiple dispersions have been observed for mixtures of gases in ultrasonic experiments.[1] The present pace of experimental development ensures that our knowledge of V ⟶ V energy transfer will be expanded to many new systems. At the same time it is likely that our experimental resolution will be improved to the point that the rotational quantum number dependence of some V ⟶ V cross sections may be studied. For some systems angular distributions for V ⟶ V scattering might be obtained from molecular beam experiments.

A. Laser Experiments

Since the first laser experiments in vibrational relaxation were reported in 1966,[27–29] many new energy transfer cross sections have been reported. The variety of methods and the number of molecules and sorts of energy transfer processes to which these methods may be applied continue to grow rapidly. Work published to date has employed fixed frequency laser sources that overlap molecular absorption transitions. In addition to the systems listed in Table I there are many others for which fortuitous overlaps between laser lines and vibrational absorption spectra exist.[22,30,31] Much more promising is the current development of tunable infrared sources.[32] The parametric oscillator[33] provides tunable laser light at frequencies higher than 2800 cm^{-1} and is thus suitable for exciting hydrogen stretching fundamentals and many combination and overtone vibrations. For high overtones dye lasers are a possibility.[34] The spin flip Raman laser may be useful in the 1900–1550 cm^{-1} range at low magnetic

TABLE I
Laser Measurements of Vibrational Relaxation Rates

Molecule (mode)	Excitation source	Detection	Processes studied	Ref.
H_2	Ruby + H_2 stokes	Refractive index	$V \rightarrow T$, R pure H_2	27, 38
HF, DF HCl, DCl HBr, DBr	HX or DX chemical laser	Fundamental emission	$V \rightarrow T$, R and $V \rightarrow V$ in many mixtures	44–46 17, 47–51 18, 52
$CH_4(\nu_3)$	He–Ne c.w. 3.39 μ	Emission 3 and 8 μ	$V \rightarrow V$ and $V \rightarrow T$, R	16, 29, 53, 54
$CO_2(00^01)$	CO_2 10.6 μ	Emission 4.3 μ	$V \rightarrow V$ in many mixtures	28, 49, 55–69
		Absorption of CO_2 laser	$V \rightarrow V$ pure CO_2	39
$N_2O(00^01)$	N_2O	Emission 4.5 μ	$V \rightarrow T$ in mixtures	49, 65, 70, 71
$SF_6(\nu_3)$	CO_2 10.6 μ	Emission 10 and 16 μ	$V \rightarrow V$ and $V \rightarrow T$ in pure gas and mixtures	43, 72
		Absorption of CO_2 laser		40
		Microphone		42
$C_2H_4(\nu_7)$	CO_2 10.6 μ	Emission 5 and 3.3 μ	$V \rightarrow V$ and $V \rightarrow T$, R	73
CH_3F	CO_2 9.6 μ	Emission 3, 5, and 7 μ	$V \rightarrow T$, R in mixtures	74
$CH_3Cl(\nu_6)$	CO_2 9.6 μ	Emission 3 μ and 13 μ	$V \rightarrow V$ and $V \rightarrow T$, R	75

fields[35] and at fields to 100 kG between 1900 and 400 cm^{-1}.[36,37] The excitation of Raman active transitions can be carried out for transitions at frequencies higher than 600 cm^{-1} by using a tunable dye laser along with a ruby as pump instead of relying on the stimulated Raman effect to generate the second frequency.[27,38] Thus we are rapidly reaching the point when it will be possible to laser-excite any desired vibrational transition.

Relatively few of the many conceivable techniques for monitoring the kinetics of vibrational relaxation in laser-excited gases have actually been used. Changes in the populations of vibrational levels are observed by infrared fluorescence[28–30] (intensity $\propto$ quanta of vibrational excitation) and by absorption of light from a probe laser[39,40] (intensity change ∝ difference in population between the upper and lower vibration–rotation levels whose frequencies overlap the laser transition). Infrared absorption monitored with a laser source operating at a molecular transition frequency can be a sensitive probe. It can allow individual rotational state populations to be obtained. Fluorescence emission is usually not sufficiently intense to be analyzed in a monochromator and detected. Changes in population induced by microwave pumping have also been detected by infrared laser absorption measurements.[41] Lasers operating on resolved rovibronic transitions could yield still better sensitivity, especially if the excited electronic state were fluorescent. It has been even feasible to use Raman scattering to monitor population versus time in a system with exceptionally high excitation densities.[27] The pressure change caused by the release of energy to translation has been monitored by phase contrast detection of the change in refractive index[38] and by sensitive pressure transducers (microphones).[42] Translational temperature changes are also observed through infrared emission.[29,43,51] The discussion of a few specific experiments, below, is intended to illustrate some of the promise and pitfalls of laser methods. It must be remarked that laser experiments suffer equally from the usual problems of trace impurities, absorption of samples on vessel walls, etc., which have always plagued the field of vibrational relaxation.

1. Laser-Excited Vibrational Fluorescence in Hydrogen Halides

Hydrogen halide pulsed chemical sources operating on the $v = 1$ to $v = 0$ transition are now available for HF, DF, HCl, DCl, HBr, and DBr.[32] Thus each of these molecules is resonantly excited from the ground vibrational level to the first vibrationally excited state. A typical experimental apparatus is shown in Fig. 1. The recently developed transverse electrical discharge chemical lasers are the best and simplest sources now available.[76] At modest laser powers and pressures the density of excited molecules is given by the product of the laser energy per cm^2 per pulse and the absorption coefficient at line center for the laser transition. At sufficiently high laser powers or low pressures the sample absorption may become saturated. This occurs whenever the rate at which molecules are pumped from the absorbing rotational level of the lower vibrational state becomes comparable to the rotational relaxation rate. Excess population of the rotational level excited in the upper vibrational state may likewise diminish the amount of absorption. For experiments in which the duration of the laser pulse is less than the time between collisions, lasers of quite modest pulse energy are sufficient to excite all of the molecules which can absorb.

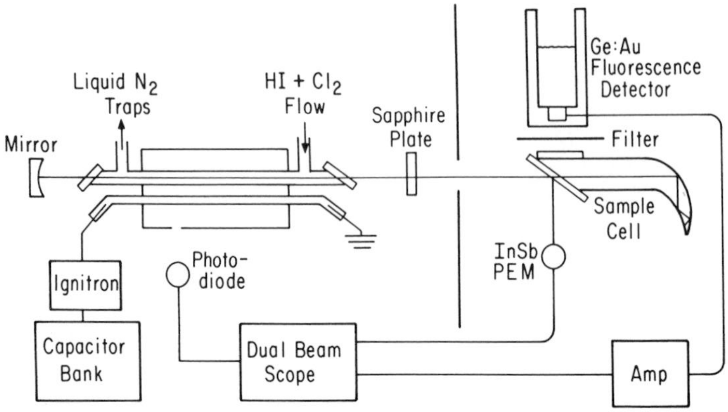

Fig. 1. HCl laser-excited vibrational fluorescence apparatus of Chen and Moore. Laser pulses are of 10^{-5} sec duration and 10^{-2} J energy. A sample cell for DCl fluorescence detection is shown. Fluorescence is observed through interference filters with 30 to 100 cm^{-1} transmission bands. A detailed description is given in Ref. 17.

Ideally, when one observes the vibrational fluorescence following laser-excitation of a hydrogen halide gas, one finds a single exponential decay for which the product of lifetime and pressure is constant over a wide pressure range. Deviations from this behavior often arise from heating of the sample as energy is transferred from vibration into translation and rotation. Samples may be heated easily by several hundred degrees. The effect on observed decay rates can be substantial.* If the laser is saturating the molecular absorption, a decrease in laser power or an increase in sample pressure may have relatively little effect on the temperature rise. It is often necessary to buffer the hydrogen halide with a second gas, such as argon, which has little or no effect on the relaxation rates but decreases heating by adding translational heat capacity and by lowering the absorption coefficient through pressure broadening. The observation of non-exponential decays could be interesting if rotational relaxation were not very fast compared to the time scale of observation for vibrational relaxation. Unfortunately, such work has not yet been carried out. Hopefully it will become possible to observe the dependence of vibrational energy transfer rates on rotational quantum number by studying very fast $V \rightarrow V$ exchanges and by observing vibrational deactivation during the first few collisions. At sufficiently low pressures the pressure–relaxation time product decreases due to vibrational deactivation by fluorescence emission and by diffusion to and deactivation at the cell walls. This situation was first

* See Section III.A for a more detailed discussion of heating effects.

studied by Kovacs and Javan[58] for CO_2 and later by Henry and co-workers[49] for HCl, CO_2, and N_2O. The cross sections for diffusion of vibrational excitation and the probabilities for vibrational deactivation on surfaces have been reported for these molecules.

Laser-excited vibrational fluorescence has been used to measure rates of vibrational deactivation in pure hydrogen halides and in mixtures of hydrogen halides with other gases.[17,18,44–50] Craig and Moore have studied vibrational relaxation of HCl by Cl atoms[50] by photolytically producing Cl atoms (a 20 nsec duration pulse of double ruby, 3472 Å) shortly after the infrared excitation pulse. V → V energy transfer from hydrogen halides to a variety of diatomic and simple polyatomic molecules has been studied. Fig. 2 illustrates an experiment in which HCl is laser-excited. The rapid V → V transfer of excitation from HCl to DCl is observed via the decay of fluorescence from HCl or the rise of fluorescence from DCl. By studying the decay of DCl fluorescence the vibrational deactivation of DCl may be studied. Thus through V → V transfer it is possible to study the relaxation of molecules which cannot be laser-excited directly. In this way Zittel and Moore[51] have studied vibrational relaxation of CH_4 and CD_4 by a variety of collision partners. In mixtures of HCl–CH_4–M or DCl–CD_4–M the hydrogen chloride was laser-excited. V → V energy transfer to the methane stretches followed by transfer to the methane bends is rapid. The decay of fluorescence from the methane bending vibrations then gives the CH_4–M relaxation rate.

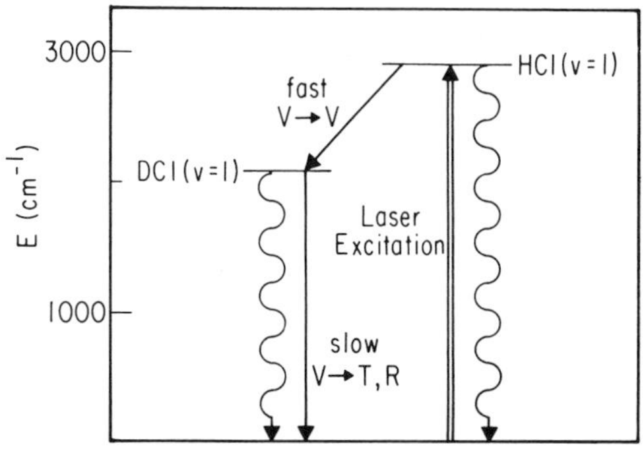

Fig. 2. Kinetic diagram for laser-excited vibrational fluorescence in HCl–DCl mixtures. HCl is vibrationally excited by a laser pulse. Vibrational energy is transferred rapidly from HCl to DCl by V → V energy transfer. Subsequently, DCl decays more slowly by V → T,R energy transfer. The concentrations of excited species are proportional to the observed fluorescence intensity, but the optical transition rates make a negligible contribution to the observed kinetics.

2. *Infrared Double Resonance in CO_2*

Laser-excited vibrational fluorescence studies of the relaxation of the asymmetric stretching vibration of CO_2 have been made in pure CO_2 and in many mixtures of CO_2 with other gases.[28,49,55–69] Ultrasonic studies of the relaxation of the bending mode of CO_2 have been carried out in many systems as well.[2,3,77] Data as a function of temperature are available from shock tubes,[3,77–79] laser,[59–64,67–69] and ultrasonics.[3,4] The kinetics of the energy levels of CO_2 between the asymmetric stretch (00^01) at 2349 cm^{-1} and the bend (01^10) at 667 cm^{-1} has not been experimentally studied in such detail.

To examine the kinetics of energy transfer among the levels of the symmetric stretch and bend vibrations (Fig. 3), Rhodes, Kelly, and Javan[39] carried out an experiment using one laser source to create population changes and a second to monitor level populations through transient absorption. A 9.6 μ pulsed CO_2 laser pumped molecules from the (02^00) level to the (00^01) level. The intensity was sufficient to equalize the populations of these two levels. A second CO_2 laser operating at 10.6 μ on the (10^00)–(00^01) transition probes absorption of the sample with a power level too low to affect the level populations. During the pump laser pulse, the transmission of the probe increased due to population of the upper laser state (00^01). Following the pulse, a further increase in transmission occur-

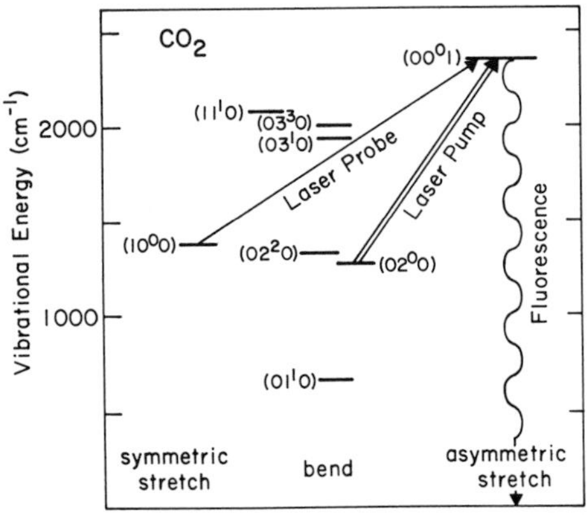

Fig. 3. Infrared laser "double resonance" experiment for CO_2. The sample is strongly excited at 9.6 μ and weakly probed at 10.6 μ. This technique allows the rapid V → V energy transfers coupling the (10^00), (02^20), (02^00), and (01^10) levels to be studied.[39] Rotational relaxations can also be observed. The levels labeled (10^00) and (02^00) are completely mixed by Fermi resonance. The wave function for each level is about half symmetric stretch and half bending overtone.

red with a rise time which was barely resolved with a 1 μsec per channel signal averager at pressures of a few tenths of a torr. Next, transmission of the probe decreased with a relaxation rate observed between 0.1 and 1.6 torr to be $(4 \pm 1) \times 10^5$ sec^{-1} torr^{-1}. Finally, on a much longer time scale, known from fluorescence measurements, the probe transmission decreased further as the (00^01) state was depopulated. A lower limit on the rate of the first rapid relaxation is set at 10^6 sec^{-1} torr^{-1}, or five kinetic collisions. This rate is assigned to the transfer of molecules from the (10^00) level (lower level of the probe laser) to the (02^00) level (depleted by the pump laser)

$$CO_2(10^00) + CO_2(00^00) \rightarrow CO_2(02^00) + CO_2(00^00) + \Delta E = 103 \text{ cm}^{-1} \tag{2.1}$$

The rate of 4×10^5 sec^{-1} torr^{-1} was assigned to the sum of processes such as

$$2CO_2(01^10) \rightarrow CO_2(10^00) + CO_2(00^00) + \Delta E = -54 \text{ cm}^{-1} \tag{2.2}$$

$$\rightarrow CO_2(02^00) + CO_2(00^00) + \Delta E = 50 \text{ cm}^{-1} \tag{2.3}$$

$$\rightarrow CO_2(02^20) + CO_2(00^00) + \Delta E = -1 \text{ cm}^{-1} \tag{2.4}$$

which couple the three levels near 1300 cm^{-1} to the (01^10) level. Sharma argues that the third of these processes, the most closely resonant, has a rate greater than 10^6 sec^{-1} torr^{-1} and that the rate limiting process should be

$$CO_2(02^20) + CO_2(00^00) \rightarrow CO_2(10^00) + CO_2(00^00) \tag{2.5}$$

This proposal is consistent with the data but rests on the assumption that long range forces are dominant in (2.2)–(2.4).

The double resonance method has been applied to "rotational" relaxation in the asymmetric stretching level of CO_2.[80] The sensitivity to observe such fast processes was achieved by using short pumping pulses (20 nsec) and by using a dc electrical discharge to create a large population of the $CO_2(00^01)$ level. (The experiment of Rhodes, Kelly, and Javan[39] on room temperature equilibrium gas was a real tour de force.) Cheo and Abrams[80] found that the cross section for rotational relaxation was about twice the kinetic cross section and that changes in J from -2 to -18 were about equally probable (positive changes in J were not studied). It seems very likely that the *vibrational* transfer process, in which a quantum of asymmetric stretching energy is transferred resonantly fron one CO_2 molecule to another, is largely responsible for these observations.

$$CO_2\,(00^01, J = 19) + CO_2(00^00, J_i) \rightarrow CO_2(00^00, J = 19 \pm 1) + CO_2(00^01, J_i \pm 1) \tag{2.6}$$

The observation of comparable rates for CO_2–N_2 and CO_2–He collisions[81] unsettles this conclusion a bit. Unfortunately, changes in *J* were not resolved in the He and N_2 work. This resonant dipole–dipole process is expected to have a cross section greater than kinetic[57,58,82] and to be responsible for the observation that the diffusion cross section for $CO_2(00^01)$ is some 80% larger than that for purely molecular diffusion of ground state CO_2 molecules.[58]

Recently an alternative mechanism for the results of Rhodes, Kelly, and Javan has been proposed.[83] It is suggested that the rate $k = 4 \times 10^5$ sec^{-1} torr^{-1} arises from rotational relaxation in the (00^01) level. However, Cheo and Abrams [80] show definitively that this rate is 10^7 sec^{-1} torr^{-1}. Seeber[83] suggests that the fast process ($k > 10^6$ sec^{-1} torr^{-1}) actually occurs during the laser pulse. The implication is that Rhodes, Kelly, and Javan are mistaken in their explicit statement that this relaxation occurs after the laser pulse. It is stated that a two-photon process which occurs during the laser pulse is responsible for the observation. The presence of such a process can alter the saturation limit at the probe wavelength but should have no effect on the absorption of a weak probe under the experimental conditions. It is not surprising that the SSH calculations of Seeber for the V $\rightarrow$ V processes coupling the symmetric stretching and bending levels badly underestimate the observed cross sections. The data to which these calculated cross sections compare favorably are from experiments that give cross sections more closely related to the V $\rightarrow$ T deactivation of (01^10) or to V $\rightarrow$ V energy transfer between $CO_2(00^01)$ and $N_2(v = 1)$.

In principle a great deal more may be learned about CO_2 through double resonance experiments. The rate which couples (02^00) to (10^00) could actually be measured. Considerably more detail on rotational relaxation and on the rotational quantum number dependence of some of the vibrational relaxation cross sections could be obtained. Since the laser sources are polarized, changes in angular momentum direction could be observed as well. For these experiments to be possible without an electrical discharge in the sample, improvements in signal to noise of one or two orders of magnitude above Rhodes, Kelly, and Javan[39] would be required. This might be accomplished by heating the sample to increase the thermal (02^00) population, by constructing an extremely stable probe laser, and by improving the time resolution.

The infrared double resonance method has been applied to vibrational relaxation in SF_6 by Steinfeld and co-workers.[40] The results of these studies have already been well summarized by him in a recent review.[84] Absorption of infrared laser light whose frequency matches the center of a vibration–rotation spectral line is a particularly sensitive detection technique for short time resolution and long path lengths. It makes possible the resolution of rotational levels where this would be hopeless in fluorescence detection. The development of practical, tunable, high-resolution infrared laser sources may make the infrared double resonance technique one of considerable general value.

3. Stimulated Raman Excitation of H_2

The generation of Stokes Raman laser radiation by a high-power laser simultaneously creates a high density of vibrationally excited molecules. Ducuing and DeMartini[27] produced vibrationally excited H_2 at the focus of a ruby laser in high-pressure H_2 gas. They probed the concentration of excited molecules as a function of time by observing the spontaneous Raman scattering of a weak probe laser at right angles to the excitation. Ducuing, Joffrin, and Coffinet[38] have recently improved the sensitivity and flexibility of this method (Fig. 4). Rather than study energy transfer

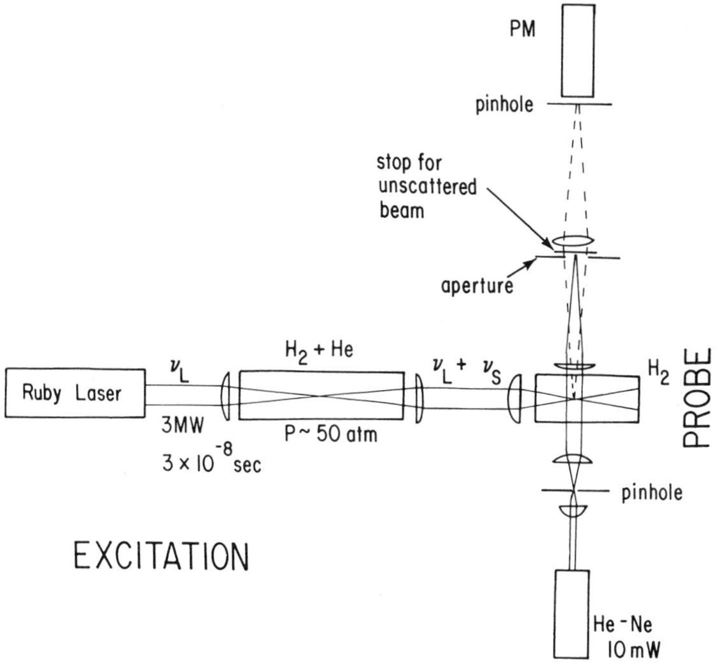

Fig. 4. Stimulated Raman excitation of hydrogen.[38] V → T energy transfer is studied by monitoring the refractive index changes as vibrational excitation is transferred to translation. The operating pressure range in the experimental cell ranges from 1 to 50 atm.

kinetics in the cell where stimulated Raman is produced (this is possible only at rather high pressures and excitation densities), the stimulated Stokes Raman line from a cell optimized for Raman generation was refocused along with the ruby laser into a second cell. In this second cell the Raman transition is stimulated by the laser and Stokes radiations. The density of excited molecules produced is proportional to the laser intensity, the Stokes intensity, and the pressure in the cell. A large fraction of the molecules at the focus of the beams may be excited. Vibrational relaxation is detected in this system by monitoring the light scattered by the change in refractive index of the focal zone as it expands due to the increase in translational temperature which in turn results from the transfer of vibrational energy into translation. This detection technique is orders of magnitude more sensitive than the spontaneous Raman scattering used in the earlier work.

Although the only result published to date is the relaxation time of pure H_2 at room temperature, the method promises to be powerful and useful for many molecules.[85] Stimulated Raman scattering has been produced in many simple molecules using picosecond laser pulses.[85,86] It has also been suggested that the Stokes beam be replaced by a tunable dye laser.[38] In this way any Raman active transition could be excited, rather than only those for which stimulated scattering could be produced. For excitation in gases this would best be done with a highly monochromatic laser; in condensed phases standard dye lasers would be excellent. Resonant Raman scattering with a dye laser source as probe could prove useful in monitoring level populations in some experiments. In many systems infrared detection may also be useful. For infrared inactive transitions and for transitions outside the tuning ranges of infrared lasers, stimulated Raman excitation and spontaneous Raman detection may become important experimental methods.

III. RELAXATION MODELS

In most vibrational energy transfer experiments a system of molecules at thermal equilibrium is subjected to a perturbation and the return of the system to equilibrium is monitored by a detector sensitive to some property of the system. In order to deduce the bimolecular rate constants for vibrational energy transfer processes four things must be known: (1) the energy levels or reservoirs of the system; (2) the changes in level populations of shifts of energy content from equilibrium caused by the perturbation; (3) the kinetic processes by which molecules change energy level or by which energy flows from one reservoir to another; and (4) the response of the detector to the population in each level or to the energy content of

B. Distributions for Relaxing Anharmonic Oscillators

The distribution of molecules among the highly excited vibrational levels of a relaxing system which is vibrationally hot and translationally cold has been the subject of active experimental and theoretical investigation over the past few years. These studies are of great practical interest now for chemical and molecular lasers. Much of the original motivation for this work came from the curious result that vibrational relaxation of a gas upon cooling from a high temperature appeared faster than would be predicted from the application of microscopic reversibility to rates measured in a standard shock tube experiment.

For a gas containing a single species of diatomic harmonic oscillators[88] the V → V energy transfer processes

$$\mathrm{AB}(v) + \mathrm{AB}(v') \leftrightarrows \mathrm{AB}(v \pm 1) + \mathrm{AB}(v' \mp 1) + \Delta E = 0 \quad (3.1)$$

are exactly resonant and serve to maintain a "Boltzmann" distribution among the vibrational levels at a vibrational temperature, T_v, given by

$$\frac{n_{v+1}}{n_v} = \exp\left(\frac{-h\nu_{\mathrm{vib}}}{kT_v}\right) \quad (3.2)$$

When the vibrational energy spacings of the colliding molecules are different, as with a mixture of two different molecules or a pure gas with anharmonic levels, then the exchange is no longer resonant:

$$\mathrm{AB}(v) + \mathrm{CD}(v') \rightarrow \mathrm{AB}(v + 1) + \mathrm{CD}(v' - 1) + \Delta E_{v,v'} \neq 0 \quad (3.3)$$

Whereas the equilibrium constant for reaction (3.1) is unity and gives distribution (3.2), the equilibrium constant for (3.3) is exp $(\Delta E_{v,v'}/kT)$. Here $\Delta E_{v,v'}$ is the energy released to translation and rotation and T is the translational (and rotational) temperature. In this way the quanta of vibrational excitation are shifted to the molecules with smaller energy level differences (i.e., highly excited levels of an anharmonic oscillator or the molecule of lower frequency in mixtures). Thus if V → V processes come to equilibrium in a system with a substantial excess of vibrational energy (low translational temperature), the vibrational temperature as defined by (3.2) increases as vibrational quantum number increases for anharmonic oscillators. Level populations may even be inverted, $n_{v+1} > n_v$. For a mixture of two diatomics the one of lower vibrational frequency will have a higher vibrational temperature. Treanor, Rich and Rehm[89] have given an expression for the quasiequilibrium distributions reached when V → V equilibration is rapid compared to V → T transfer (i.e., distributions for a fixed number of vibrational quanta at fixed translational temperature).

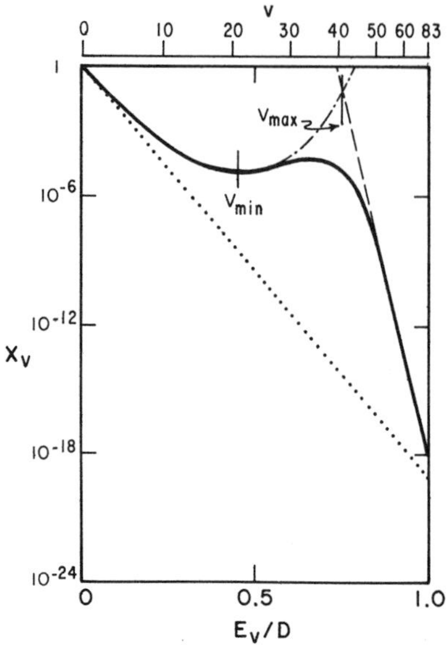

Fig. 6. Vibrational distributions for anharmonic oscillators. X_v is population of level v relative to level $v = 0$. The ratio of $v = 1$ to $v = 0$ populations defines a vibrational temperature of 3000°K. · · · ·, $T_{translation} = 3000$°K. ·—·—, $T_{translation} = 800$°K, no V → T transfer, Treanor distribution. ——, $T_{translation} = 800$°K, V → T transfer included, Brey. – – –, with slope $T_{vibration} = 800$°K. For very high v levels, V → T transfer is faster than V → V and the vibrational distribution is tied to the translational temperature. Curves are from Bray.[90]

Figure 6 shows such a distribution (—·—·) for CO with a translational temperature of 800°K and the ratio of the $v = 0$ and $v = 1$ populations defining a vibrational temperature for that pair of levels as 3000°K. For comparison the Boltzmann distribution for the anharmonic oscillator, CO, at thermal equilibrium at $T = 3000$°K is also shown (· · · · ·). The large populations of high v levels are readily evident at the low translational temperature.

Bray[90] has shown that the steep increase in population for the high v levels in the Treanor distribution is limited by V → T energy transfers. At high v the energy discrepancies for V → V transfers become large and the transfer probabilities decrease correspondingly. On the other hand V → T transfers increase in rate as the size of the quantum decreases and at sufficiently high v they can limit the populations. In the limit or rapid V → T and slow V → V processes the distribution at high v will conform to a Boltzmann distribution at the translational temperature.[90] The calculations of Bray[90] are shown in Fig. 6 for a reasonable set of V → V and V → T energy transfer probabilities.

Calculations have also been done[90] appropriate to the nozzle expansion system, in which the translational temperature decreases as a function of time. There are two phenomena which produce a faster relaxation in this system than for harmonic oscillators. First, the overpopulation of levels with low vibrational frequencies leads to a faster conversion of vibrational quanta to translation. Second, as the translational temperature drops the V → V transfers shift quanta from low v to high v. This lowers the total vibrational energy a bit since the quanta are getting smaller, and it decreases substantially the vibrational temperature as measured by the $v = 1$ to $v = 0$ population ratio.

The most complete experimental illustrations of these effects are the distributions of vibrational energy observed in CO lasers and CO–N_2 mixtures.[91–93] An immediate explanation is provided for the fact that CO lasers operate on high v transitions but not on low v transitions. Curves qualitatively similar to the solid line in Fig. 6 have been observed experimentally and can be matched with curves calculated from reasonable sets of vibrational relaxation rates.[94,95] In CO at pressures below 1 torr spontaneous radiative decay may limit the populations of high v levels.[94] A crucial element in these calculations has been the experimental results of Hancock and Smith[96] on V → V transfer in highly excited states. The basic principles have also been demonstrated in an expanding flow of an N_2–CO mixture by Teare, Taylor, and von Rosenberg.[97] The vibrational temperature of CO ($\nu = 2143$ cm^{-1}) actually increased as the translational temperature decreased upon expansion with N_2 ($\nu = 2331$ cm^{-1}).

The unusually rapid relaxations observed in expansion nozzles have found their explanations only partly with the considerations above.[98] These experiments have been carried out by shock heating a gas to a high stagnation temperature and then allowing it to flow through a nozzle orfice and cool upon expanding. While the gas remains at the stagnation temperature, impurity molecules can be dissociated into reactive fragments which are highly efficient at vibrational deactivation. von Rosenberg, Taylor, and Teare[99] demonstrated this effect and measured the efficiency of deactivation of CO by H atoms. At the lowest attainable impurity levels they found an apparent acceleration of the relaxation rate of a factor of five. This is within the range that can be calculated from reasonable sets of relaxation rates for the individual vibrational levels.[90] McLaren and Appleton have found virtually no acceleration for CO at 0.5% in Ar.[98] Similar results have been found by Just and Roth.[100]

The striking effects of anharmonicity and of energy pumping between diatomics of slightly different frequencies have already explained several interesting experimental phenomena. The effects are the basis of some methods of obtaining laser gain.[97,101] These phenomena may be expected to play an important role in other systems as well. For example, dissociation rates in gases which are vibrationally heated by a laser should be strongly influenced by V → V coupling rates.[102]

C. Rate Constants from Data on Multilevel Systems

Relaxation rate data for gases with two or more vibrational degrees of freedom are often insufficient to independently determine the rate constants for each of the significant energy transfer processes. In systems with a modest number of levels, algebraic equations giving the observed relaxation rates in terms of the rate constants for each of the possible energy transfer processes may be written down. Usually the number of unknown rate constants is greater than the number of equations. In this unsatisfying situation assumptions are often made, explicitly or implicitly, about the relative rates of the various relaxation processes. Often some or all of the $V \rightarrow V$ rates may be assumed to be very fast by comparison to the $V \rightarrow T,R$ rates or some $V \rightarrow T,R$ processes may be neglected since they involve much larger energy transfers than others. These assumptions are often based on observed rates for processes which appear to be similar. Occasionally these assumptions are quite unjustifiable and lead to false conclusions. If a valid set of assumptions is made, an accurate simplified set of equations is obtained. In the application of these equations to the experimental data, it is imperative to explore all sets of rate constants which satisfactorily fit the data within the experimental uncertainty. In evaluating published data a useful but not sufficient test is to discover whether there are more rate constant values given than independent pieces of experimental data determined. Some of the difficulties in these analyses are illustrated below for the data on CH_4–H_2O mixtures.

The methane molecule has four modes of vibration and the water molecule three. For ultrasonic measurements at room temperature the stretching vibrations make a negligible contribution to the heat capacity. Likewise, combination and overtone levels of the bending vibrations are unimportant. For the low levels of excitation involved the considerations of anharmonicity introduced in the previous section are not a problem.

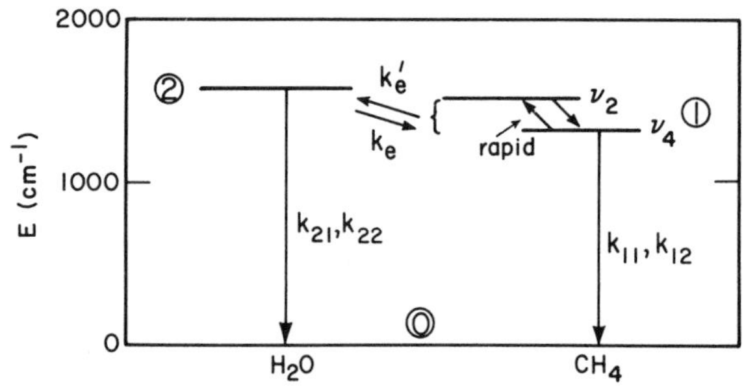

Fig. 7. Energy levels for ultrasonic experiments on CH_4–H_2O mixtures (see text).

Thus in CH_4–H_2O mixtures we are concerned with the three excited vibrational levels shown in Fig. 7. The further simplifying assumption is made that the two methane levels are in rapid equilibrium during the relaxation process. It is known[54] that energy transfer between these two levels requires less than 100 methane–methane collisions, but even this limit leaves the hypothesis open to suspicion. By treating the two methane levels as a single reservoir of vibrational excitation, the problem is reduced to the five processes (3.4)–(3.8).

$$CH_4^* + CH_4 \xrightarrow{k_{11}} CH_4 + CH_4 \tag{3.4}$$

$$CH_4^* + H_2O \xrightarrow{k_{12}} CH_4 + H_2O \tag{3.5}$$

$$H_2O^* + CH_4 \xrightarrow{k_{21}} H_2O + CH_4 \tag{3.6}$$

$$H_2O^* + H_2O \xrightarrow{k_{22}} H_2O + H_2O \tag{3.7}$$

$$CH_4^* + H_2O \underset{k_e'}{\overset{k_e}{\rightleftarrows}} CH_4 + H_2O^* \tag{3.8}$$

The reverse directions of the first four reactions are neglected since the vibrational quanta are six to eight times kT. Detailed balancing tells us that (k_e/k_e') is given by the ratio of the number of excited H_2O molecules to the number of excited CH_4 molecules at thermodynamic equilibrium. This ratio is 0.07 when one takes account of the threefold degeneracy of the 1306 cm^{-1} mode and the twofold degeneracy of the 1533 cm^{-1} mode. The experimental data on CH_4–H_2O mixtures from pure CH_4 to CH_4 with 30 mole-percent H_2O is fit by the expression[103]

$$(p\tau)^{-1} = (0.307 + 15.0X_{H_2O} + 98X_{K_2O}{}^2) \times 10^6 \text{ sec}^{-1} \text{ atm}^{-1} \tag{3.9}$$

Three independent numbers are experimentally measured, and there are five unknown rate constants. Many different sets of rate constants will be consistent with the quardratic form of the rate and the three measured parameters. Monkewicz[103] presents an algebraic solution of the problem which gives "unique" values for each of the five rate constants. By assuming $k_{12} = 0$ (or more precisely $k_{12} \ll .07\, k_{21}$) and by assuming that extension of (3.9) to a cubic gave exactly zero as the coefficient, Bauer[104] was able to derive four rate constants. A plausible model which illustrates the limitations placed on valid interpretations of this data is that the V → V rates are much faster than any of the V → T rates. In this situation no knowledge of the V → V rate can be obtained except that it must be larger than some lower limit set by the V → T rates. In this approximation the relaxation rates are given by

$$(p\tau)^{-1} = [X_{CH_4} + 0.07X_{H_2O}]^{-1} \\ [X_{CH_4}{}^2k_{11} + X_{CH_4}X_{H_2O}(k_{12} + 0.07\, k_{21}) + 0.07\, X_{H_2O}{}^2\, k_{22}] \tag{3.10}$$

Comparing with (3.9) we see that the quadratic term comes primarily from k_{22} and the constant term entirely from k_{11}. These conclusions follow for almost any imaginable set of rate constants. In the fast V → V approximation the linear term allows the sum $[k_{12} + (k_e/k_e')k_{21}] = [k_{12} + 0.07\ k_{21}]$ to be determined. Although this set of constants substituted into (3.10) does not precisely match (3.9), it does match the original data. There is no compelling reason to believe that the rate constants derived from this model are the true set. Consideration of the model does illustrate that in the usual situation where V → V rates are faster than V → T rates it is almost impossible to determine the magnitude of the V → V rates from acoustic data. It is similarly difficult to independently determine the rates k_{12} and k_{21} for deactivation in collisions between type 1 and type 2 molecules. Unfortunately, these simple physical facts may quickly become lost in complicated, nearly singular algebraic equations. Detailed treatments of specific multilevel problems may be found in Refs. 26, 53, 54, 67, 78, 105, and 106.

D. Interpretation of Spectrophone Data

In the spectrophone method of studying vibrational relaxation a single vibrational mode is optically excited and the kinetics of transfer of energy from this mode into translation is monitored with a microphone. In contrast to purely acoustic methods, such as shock waves and ultrasonics, the spectrophone allows vibrational levels of low heat capacity to be studied and gives some information on the kinetics of V → V transfers when high frequency modes are excited. The method also gives information which often cannot be obtained in purely optical experiments. Since energy liberated into translation is detected, the energy discrepancy in a V → V process and thus the V → V mechanism in a polyatomic gas or gas mixture may be studied.

Until recently the spectrophone had not yielded any reliable energy transfer rates. Read[107] reviewed progress in the field from 1880 to 1968. Systematic errors in the experimental method (spurious apparatus phase shifts), impurities in samples, and incorrect data analysis of results have been constant plagues. Experiments on the hydrogen halides[107] gave relaxation times several orders of magnitude too long.[17,18] Recent data on CO_2,[108,109] on N_2O,[108] and on mixtures of CO_2 or N_2O with other gases[110,111] appear to be correct. These data, when analyzed using a sufficiently detailed kinetic model, are in agreement with laser and ultrasonic relaxation measurements and give new information on V → V energy transfer mechanisms in these mixtures.[26]

The possibility of obtaining new information by combining the results of purely optical and purely acoustic experiments with those of spectro-

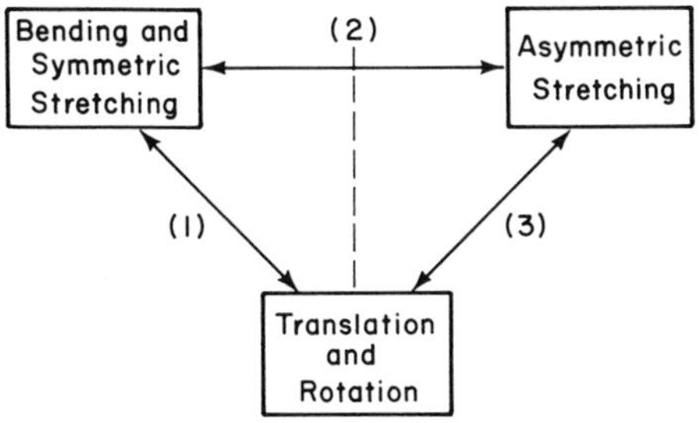

Fig. 8. Energy reservoirs for analysis of vibrational relaxation in CO_2. In V → V transfer from asymmetric stretching to the combined bending–symmetric stretching manifolds a fraction of the vibrational energy, *G*, is transferred to translation and rotation. The bulk of the V → T,R transfer occurs through (1) since (3) involves much larger vibrational quanta. See text and Ref. 26.

phone experiments appears to have gone unnoticed before the work of Huetz-Aubert and collaborators.[112] A treatment of CO_2 sufficiently complete for accurate treatment of results from room temperature to a few hundred degrees Centigrade has been published by Tripodi and Vincenti,[113] and is generally accepted.[114] The model for CO_2 treated by these authors is illustrated in Fig. 8. Three reservoirs of energy are considered: the asymmetric stretching energy levels; the combination of the bending and symmetric stretching levels which is taken to be a single reservoir with rapid equilibration among its entire manifold of levels; and the combination of rotational and translational degrees of freedom. Thus, in accord with rate data from laser experiments,[39] it is assumed that excitation transferred to the bending and symmetric stretching manifold rapidly equilibrates with the lowest excited bending vibrational level through proceeses such as

$$CO_2(11^10) + CO_2(00^00) \rightarrow CO_2(10^00) + CO_2(01^10) \qquad (3.11)$$

$$CO_2(10^00) + CO_2(00^00) \rightarrow 2CO_2(01^10) \qquad (3.12)$$

$$CO_2(02^00) + CO_2(00^00) \rightarrow 2CO_2(01^10), \text{ etc.} \qquad (3.13)$$

Transfer from the asymmetric stretch via processes such as

$$CO_2(00^01) + CO_2(00^00) \begin{cases} \rightarrow CO_2(10^00) + CO_2(01^10) & (3.14) \\ \rightarrow CO_2(11^10) + CO_2 & (3.15) \\ \rightarrow CO_2(03^30) + CO_2 & (3.16) \end{cases}$$

would give as a net result three molecules in the (01^10) level. Transfer to other levels can lead to one, two, three, or four quanta of bending excitation for each quantum of asymmetric stretching energy transferred. The amount of energy transferred into translation in the V → V process is then the difference between the energy of an asymmetric stretching quantum and that of the multiple of bending quanta excited. If the fraction of energy transferred into translation is denoted by G, then the spectrophone relaxation time is given by[113]

$$\tau_{sp} = (G - 0.28)\tau_{V\to T}' + \tau_{V\to V} \tag{3.17}$$

where $\tau_{V\to T}$ and $\tau_{V\to V}$ are the relaxation times measured in ultrasonic and laser fluorescence experiments. The value of G determined by correct analysis of the experimental data is between 0.75 and 0.90. If three quanta of the bend are always excited, then $G = 0.85$. Thus processes such as those in (3.14)–(3.16) are dominant in the relaxation of $CO_2(00^01)$. This method of analysis has been applied to CO_2–N_2 and to CO_2–rare gas mixtures as well.[26] Although the experimental data available are not sufficiently accurate to determine good values for G in these experiments, it has been shown[26] that the apparent inconsistencies between spectrophone results[111] and the combination of ultrasonic and laser data[56] were the result of an incorrect analysis of the spectrophone data. Accurate spectrophone data, carefully analyzed and combined with accurate acoustical and optical data, can yield significant new information on V → V energy transfer mechanisms.

IV. VIBRATION → VIBRATION ENERGY TRANSFER

A. Vibration → Vibration Energy Transfer Processes

Historically, V → V energy transfer processes have been very difficult if not impossible to study. Since these processes are usually much faster than V → T,R energy transfer processes, their influence on the translational degree of freedom is small. Only in a few special and carefully studied systems has it been possible to deduce accurate V → V energy transfer rates from ultrasonic measurements.[1,115] Vibrational excitation in shock tubes monitored by infrared fluorescence has provided good V → V rates on several systems.[24,79,105,116,117] The specificity of flash photolysis and especially laser techniques has made these methods the most generally useful and powerful ones. Since the work reviewed by Callear and Lambert in 1969[1] many new sorts of V → V energy transfer processes have been observed and a great deal of new information is available.

A sampling of the types of processes which are responsible for energy transfer among the vibrational degrees of freedom of a gas are as follows:

I. Intramolecular single-quantum V → V transfer

$$CH_4(\text{sym. bend}) + M \rightarrow CH_4(\text{asym. bend}) + M + \Delta E = 227\ \text{cm}^{-1} \quad (4.1)$$

II. Intermolecular single-quantum V → V transfer

Fundamental levels

$$HCl(v = 1) + HBr(v = 0) \rightarrow HCl(v = 0) + HBr(v = 1) + \Delta E = 327\ \text{cm}^{-1} \quad (4.2)$$

$$CO_2(00^01) + N_2(v = 0) \rightarrow CO_2(00^00) + N_2(v = 1) + \Delta E = 19\ \text{cm}^{-1} \quad (4.3)$$

$$CO_2(00^01) + CO_2(00^00) \rightarrow CO_2(00^00) + CO_2(00^01) + \Delta E = 0 \quad (4.4)$$

Combinations and overtones

$$HCl(v = 2) + HCl(v = 0) \rightarrow HCl(v = 1) + HCl(v = 1) + \Delta E = -103\ \text{cm}^{-1} \quad (4.5)$$

$$CO_2(10^01) + CO_2(00^00) \rightarrow CO_2(10^00) + CO_2(00^01) + \Delta E = -22\ \text{cm}^{-1} \quad (4.6)$$

III. Intramolecular multiquantum V → V transfer

$$CO_2(00^01) + M \rightarrow CO_2(11^10) + M + \Delta E = 272\ \text{cm}^{-1} \quad (4.7)$$

IV. Intermolecular multiquantum V → V transfer

$$HCl(v = 1) + {}^{18}O_2(v = 0) \rightarrow HCl(v = 0) + {}^{18}O_2(v = 2) + \Delta E = -24\ \text{cm}^{-1} \quad (4.8)$$

$$CO(v = 1) + CO_2(00^00) \rightarrow CO(v = 0) + CO_2(11^10) + \Delta E = 65\ \text{cm}^{-1} \quad (4.9)$$

V. Intermolecular V → V energy sharing

$$CO_2(00^01) + SF_6(\nu_i = 0) \rightarrow CO_2(10^00) + SF_6(\nu_3 = 1) + \Delta E = 13\ \text{cm}^{-1} \quad (4.10)$$

The intermolecular forces which cause V → V transfers vary widely according to the type of process, the type of molecules involved, and structural parameters such as vibrational energy discrepancy, moments of inertia, reduced masses, and transition dipole moments. For a particular type of process it is possible to make empirical correlations of V → V cross sections for families of similar molecules. However, no generally satisfactory empirical correlations have been made which include most of the data now available. Good quantitative theoretical treatments are available for few cases indeed.

B. Long-Range Interactions in Nearly Resonant Energy Transfer

The probability that vibrational energy transfer will occur during a collision described by a classical trajectory $R(t)$ may be written as

$$P = \frac{1}{\hbar^2}\left| \int_{-\infty}^{\infty} e^{i\omega t} \langle f | V[R(t)] | i \rangle \, dt \right|^2 \tag{4.11}$$

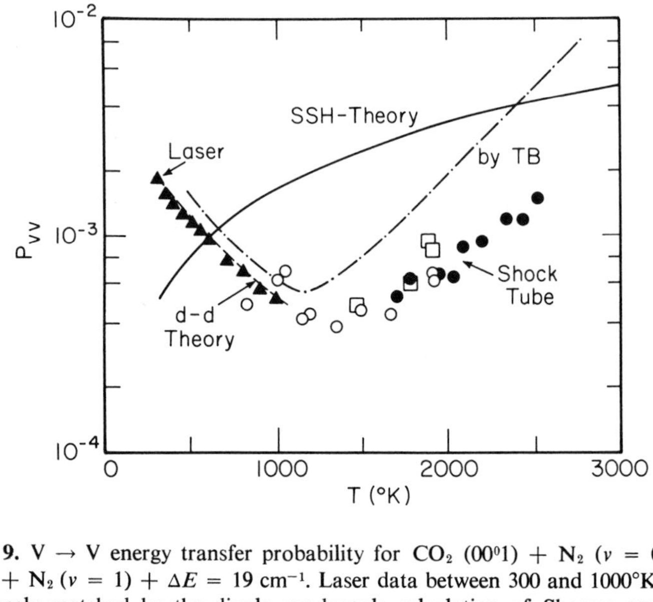

Fig. 9. V → V energy transfer probability for CO_2 (00^01) + N_2 $(v = 0)$ → CO_2 (00^00) + N_2 $(v = 1)$ + ΔE = 19 cm^{-1}. Laser data between 300 and 1000°K (Ref. 59) are closely matched by the dipole–quadrupole calculation of Sharma and Brau.[119] Two sets of shock tube data are shown. The earlier results (Taylor and Bitterman[121]) do not match quite so well with the laser data as do the more recent results.[79] Taylor and Bitterman's work went to lower temperature and clearly demonstrated a minimum in the probability near 1200°K. The increase in probability with temperature is presumably relaxation via short range interactions. However, the SSH calculation provides no useful guidance.

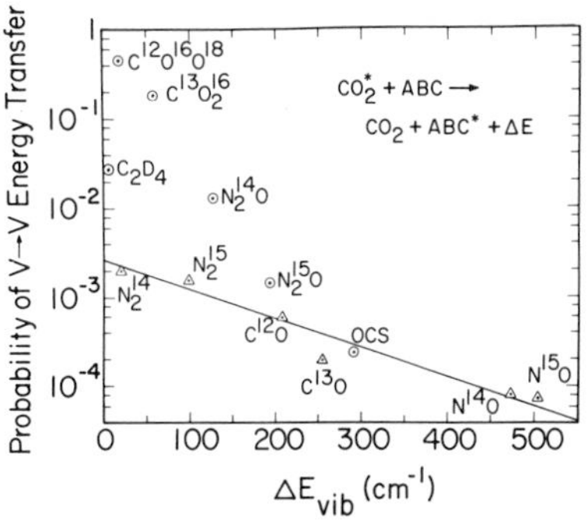

Fig. 10. Probabilities for nearly resonant V ⟶ V transfers from the asymmetric stretch of CO_2.[57] The isotopic CO_2 molecules, N_2O, and OCS all have comparable transition dipole moments. The curve traced by these points shows roughly the dependence of dipole–dipole transfer cross section on ΔE. The cross section for CO_2–N_2, probably dipole–quadrupole, is a factor of 200 less than the corresponding dipole–dipole process for CO_2–$CO^{16}O^{18}$. The data for NO are upper limits since processes of type (4.7) may contribute to the rate.

in the first order of perturbation. Here $\hbar\omega$ is the amount of energy transferred into translation (the net change in vibration–rotation energy during the collision, $|i\rangle$ and $|f\rangle$ are the initial and final vibration–rotation wave functions, and $V[R(t)]$ is the intermolecular potential function including its dependence on all the vibrational and rotational coordinates. The long-range parts of the intermolecular potential may give significant probabilities for nearly resonant energy transfers but are rapidly averaged to zero as ω increases. Mahan[118] suggested that the R^{-3} interaction between the vibrational transition dipole moments of the collision partners could lead to resonant V ⟶ V energy transfer. Sharma and Brau[119] calculated energy transfer cross sections for the process (4.3) due to the interaction of the N_2 transition quadrupole with the CO_2 transition dipole, $V \propto R^{-4}$. Their calculations were in excellent agreement with the magnitude and temperature dependence of data (Fig. 9) between 300 and 1000°K.* These were the first vibration relaxation cross sections observed[59] to be proportional to T^{-1}. Shock tube data show that the cross section goes through a minimum near 1200°K and then increases with temperature.[79,121] Cross sections for a series of nearly resonant transfers from $CO_2(00^01)$ are shown in Fig. 10. Those for transfer to isotopic CO_2 molecules or to N_2O molecules whose transition dipole moments are large show cross sections much larger than those for molecules such as N_2 or CO, with zero or small transition dipole moments.[57] The rapid decrease of these cross sections with increasing energy discrepancy and the agreement of calculated and observed cross sections for small ΔE indicate that the dipole–dipole interaction dominates these results.[57,61,67] The transition dipole moment of C_2D_4 is less

* This theory has recently been criticized and an alternative calculation proposed.[120] The arguments appear not to be valid since only zero-impact parameter collisions were treated.

than for CO_2. The calculated dipole–dipole cross sections are in excellent agreement with experiment over the temperature range 300–800°K.[68] Results on N_2O–CO and N_2O–N_2 fit the same general pattern of comparison between experiment and theory.[65] Morely and Smith[19] have observed the resonant deactivation of CS by CO_2 and N_2O.

$$CS(v = 1) + CO_2(00^00) \rightarrow CS(v = 0) + CO_2(02^00) + \Delta E = -13 \text{ cm}^{-1}$$

$$CS(v = 1) + N_2O(00^00) \rightarrow CS(v = 0) + N_2O(100) + \Delta E = -13 \text{ cm}^{-1}$$

The rate with N_2O was 100 times faster than with CO_2. The transition in N_2O is a dipole allowed fundamental whereas the transition in CO_2 is a fundamental with zero transition dipole [(02^00) is in Fermi resonance with (10^00) and is thus one half symmetric stretch in character]. As in Fig. 10 this indicates the importance of the dipole–dipole interaction. A value of

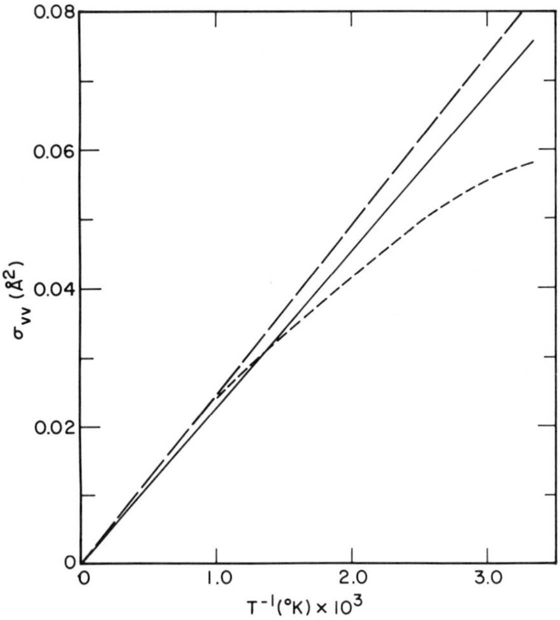

Fig. 11. V → V energy transfer cross section for CO_2–N_2 vs. reciprocal temperature. - - -, CO_2–$^{14}N_2$, $\Delta E = 19 \text{ cm}^{-1}$, experiment[59] and dipole–quadrupole theory[119] coincide. ——, CO_2–$^{15}N_2$, $\Delta E = 99 \text{ cm}^{-1}$, experiment.[67] - - - -, CO_2–$^{15}N_2$, dipole–quadrupole theory.[119] First-order theory predicts cross sections will decrease more rapidly with increasing ΔE than is observed. Results above 1000°K are not included here. Lines drawn to the origin show an extrapolation of low-temperature data. Fig. 9 gives actual behavior at high temperature.

the CS transition dipole moment was calculated from the observed cross section with N_2O and an assumed hard sphere radius for CS of 4.7 Å.[19] Figure 11 shows that for resonant processes V → V cross sections are proportional to T^{-1}. As ΔE increases cross sections decrease less rapidly with increasing temperature, become more or less temperature insensitive, and finally increase with temperature (Fig. 12). This transition is made more rapidly for the dipole–dipole processes than for those involving higher-order multipole moments (Fig. 11). Calculations using first-order perturbation theory predict that cross sections should be far more sensitive to ΔE than are the observed cross sections.[57,67,122,123] Qualitatively, it appears that

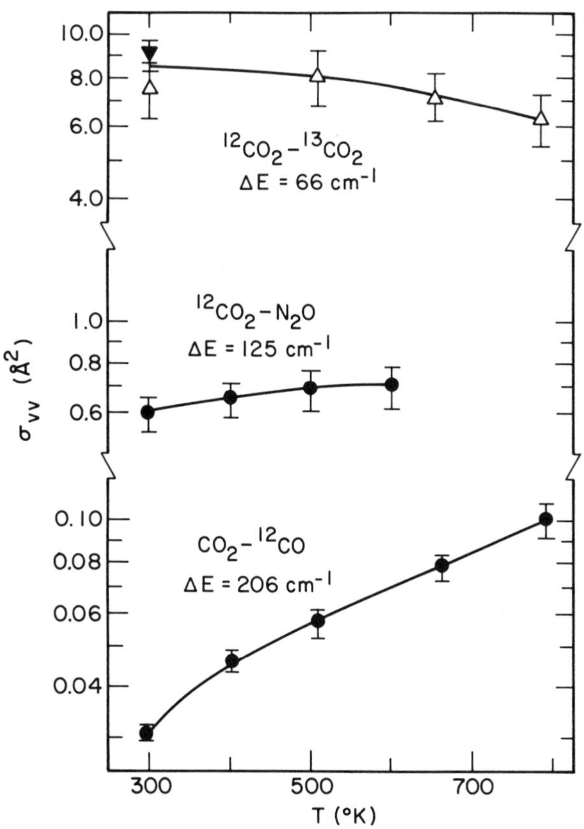

Fig. 12. Temperature dependence of cross sections for nearly resonant V → V single-quantum transfers. As the energy discrepancy increases σ_{vv} decreases and the slope changes from decreasing to increasing with temperature.[67]

forces which couple rotation and translation act simultaneously with the transition multipolar potential to transfer larger amounts of the vibrational energy discrepancy into rotation than are allowed in first-order perturbation theory.[67] The mechanism of these energy transfers appears to be on solid ground qualitatively. For the very nearly resonant cases reasonable quantitative agreement between theory and experiment is obtained. For not-so-nearly resonant cases calculations are needed which consider accurate trajectories and account carefully for the rotation–translation interaction throughout the collision.

Cross sections for nearly resonant multiquantum V → V processes have also been observed. For process (4.10) the dipole–dipole mechanism gives a calculated cross section at room temperature of 3.6 Å^2 compared to an experimental value of 4.3 Å^2 for the deactivation of $CO_2(00^01)$ by SF_6.[61] Both experiment and theory give cross sections which are linear with T^{-1}. Although the data do not extrapolate to zero cross section at infinite temperature, the agreement between experiment and theory is remarkable and strongly supports the hypothesis that CO_2 is deactivated by process (4.10) as well as the conclusion that dipole–dipole forces are responsible for the observed cross section. The methyl halides all deactive $CO_2(00^01)$ with cross sections between 0.28 and 0.42 Å^2 at room temperature. These cross sections are also linear with T^{-1}.[61] The dipole–dipole calculations for process (4.10) with CH_3F are consistent with the experimental data if one allows ΔJ to be larger than the ± 1 allowed by first-order perturbation theory. For the other hydrogen halides the transition dipole moments are an order of magnitude too small to explain the observed cross sections. It is possible for CH_3Cl, CH_3Br, and CHI_3 that other processes such as

$$CO_2(00^01) + CH_3Cl \rightarrow CO_2(00^00) + CH_3Cl(\nu_2 + \nu_6) + \Delta E = -19 \text{ cm}^{-1} \quad (4.12)$$

are responsible for the deactivation.[68] Unfortunately, the transition dipole moments for these combination bands are not known.

Long-range multipole moment calculations have been carried out for several systems in which near-resonant vibrational energy transfer to rotation is possible. Sharma[13] calculated the cross section for relaxation of CO_2 in ortho H_2,

$$CO_2(01^10) + H_2(J = 1) \rightarrow CO_2(00^00) + H_2(J = 3) + \Delta E = 81 \text{ cm}^{-1} \quad (4.13)$$

using the interaction between the CO_2 transition dipole and the H_2 permanent quadrupole. The rate for para H_2,

$$CO_2(01^10) + H_2(J = 2) \rightarrow CO_2(00^00) + H_2(J = 4) + \Delta E = -146 \text{ cm}^{-1} \tag{4.14}$$

was about one order of magnitude smaller due to increased ΔE and decreased fractional population of $J = 2$. He therefore predicted that o-H_2 would be an order of magnitude more efficient than p-H_2 for deactivating the bending vibration of CO_2. In fact o-H_2 is more efficient but the ratio is a factor of two at room temperature and about 50% at 1000°K (Fig. 13).[14,15,15a] It is most probable that V → T energy transfer contributes to the relaxation in both cases, thus decreasing the difference in observed relaxation rates. It is also likely that the rate for process (4.14) is badly under-estimated by the first-order theory. A similar calculation by Sharma and Kern[12] for the process

$$CO(v = 1) + H_2(J = 2) \rightarrow CO(v = 0) + H_2(J = 6) + \Delta E = 88 \text{ cm}^{-1} \tag{4.15}$$

gives an explanation for the observed difference in relaxation rate of CO by ortho and para H_2.[11] The authors point out the approximate nature of

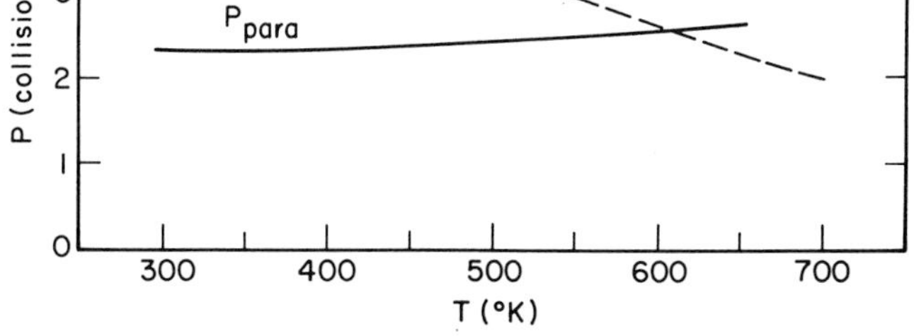

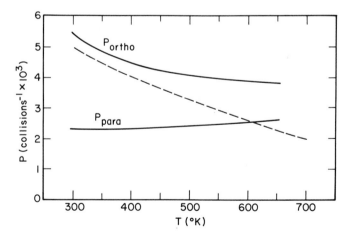

Fig. 13. Probability per collision for vibrational relaxation of the bending vibration of CO_2 by ortho H_2 and para H_2 vs. temperature. Solid curves are experiment.[15] The broken curve is the V → R calculation of Sharma for ortho H_2.[13] The calculation for para H_2 gives probabilities about one order of magnitude smaller.

their calculation.[12] The fact that the magnitude and temperature dependence of the effect are reproduced is convincing evidence for the importance of (4.15). This same model has been applied to the deactivation of CO_2 by H_2O,[124] but here the approximations of straight-line constant-velocity trajectories and of interactions due only to noninteracting charge distributions seem particularly dangerous, and the comparison to data uncompelling. Morley and Smith[19] found that para H_2 deactivates $CS(v = 1)$ with a cross section nearly double that for ortho H_2. The difference in cross section was nicely matched by a first-order perturbation calculation using the transition dipole calculated from the N_2O data along with the assumed hard sphere radius. The probability of

$$CS(v = 1) + H_2(J = 4) \rightarrow CS(v = 0) + H_2(J = 6) + \Delta E = 28 \text{ cm}^{-1} \quad (4.16)$$

is calculated as 1.1×10^{-2}.[19]

Near-resonant dipole–dipole energy transfers have recently been observed in molecular electronic energy transfer as well.[125,126]

All these results taken together prove beyond any reasonable doubt that the interaction of electric transition multipole moments can give large cross sections for nearly resonant energy transfer processes. For long-range interactions to play an important role in a $V \rightarrow V$ process the amount of energy transferred into translation must be quite small. The integral in (4.11) decreases rapidly for more than 15 cm^{-1} of energy transferred into translation for CO_2 at room temperature. Larger vibrational energy discrepancies can be accommodated to some extent by the changes in rotational energy allowed in first-order perturbation theory. The resonance requirement is further weakened by the operation of rotation–translation coupling interactions simultaneously with the multipolar forces. Empirically, for CO_2 at room temperature (Fig. 10) energy transfer cross sections decrease an order of magnitude for each 60 cm^{-1} beyond the first 20 cm^{-1} away from resonance. This sensitivity to energy discrepancy decreases slowly as temperature increases. For higher-order multipole moments, such as dipole–quadrupole (R^{-4}) the forces are shorter range and the cross sections are not so sensitive to energy discrepancy. However, the magnitudes of the cross sections at resonance decrease by powers of ten for higher-order moments. Even if the vibrational energy discrepancy is small enough for long-range forces to be operative, it is still possible that the observed energy transfer rates will be dominated by other interactions. It should be pointed out that most of the cases cited above have larger than average transition multipole moments. A slide rule estimate of the multipolar cross section at exact resonance sets an upper limit on multipolar effects and is often helpful in deciding the importance of various multipolar interactions. In considering the temperature dependence of observed cross sections it should be borne in mind that the impact parameter, constant-velocity, straight-line form of the Born approximation will give resonant cross sections proportional to T^{-1} for any dependence of the potential on R. Thus it is often difficult to establish whether multipolar forces, medium-range exchange-type attractive forces or short-range repulsive forces are responsible for a particular observed process.

C. Single-Quantum Intermolecular V → V Transfers

Most of the data presently available on single-quantum V → V energy transfer are assembled in Table II. The majority of these processes cannot be accounted for in terms of the multipolar interactions discussed in the previous section. For the most part these V → V energy transfer probabilities decrease monotonically with increasing vibrational energy discrepancy within a restricted family of molecules for which the intermolecular forces, the moments of inertia, the molecular masses, and the vibrational frequencies are nearly the same.[1,22,23,25,48,52,57,68,69] It is not surprising that a satisfying empirical correlation of the entire body of data in Table II has not been possible. The ranges of intermolecular forces, types and frequencies of vibration, and molecular structures, masses, and moments of inertia are just too great to be accounted for by any simple single theory or mechanism of energy transfer.

For some groups of molecules and for the temperature dependence of some energy transfer rates simple theoretical treatments[127] of energy transfer due to the repulsive part of the intermolecular potential compare favorably with experimental data. Fig. 14 shows energy transfer probabilities for several diatomic–diatomic systems. In all these cases the probabilities of energy transfer increase with increasing temperature. The slope of this increase becomes steeper as the energy discrepancy increases. The probabilities generally decrease as ΔE increases. For CO_2–N_2 (Fig. 9) the situation is dramatically different. At low temperature the data are accurately fit by the multipole moment calculations of Sharma and Brau.[119] At higher temperatures the behavior is similar to that for the diatomics at high temperatures. It appears that the repulsive interactions dominate the transfer at the higher temperatures. The most recent shock tube data[79] for CO_2–N_2 agree well with the laser results in the temperature range where they overlap. However, earlier shock tube results exhibit considerable scatter and are systematically higher.[121] A recent shock tube study of CO_2–CO mixtures[130] compares very poorly with accurate laser work.[62,67] Data are also available for N_2O–N_2,[131] HI–N_2, and DI–N_2.[132]

TABLE II

Single-Quantum V → V Energy Transfers,

AB (v = 1) + CD → AB + CD (v = 1) + ΔE

Molecule		ν_{AB} (cm^{-1})	ΔE[a] (cm^{-1})	T[b] (^0K)	$\sigma_{V \to V}$[c] ($Å^2$)	P[c]	Ref.
AB	CD						
CO	O_2	2143	587	286	8.6(−6)	2.2(−7)	128[d]
	DBr	2143	303	295		5.9(−4)	51
	DI	2143	548	295		3.1(−5)	51
	$CD_4(\nu_1)$	2143	58	295		1.0(−3)	51
	(ν_3)		−115				
	$CH_4(\nu_2)$	2143	610	303	1.3(−3)	3.0(−5)	134
				363	1.8(−3)	4.2(−5)	134
N_2	O_2	2331	775	1000	4(−5)	1.0(−6)	116[d]
N_2	CO	2331	188	295		5.1(−5)	51[d]
O_2	$CH_4(\nu_2)$	1556	23	297		2.0(−3)	53
	(ν_4)		250				
NO*	N_2	2341	10	RT		1.3(−3)	23[e]
NO	CO	1876	267	RT		1.0(−4)	23
	N_2	1876	455	298		6(−6)	129
				433		9(−6)	129
	$CO_2(10^00)$	1876	491	RT		~1.7(−4)	135[f]
	D_2S	1876	−16	RT	5(−1)	1.1(−2)	136
	H_2O	1876	281	RT	1.6(−1)	6(−3)	136
	H_2S	1876	586	RT	1.4(−1)	3.2(−3)	136
	D_2O	1876	697	RT	2.5 (−2)	1.0(−3)	136
	$CH_4(\nu_2)$	1876	343	RT	4.1(−2)	9(−4)	136
CS	N_2O	1272	−13	312	2.3	4.0(−2)	19
	NO_2	1272	−48	325	1.1(−1)	1.6(−3)	137
	$CO_2(02^00)$	1272	−13	312	2.3(−2)	3.9(−4)	19[g]
	$CS_2(00^01)$	1272	−251	325	5.5(−3)	5.5(−5)	137
	$CS_2(030)$		70				
	H_2O	1272	−323	312		4.9(−2)	19[i]
	D_2O	1272	93	312		1.8(−2)	19
	H_2S	1272	89	312		1.6(−2)	19
	D_2S	1272	398	312		1.7(−3)	19
HF	H_2	3958	−201	294		4.2(−3)	45[j]
	N_2	3958	1627	294		1.3(−5)	45[i]
	D_2	3958	964	294		3.1(−4)	45
DF	CO_2	2907	558	400	.5−2		46[j]
HCl	D_2	2886	−108	295	2.1(−1)	7.0(−3)	48
	HBr	2886	327	295	2.1(−1)	6.0(−3)	48, 138
	N_2	2886	555	295	4.2(−3)	1.1(−4)	48
	HI	2886	656	295	3.5(−2)	8.0(−4)	48

(Continued)

TABLE II (continued)

Molecule AB	Molecule CD	ν_{AB} (cm^{-1})	ΔE[a] (cm^{-1})	T[b] (°K)	$\sigma_{V \to V}$[c] ($Å^2$)	P[c]	Ref.
	CO	2886	743	295	1.3(−2)	3.4(−4)	48
	DCl	2886	795	295	1.66(−2)	5.0(−4)	48
	NO	2886	1010	295		4.0(−4)	51
	CH_4	2886	−30	295	3.7(−1)	9.4(−3)	48[k]
			−133				
DCl	^{13}CO	2091	−5	295		8.8(−3)	51
	CO	2091	−52	295		6.4(−3)	51
	NO	2091	215	295		5.1(−3)	51
	DBr	2091	251	295		2.8(−3)	51
	DI	2091	496	295		2.7(−4)	51
	N_2	2091	−240	295		2.2(−4)	51
	O_2	2091	535	295		8.6(−5)	51
	CD_4	2091	6	295		5.5(−3)	51
			−167				
	CH_4	2091	558	295		2.8(−3)	51
			785				
HBr	N_2	2559	228	295	1.8(−2)	4.5(−4)	52, 138
	HI	2559	329	295	1.9(−1)	4.2(−3)	52
	CO	2559	416	295	5.5(−2)	1.4(−3)	52
						7(−4)	138
	D_2	2559	−435	295	9.1(−3)	2.9(−4)	52
	O_2	2559	1003	295	9.3(−4)	2.5(−5)	52
	$CH_4(\nu_2)$	2559	1026	300		1.1(−3)	138[i]
	(ν_1)		1253				
CO_2	$CO^{18}O(00^01)$	2349	18	298	24		57
	$^{13}CO_2(00^01)$	2349	66	298	8.5		67[l]
	$N_2O(00^01)$	2349	125	298	6.0(−1)		67[l]
	$^{15}N_2O(00^01)$	2349	193	298	7.2(−2)		57
	CO	2349	206	298	3.0(−2)	6.6(−4)	67[l]
	^{13}CO	2349	256	298	8(−3)	1.8(−4)	67
	$OCS(00^01)$	2349	280	298	1.5(−2)		68
	NO	2349	473	298		8.5(−5)	57[i,m]
	^{15}NO	2349	502	298		8(−4)	57[i]
	N_2	2349	19	298	8.2(−2)		55, 59[n]
	$^{15}N_2$	2349	99	298	7.6(−2)		67[n]
	$(CN)_2$	2349	27	298	1.1		68[i]
			200				
	CH_3CN	2349	100	298	6.3(−1)		68[i]
	HCN	2349	260	298	5.2(−2)		68[i]
	HCl	2349	−537	298	4.8(−1)		69
				500	3.8(−1)		69
	HBr	2349	−209	298	1.9		69

(*Continued*)

TABLE II (continued)

Molecule AB	Molecule CD	ν_{AB} (cm^{-1})	ΔE[a] (cm^{-1})	T[b] (°K)	$\sigma_{V\to V}$[c] (Å^2)	P[c]	Ref.
	HI	2349	116	298	1.5		69
	DCl	2349	258	298	5.5(−1)		69
				500	3.8(−1)		69
	DBr	2349	540	298	6.0(−2)		69
	DI	2349	737	298	8.5(−3)		69
	H_2Se	2349	0	298	2.4		68
	C_2D_4	2349	5	298	1.5		68
				468	9.2(−1)		68
				785	4.9(−1)		68
	CD_2CH_2	2349	16	298	9.0(−1)		68
	CD_3I	2349	50	298	1.0		68
	CD_3Cl	2349	63	298	7.4(−1)		68
	C_2D_2	2349	−78	298	5(−1)		68
				392	3(−1)		68
				509	3(−1)		68
	CD_2H_2	2349	94	298	3.9(−1)		68
	CD_4	2349	92	298	7.6(−1)		68
	D_2	2349	−645	298	2.5(−2)	6.6(−4)	67
N_2O	N_2	2224	−107	298	3.9(−2)	8.7(−4)	65, 71[o]
	CO	2224	81	RT	1.0	2.3(−2)	65[o]
	HBr	2224	−335	RT	2.9(−1)	6.4(−3)	65
	HI	2224	−6	RT	1.4	2.8(−2)	65

[a] $\Delta E = \nu_{AB} - \nu_{CD}$.

[b] RT denotes that room temperature was not specified.

[c] P and σ_{VV} are given for the process occurring in the exothermic direction. $P = \sigma_{VV}/\sigma_{kin}$ where $\sigma_{kin} = \pi\,(d_{AB} + d_{CD})^2/4$ and d_{AB} and d_{CD} are the collision diameters for an L–J 6–12 potential as given in J. O. Hirschfelder, C. F. Curtiss, and R. B. Bird, *Molecular Theory of Gases and Liquids*, Wiley, New York, 1964.

[d] See Fig. 14 for high-temperature data.

[e] NO* is electronically excited NO (A $^2\Sigma^+$).

[f] May not be single quantum V→V process.

[g] $CO_2(02^00)$ and $CO_2(10^00)$ are strongly coupled by Fermi resonance. This level is roughly half symmetric stretching fundamental in character.

[h] The (030) levels are at about 1200 cm^{-1} and may well contribute to this deactivation. P and σ_{VV} correspond to the net observed rate for deactivation of CS.

[i] Upper limit on the single quantum V→V rate. The mechanism is not established.

[j] Analysis of data is not clear.

[k] P and σ_{VV} are for deactivation of CH_4 (endothermic).

[l] See Fig. 12 for high-temperature data.

[m] See ref. 64 for high-temperature data.

[n] See Figs. 9 and 11 for high-temperature data.

[o] See ref. 65a for high-temperature data.

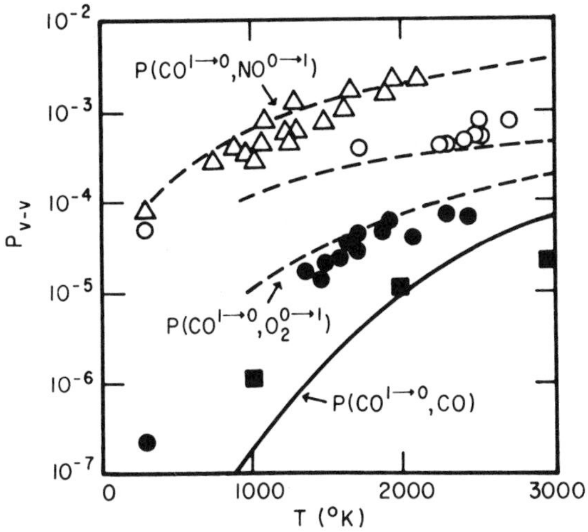

Fig. 14. Single-quantum V → V energy transfer probabilities as a function of temperature.

△	$CO \rightarrow NO$ $\Delta E = 267\ cm^{-1}$		(Refs. 23, 24)
○	$N_2 \rightarrow CO$ $\Delta E = 188\ cm^{-1}$		(Refs. 51, 105)
●	$CO \rightarrow O_2$ $\Delta E = 587\ cm^{-1}$		(Refs. 105, 128)
■	$N_2 \rightarrow O_2$ $\Delta E = 775\ cm^{-1}$		(Ref. 116)

The solid line is for V → T relaxation of CO. The dashed lines are SSH calculations.[105]

Hancock and Smith[25] have recently measured V → V energy transfer rates for highly excited levels of CO, $4 \leq v \leq 13$, in collisions with many species.

$$CO(v) + AB(v = 0) \rightarrow CO(v - 1) + AB(v = 1) + \Delta E_v \quad (4.17)$$

Figure 15 shows results for a number of collision partners. When the observed probabilities are divided by the quantum number they give smooth correlations of $\log(P/v)$ vs. ΔE. The results correlate quite well with data on $CO(v = 1)$ from other experiments on NO,[23] N_2,[51] and O_2.[128] Recent results on CO–N_2 at room temperature[51] show two earlier investigations[92,133] to be in error and yield a rate which correlates quite well both with high-temperature shock tube data (Fig. 14)[105] and with Ref. 25 (Fig. 15). Hancock and Smith[25] point out that for a given pair of collision partners the normalized probabilities appear constant for ΔE less than 60 cm^{-1} and that for ΔE greater than 60 cm^{-1} each set of points shows an approximately linear decrease of $\log(P/v)$ with a slope similar to that of Callear s plot for second-row diatomics. A more complicated picture seems some-

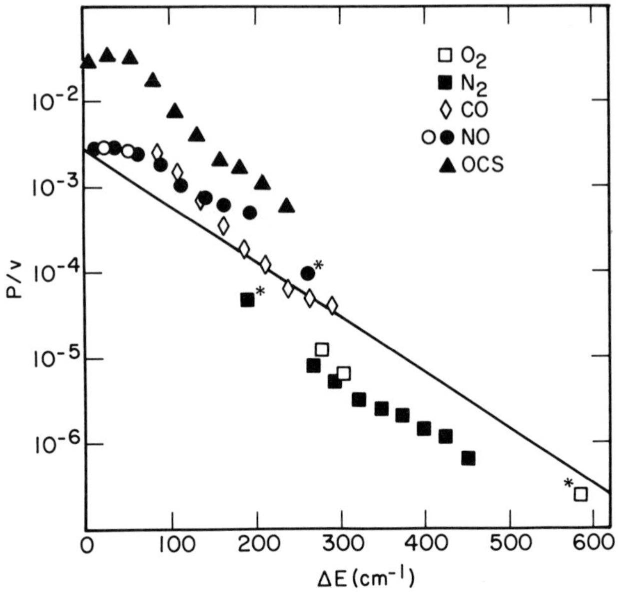

Fig. 15. Single-quantum V → V transfers from CO(v) to O_2, N_2, CO, NO, and OCS from Handcock and Smith.[25]. Data from other sources for CO(v = 1) are marked *: O_2 (Ref. 128), N_2 (Ref. 51), and NO (Ref. 23). Probabilities per collision are given for the exothermic direction. The harmonic oscillator vibrational matrix elements are proportional to the initial vibrational quantum number, v, of CO. The normalized probability, P/v, is plotted. The line is from Callear's correlation of heavy atom diatomic V → V rates.[23]

what more attractive from a theoretical point of view and appears to be equally compatible with the data. Between about 60 and 200 cm^{-1} the slopes are steeper than Callear's line. This is especially clear if one ignores the last few points in each series. These points carry large experimental uncertainties in the method of Hancock and Smith. They calculate that the dipole–dipole interaction accounts for the observed probabilities at small values of ΔE for CO–NO and CO–CO and surely for OCS as well. The variation in $\Delta E = 0$ intercepts parallels the magnitudes of the transition dipole moments. Thus a slope similar to the 60 cm^{-1} per decade of Fig. 10 might be expected in the range of ΔE where dipolar coupling is important. For ΔE sufficiently large, shorter-range forces will be responsible for the energy transfer and the slope should be closer to the 150 cm^{-1} per decade of Callear's[23] correlation. The data for CO–N_2 and CO–O_2 illustrate this (Fig. 15).

1 PRIGOGINE ADVANCES V23 Y
SBN# 0471699276, 02443A 22.50

PACKING LIST

NET WEIGHT 1LBS 5OZS
NET AMOUNT
SHIPPING AND HANDLING CHARG

1 ← TOTAL QUANTITY SHIPPED R 72166

IF SHIPMENT ARRIVES SHORT OR DAMAGED,
ADVISE WITHIN 30 DAYS

JOHN WILEY & SONS, INC.

ONE WILEY DRIVE, SOMERSET, N. J. 08873

The good correlation between single-quantum V → V rates for molecules in $v = 1$ and molecules in much higher vibrational states is comforting from both a theoretical and an experimental point of view. Empirical correlations can now be used to predict rates for these higher levels, rates which are experimentally difficult to measure. It is possible, however, that this pleasant situation will not prevail in cases where strongly attractive forces lead to orbiting or wrestling collisions. Tunable infrared lasers may now be used to excite overtone and combination vibrations. Thus more results on highly excited states should be available soon. A recently reported experiment[34] on relaxation of HCl($v = 3$) in HCl gives a relaxation time of 0.3 μsec-atm. This is apparently a rate for the overall V → T,R relaxation which follows the V → V exchange processes.

There is a considerable quantity of ultrasonic data which relates to single-quantum V → V transfers. This has been reviewed recently and in detail.[1,5,115] Though it is probable that some of these data have been overinterpreted, as in the CH_4–H_2O case described above, some of these rates will surely stand the test of time. The rates for vibrations below 600 cm^{-1} will tend to complement data from the optical experiments which are more easily performed at higher frequencies. Some recent work has been done by Bauer,[139,140] Shields,[141] and their co-workers.

D. Other V → V Energy Transfer Processes

Although examples of most of the other types of energy transfer processes listed in Section IV.A above have been reported, there is no large representative body of data available for them. For intramolecular processes involving the vibrational degrees of freedom of only one of the collision partners there are many data for the deactivation of the asymmetric stretching vibrations of CO_2 and N_2O. Stephenson, Wood, and Moore[63] have discussed this work, but some more data have become available recently for the hydrogen halides as collision partners.[65,66,69] For processes for which several quantum number changes occur it is important to consider the mixing of the harmonic oscillator states of the free molecule by Coriolis and anharmonic effects. As the power of our experimental tools increases it will become possible to study the rates of these more complex processes in a variety of systems and to identify measured rates with particular processes in more and more complicated sets of energy levels.

V. CONCLUSION

During the past ten years our knowledge of V → V energy transfer processes has developed from essentially none into a substantial body of data. For the most part these data fit into reasonable qualitative patterns; however, we are far from an understanding of the relative importance of each of the parameters which influence V $\rightarrow$ V energy transfer cross sections. Reliable prediction of rates is confined to situations for which there are data available on a number of closely analogous systems. For a few systems in which known long-range interactions are responsible for nearly resonant energy transfers a quantitative comparison between experiment and theory has been made.

Most of our experimental information to date gives rate constants for energy transfers between particular vibrational levels or groups of vibrational levels. As detailed and specific as this information may seem by contrast to the measurements of bulk thermal relaxation times, it does not represent the ultimate possible or desirable level of detail. In spectroscopic experiments it is becoming possible to resolve individual vibration–rotation levels. The rotational quantum number dependence of vibrational relaxation could yield much information about detailed mechanisms of V $\rightarrow$ V transfers and about the shapes of the intermolecular potentials which cause them. The possibility of combined spectroscopic–molecular beam experiments[142] is also attractive. Information on rotational distributions, velocity distributions, and angular distributions would tell directly about the importance of head-on collisions relative to grazing collisions, of the rotational degrees of freedom in absorbing the vibrational energy discrepancy, of collision complexes, and of particular states or velocities whose large cross sections are the dominant contributors to the small thermally averaged cross sections usually observed.

A complete quantitative, quantum mechanical calculation of a real V $\rightarrow$ V energy transfer cross section has never been carried out. Accurate calculations for V $\rightarrow$ T,R energy transfer are becoming feasible for the H_2–He system.[143] Perhaps the H_2–H_2 problem will become tractable as well. For the bulk of the processes which we can study and for which information is practically valuable we must be content with exact solutions of unphysical problems, with approximate solutions of realistic models, with semiempirical theories, and with empirical correlations. There has been progress in these areas in recent years as good semiclassical scattering treatments have been developed[21] and as the experimental facts have forced model calculations out of the one-dimensional rut.

Our ability to make empirical correlations of experimental data, to make honest rationalizations for these correlations, and to make clear tests of approximate theoretical treatments is severely hampered by a dearth of intermolecular potential data. Big calculations on small systems and small calculations[144] on big systems both appear to be promising at present.

Molecular beam experiments have provided a good deal of data in the past and may become more useful still. Perhaps spectroscopic measurements of gas-phase and matrix-isolated dimer molecules could be put together with spherically averaged potentials from molecular beam work and other experimental information to yield good potentials. Potential functions are needed which will not only give good trajectories but also give the forces coupling vibrational and rotational coordinates of the two molecules. The shape of the potential and its dependence on all the vibrational coordinates must be known in addition to the usual radial dependence.

V → V transfer rates determine the behavior of a considerable number of scientifically and technologically interesting systems. New opportunities to apply our knowledge of energy transfer processes continue to arise. The use of lasers to excite strongly single vibrational modes suggests a number of interesting experiments whose feasibility and interpretation depend crucially on vibrational relaxation rates. It has been suggested that by laser-exciting a single vibration one could study dissociation at low translational temperatures,[102,145] or the chemical reactions of vibrationally hot but otherwise cold molecules.[146] Selective vibrational excitation of one isotopic form of a molecule with a laser followed by a short wavelength pulse to dissociate the vibrationally excited molecules has been suggested[147] as a method of isotopic separation.

Acknowledgments

It is a pleasure to thank the National Science Foundation, the Advanced Research Projects Agency and Army Research Office, Alfred P. Sloan Foundation, and the Berkeley Chemistry Department for their support of our own work on vibrational relaxation. I am very grateful to my collaborators and to my friends in other laboratories for sharing their thoughts, their results, and their experimental arts with me.

References

1. A. B. Callear and J. D. Lambert, "The Transfer of Energy Between Chemical Species," in *Comprehensive Chemical Kinetics*, Vol. 3, *The Formation and Decay of Excited States*, C. H. Bamford and C. F. H. Tipper, Eds., Elsevier, Amsterdam, 1969, Chap. 4, pp. 182–273.
2. R. G. Gordon, W. Klemperer, and J. I. Steinfeld, *Ann. Rev. Phys. Chem.*, **19**, 215 (1968).
3. B. Stevens, *International Encyclopedia of Physical Chemistry and Chemical Physics*, *Topic* 19, *Gas Kinetics*, Vol. 3, "Collisional Activation in Gases," Pergamon Press, Oxford, 1967.
4. T. L. Cottrell and J. C. McCoubrey, *Molecular Energy Transfer in Gases*, Butterworth, London, 1961.
5. G. M. Burnett and A. M. North, Eds., *Transfer and Storage of Energy by Molecules*, Vol. 2, *Vibrational Energy*, London, Wiley-Interscience, 1969.
6. P. Borrell, *Adv. Mol. Relaxation. Processes*, **1**, 69 (1967).

7. T. L. Cottrell and A. J. Matheson, *Trans. Faraday Soc.*, **58**, 2336 (1962).
8. K. H. Welge, S. V. Filseth, and J. Davenport, *J. Chem. Phys.*, **53**, 502 (1970).
9. E. H. Fink, D. L. Akins, and C. B. Moore, *J. Chem. Phys.*, **56**, 900 (1972).
10. C. B. Moore, *J. Chem. Phys.*, **43**, 2979 (1965).
11. R. C. Millikan and E. Switkes, to be published, and refs. cited therein.
12. R. D. Sharma and C. W. Kern, *J. Chem. Phys.*, **55**, 1171 (1971).
13. R. D. Sharma, *J. Chem. Phys.*, **50**, 919 (1969).
14. S. W. Behner, H. L. Rothwell, and R. C. Amme, *Chem. Phys. Lett.*, **8**, 318 (1971).
15. C. J. S. M. Simpson and J. M. Simmie, *Proc. Roy. Soc.* (*London*) *SER. A*, **325**, 197 (1971).
15a. M. Huetz-Aubert and P. Chevalier, *Compt. Rend.*, **274B**, 1305 (1972).
16. J. T. Yardley, M. N. Fertig, and C. B. Moore, *J. Chem. Phys.*, **52**, 1450 (1970).
17. H. L. Chen and C. B. Moore, *J. Chem. Phys.*, **54**, 4072 (1971).
18. H. L. Chen, *J. Chem. Phys.*, **55**, 5551 (1971).
19. I. W. M. Smith and C. Morley, *Trans. Faraday Soc.*, **67**, 2575 (1971).
20. G. A. Kapralova, E. E. Nikitin, and A. M. Chaikin, *Chem. Phys. Lett.*, **2**, 581 (1968).
21. W. H. Miller and T. F. George, *J. Chem. Phys.*, **56**, 5668 (1972).
22. C. B. Moore, *Acc. Chem. Res.*, **2**, 103 (1969).
23. A. B. Callear, *Appl. Opt.*, *Suppl.*, **2**, 145 (1965).
24. R. L. Taylor, M. Camac, and R. M. Feinberg, *Symp. Combust.* **11***th*, *Berkeley, Calif.*, **1966**, (**1967**), p. 49.
25. G. Hancock and I. W. M. Smith, *Appl. Opt.*, **10**, 1827 (1971).
26. M. Huetz-Aubert and R. Tripodi, *J. Chem. Phys.*, **55**, 5724 (1971).
27. F. DeMartini and J. Ducuing, *Phys. Rev. Letters*, **17**, 117 (1966); J. Ducuing and F. DeMartini, *J. Chim. Phys.*, **64**, 209 (1967).
28. L. O. Hocker, M. A. Kovacs, C. K. Rhodes, G. W. Flynn, and A. Javan, *Phys. Rev. Letters*, **17**, 233 (1966).
29. J. T. Yardley and C. B. Moore, *J. Chem. Phys.*, **45**, 1066 (1966).
30. C. B. Moore, *Fluorescence*, G. G. Guilbault, Ed., Marcel Dekker, New York, 1967, p. 133.
31. R. T. Menzies, N. George, and M. L. Bhaumik, *IEEE J. Quant. Electron.*, **QE6**, 800 (1970).
32. C. B. Moore, *Ann. Rev. Phys. Chem.*, **22**, 387 (1971).
33. S. E. Harris, *Proc. IEEE*, **57**, 2096 (1969); R. W. Wallace, *Appl. Phys. Lett.*, **17**, 497 (1970).
34. R. V. Ambartzumian, V. S. Letokhov, G. N. Makarov, and N. V. Chekalin, *Chem. Phys. Lett.*, **13**, 49 (1972).
35. C. K. N. Patel, *Appl. Phys. Lett.*, **19**, 400 (1971).
36. C. K. N. Patel and E. D. Shaw, *Phys. Rev. Lett.*, **24**, 451 (1970).
37. A. Mooradian, S. R. J. Brueck, and F. A. Blum, *Appl. Phys. Lett.*, **17**, 481 (1970).
38. J. Ducuing, C. Joffrin, and J. P. Coffinet, *Opt. Commun.*, **2**, 245 (1970).
39. C. K. Rhodes, M. J. Kelly, and A. Javan, *J. Chem. Phys.*, **48**, 5730 (1968).
40. I. Burak, A. V. Nowak, J. I. Steinfeld, and D. G. Sutton, *J. Chem. Phys.*, **51**, 2275 (1969); **52**, 5421 (1970).
41. T. Shimizu and T. Oka, *Phys. Rev. A*, **2**, 1177 (1970)
42. I. Burak, P. Houston, D. G. Sutton, and J. I. Steinfeld, *J. Chem. Phys.*, **53**, 3632 (1970).

43. R. D. Bates, G. W. Flynn, J. T. Knudtson, and A. M. Ronn, *J. Chem. Phys.*, **53**, 3621 (1970).
44. J. R. Airey and S. F. Fried, *Chem. Phys. Lett.*, **8**, 23 (1971).
45. J. K. Hancock and W. H. Green, *J. Chem. Phys.*, **56**, 2474 (1972).
46. G. K. Vasil'ev, E. F. Makarov, V. G. Papin, and V. L. Tal'roze, *Sov. Phys. JETP*, **34**, 51 (1972) .*Zh. Eksperimi-i Teor. Fiz.*, **61**, 97 (1971)..
47. H. L. Chen, J. C. Stephenson, and C. B. Moore, *Chem. Phys. Lett.*, **2**, 593 (1968).
48. H. L. Chen and C. B. Moore, *J. Chem. Phys.*, **54**, 4080 (1971).
49. M. Margottin Maclou, L. Doyennette, and L. Henry, *Appl. Opt.*, **10**, 1768 (1971) and refs. cited therein.
50. N. C. Craig and C. B. Moore, *J. Phys. Chem.*, **75**, 1622 (1971).
51. P. F. Zittel and C. B. Moore, to be published.
51a. P. F. Zittel and C. B. Moore, *Appl. Phys. Lett.*, **21**, 81 (1972).
52. H. L. Chen, *J. Chem. Phys.*, **55**, 5557 (1971).
53. J. T. Yardley and C. B. Moore, *J. Chem. Phys.*, **48**, 14 (1968); **49**, 3328 (1968). See also L. B. Evans and T. G. Winter, *J. Acoust. Soc. Am.*, **45**, 515 (1969).
54. J. T. Yardley and C. B. Moore, *J. Chem. Phys.*, **49**, 1111 (1968).
55. C. B. Moore, R. E. Wood, B. L. Hu, and J. T. Yardley, *J. Chem. Phys.*, **46**, 4222 (1967).
56. J. T. Yardley and C. B. Moore, *J. Chem. Phys.*, **46**, 4491 (1967).
57. J. C. Stephenson, R. E. Wood, and C. B. Moore, *J. Chem. Phys.*, **48**, 4790 (1968).
58. M. A. Kovacs and A. Javan, *J. Chem. Phys.*, **50**, 4111 (1969).
59. W. A. Rosser, Jr., A. D. Wood, and E. T. Gerry, *J. Chem. Phys.*, **50**, 4996 (1969).
60. D. F. Heller and C. B. Moore, *J. Chem. Phys* , **52**, 1005 (1970).
61. J. C. Stephenson and C. B. Moore, *J. Chem. Phys.*, **52**, 2333 (1970).
62. W. A. Rosser, Jr., R. D. Sharma, and E. T. Gerry, *J. Chem. Phys.*, **54**, 1196 (1971).
63. J. C. Stephenson, R. E. Wood, and C. B. Moore, *J. Chem. Phys.*, **54**, 3097 (1971).
64. W. A. Rosser, Jr. and E. T. Gerry, *J. Chem. Phys.*, **54**, 4131 (1971) and **51**, 2286 (1969).
65. H. Gueguen, I. Arditi, M. Margottin-Maclou, L. Doyennette, and L. Henry, *Compt. Rend.*, **272B**, 1139 (1971).
66. R. S. Chang, R. A. McFarlane, and G. J. Wolga, *J. Chem. Phys.*, **56**, 667 (1972).
67. J. C. Stephenson and C. B. Moore, *J. Chem. Phys.*, **56**, 1295 (1972).
68. J. C. Stephenson, R. E. Wood, and C. B. Moore, *J. Chem. Phys.*, **56**, 4813 (1972).
69. J. C. Stephenson, J. Finzi, and C. B. Moore, *J. Chem. Phys.*, **56**, 5214 (1972).
70. R. D. Bates, Jr., G. W. Flynn, and A. M. Ronn, *J. Chem. Phys.*, **49**, 1432 (1968).
71. J. T. Yardley, *J. Chem. Phys.*, **49**, 2816 (1968).
72. R. D. Bates, Jr., J. T. Knudtson, G. W. Flynn, and A. M. Ronn, *Chem. Phys. Lett.*, **8**, 103 (1971). See also *J. Chem. Phys.*, Oct. 1972.
73. R. C. L. Yuan and G. W. Flynn, *J. Chem. Phys.*, **57**, 1316 (1972).
74. E. Weitz, G. W. Flynn, and A. M. Ronn, *J. Chem. Phys.*, **56**, 6060 (1972).
75. J. T. Knudston and G. W. Flynn, to be published in *J. Chem. Phys.*
76. O. R. Wood and T. Y. Chang, *Appl. Phys. Lett.*, **20**, 77 (1972).
77. R. L. Taylor and S. Bitterman, *Rev. Mod. Phys.*, **41**, 26 (1969).
78. Y. Sato and S. Tsuchiya, *J. Phys. Soc. Japan*, **30**, 1467 (1971).
79. Y. Sato and S. Tsuchiya, *Chem. Phys. Letters*, **5**, 293 (1970).
80. P. K. Cheo and R. L. Abrams, *Appl. Phys. Letters*, **14**, 47 (1969).
81. R. L. Abrams and P. K. Cheo, *Appl. Phys. Letters*, **15**, 177 (1969).

82. R. D. Sharma, *Phys. Rev.*, **177**, 102 (1969).
83. K. N. Seeber, *J. Chem. Phys.*, **55**, 5077 (1971).
84. J. I. Steinfeld, *MTP International Review of Science*, *Physical Chemistry*, Series I (A. D. Buckingham, Consult. Ed.), Vol. 9, "*Gas Kinetics*" (J. G. Polanyi, Ed.), Butterworths, London, 1972, p. 247.
85. M. E. Mack, R. L. Carman, J. Reintjes, and N. Bloembergen, *Appl. Phys. Lett.*, **16**, 209 (1970).
86. M. A. Kovacs and M. A. Mack, *Appl. Phys. Lett.*, **20**, 487 (1972).
87. R. Tripodi and M. Huetz, in press.
88. E. W. Montroll and K. E. Shuler, *Advan. Chem. Phys.*, **1**, 361 (1958).
89. C. E. Treanor, J. W. Rich, and R. G. Rehm, *J. Chem. Phys.*, **48**, 1798 (1968).
90. K. N. C. Bray, *J. Phys. B: Atom. Molec. Phys.*, **1**, 705 (1968); **3**, 1515 (1970); K. N. C. Bray and N. H. Pratt, *J. Chem. Phys.*, **53**, 2987 (1970).
91. F. Legay and N. Legay-Sommaire, *Can. J. Phys.*, **48**, 1949 (1970).
92. N. Legay-Sommaire and F. Legay, *Can. J. Phys.*, **48**, 1966 (1970).
93. K. P. Horn and P. E. Oettinger, *J. Chem. Phys.*, **54**, 3040 (1971).
94. C. A. Brau, G. E. Caledonia, and R. E. Center, *J. Chem. Phys.*, **52**, 4306 (1970).
95. J. W. Rich, *J. Appl. Phys.*, **42**, 2719 (1971).
96. G. Hancock and I. W. M. Smith, *Appl. Opt.*, **10**, 1827 (1971).
97. J. D. Teare, R. L. Taylor, C. W. von Rosenberg, Jr., *Nature*, **225**, 240 (1970).
98. T. I. McLaren, J. P. Appleton, *J. Chem. Phys.*, **53**, 2850 (1970) and refs cited therein.
99. C. W. von Rosenberg, Jr., R. L. Taylor, and J. D. Teare, *J. Chem. Phys.*, **54**, 1974 (1971).
100. T. Just and P. Roth, *J. Chem. Phys.*, **55**, 2395 (1971).
101. R. L. McKenzie, *Appl. Phys. Lett.*, **17**, 462 (1970); B. F. Gordiets, A. I. Osipov, and L. A. Shelepin, *Sov. Phys. JETP*, **33**, 58 (1971) .*Zh. Eksperim. i Teor. Fiz.*, **60**, 102 (1971)..
102. Yu. V. Afanas'ev, F. M. Belenov, E. P. Markin, and I. A. Poluentov, *JETP Lett.*, **13**, 331 (1971) and refs cited therein.
103. A. A. Monkewicz, *J. Acoust. Soc. Am.*, **42**, 258 (1967).
104. H. J. Bauer, *J. Acoust. Soc. Am.*, **44**, 285 (1968).
105. Y. Sato, S. Tsuchiya, and K. Kuratani, *J. Chem. Phys.*, **50**, 1911 (1969).
106. Y. Sato and S. Tsuchiya, *J. Chem. Phys.*, **53**, 1304 (1970).
107. A. W. Read, *Mol. Relaxation Processes*, **1**, 257 (1968).
108. T. L. Cottrell, I. M. MacFarlane, A. W. Read, and A. H. Young, *Trans. Faraday Soc.*, **62**, 2655 (1966).
109. P. J. Domnin, *Opt. Spectr.* (*USSR*), **27**, 374 (1969).
110. T. L. Cottrell, I. M. MacFarlane, and A. W. Read, *Trans. Faraday Soc.*, **63**, 2093 (1967).
111. B. J. Lavercombe, *Nature*, **211**, 63 (1966).
112. M. Huetz-Aubert and P. Chevalier, *Advan. Mol. Relaxation Processes*, **2**, 101 (1970) and *Compt. Rend.*, **B268**, 748, 965, and 1068 (1969). This work omits one important term from the relaxation equations. See ref. 114.
113. R. Tripodi and W. G. Vincenti, *J. Chem. Phys.*, **55**, 2207 (1971); R. Tripodi, *J. Chem. Phys.*, **52**, 3298 (1970); W. G. Vincenti, *Astronaut. Acta*, **15**, 559 (1970).
114. M. Huetz-Aubert, P. Chevalier, and R. Tripodi, *J. Chem. Phys.*, **54**, 2289 (1971).
115. J. D. Lambert, *Quart. Rev.*, **21**, 67 (1967).
116. D. R. White, *J. Chem. Phys.*, **49**, 5472 (1968).

117. C. W. von Rosenberg, Jr., K. N. C. Bray, and N. H. Pratt, *J. Chem. Phys.*, **56**, 3230 (1972).
118. B. H. Mahan, *J. Chem. Phys.*, **46**, 98 (1967).
119. R. D. Sharma and C. A. Brau, *Phys. Rev. Lett.*, **19**, 1273 (1967); *J. Chem. Phys.*, **50**, 924 (1969).
120. E. A. Andreev, *Chem. Phys. Lett.*, **11**, 429 (1971).
121. R. L. Taylor and S. Bitterman, *J. Chem. Phys.*, **50**, 1720 (1969).
122. R. D. Sharma, *Phys. Rev.*, **177**, 102 (1969).
123. R. D. Sharma, *Phys. Rev. A*, **2**, 173 (1970).
124. R. D. Sharma, *J. Chem. Phys.*, **54**, 810 (1971).
125. L. A. Melton and W. Klemperer, *J. Chem. Phys.*, **55**, 1468 (1971); R. G. Gordon and Y.-N. Chiu, *J. Chem. Phys.*, **55**, 1469 (1971).
126. E. H. Fink, D. L. Wallach, and C. B. Moore, *J. Chem. Phys.*, **56**, 3608 (1972).
127. D. Rapp and T. Kassal, *Chem. Rev.*, **69**, 61 (1969).
128. R. C. Millikan, *J. Chem. Phys.*, **38**, 2855 (1963).
129. J. Billingsley and A. B. Callear, *Trans. Faraday Soc.*, **67**, 257 (1971).
130. D. J. Seery, *J. Chem. Phys.*, **56**, 631 (1972).
131. J. F. Roach and W. R. Smith, *J. Chem. Phys.*, **50**, 4114 (1969).
132. W. D. Breshears and P. F. Bird, *J. Chem. Phys.*, **54**, 2968 (1971).
133. R. J. Donovan and D. Husain, *Trans. Faraday Soc.*, **63**, 2879 (1967).
134. R. C. Millikan, *J. Chem. Phys.*, **43**, 1439 (1965).
135. N. Basco, A. B. Callear, and R. G. W. Norrish, *Proc. Roy. Soc. (London) Ser. A*, **260**, 459 (1961).
136. A. B. Callear and G. J. Williams, *Trans. Faraday Soc.*, **60**, 2158 (1964).
137. I. W. M. Smith, *Trans. Faraday Soc.*, **64**, 3183 (1968).
138. R. J. Donovan, D. Husain, and C. D. Stevenson, *Trans. Faraday Soc.*, **66**, 2148 (1970).
139. H. J. Bauer, A. C. C. Paphitis, and R. Schotter, *Physica*, **47**, 58 (1970).
140. H. J. Bauer and R. Schotter, *J. Chem. Phys.*, **51**, 3261 (1969).
141. B. Anderson, F. D. Shields, and H. E. Bass, *J. Chem. Phys.*, **56**, 1147 (1972).
142. T. J. Odiorne, P. R. Brooks, and J. V. V. Kasper, *J. Chem. Phys.*, **55**, 1980 (1971).
143. W. Eastes and D. Secrest, *J. Chem. Phys.*, **56**, 640 (1972).
144. R. G. Gordon and Y. S. Kim, *J. Chem. Phys.*, **56**, 3122 (1972).
145. N. V. Karlov, Yu. B. Konev, and A. M. Prokhorov, *JETP Lett.*, **14**, 117 (1971) .*Zh. Eksperim. i Teor. Fiz.,Pis. Red.*, **14**, 178 (1971)..
146. N. G. Basov, E. P. Markin, A. N. Oraevskii, A. V. Pankratov, and A. N. Skachkov, *JETP Lett.*, **14**, 165 (1971) .*Zh. Eskperim. i Teor. Fiz., Pis. Red.*, **14**, 251 1971).; *Soviet Phys. "Doklady,"* **16**, 445 (1971) .*Dokl. Akad. Nauk. SSSR*, **198**, 1043 (1971)..
147. R. V. Ambartzumian and V. S. Letokhov, *Appl. Opt.*, **11**, 354 (1972).

ESCA*

DAVID A. SHIRLEY

Department of Chemistry and Lawrence Berkeley Laboratory,
University of California, Berkeley, California

CONTENTS

* Work performed under the auspices of the United States Atomic Energy Commission.

I. INTRODUCTION

Since Hagström, Nordling, and Siegbahn[1] discovered in 1964 that chemical shifts in the binding energies of atomic core electrons could be detected by high-resolution energy analysis of photoelectrons ejected by characteristic X-rays, a great deal of interest has developed in the study of structural problems by electron spectroscopy. There are in fact a number of ways in which electron spectroscopy—and by this term we mean experimental methods employing the energy analysis of free electrons—can be applied in chemical physics and related fields. The Uppsala group of K. Siegbahn and co-workers has given the collective name ESCA (Electron Spectroscopy for Chemical Analysis) to these methods. The subject of ESCA in the broad sense is scarcely in need of review at this time, nor could it be covered except in a very superficial manner in the space available here. There are three books available on this subject,[2–4] and the reviewer is aware of at least six more volumes presently in preparation. Instead, this article deals with ESCA in the narrow sense, that is, with X-ray photoelectron spectroscopy (XPS). The time is ripe for a comprehensive and critical review of chemical shifts in core-electron binding energies, which is by all odds the central topic of XPS. Accordingly the bulk of this article deals with these shifts. The remainder is devoted to the two special topics of valence-shell (or valence band) structure and multiplet splitting of core-orbital hole states—two areas in which, in the reviewer's opinion, the contributions of XPS are sufficiently extensive and well understood to afford a reasonably definitive review at this time. Thus this article is selective rather than general, even within the relatively narrow confines of XPS. Several important topics have been omitted either because they have been discussed adequately elsewhere or because they are too new and fragmentary to review at this time. Among these are instrumentation,[2,3] two-electron effects (Auger, "shake-up" and "shake-off"),[2,3,5] the reference level question,[6] and the recent work of Mateescu on charge distributions in norbornyl ion and related structures.[7]

The objectives of this article do not include the compilation of an exhaustive bibliography even on the topics that are discussed. Although it is hoped that no key papers have been overlooked, the reviewer hereby tenders apologies to any authors who have been slighted by omission.

II. CHEMICAL SHIFTS IN CORE-ELECTRON BINDING ENERGIES

A. General Comments

When a characteristic X-ray photon of energy $h\nu$ ejects an electron from an atomic core orbital, the electronic kinetic energy is given by the relation

$$K = h\nu - E_B \tag{2.1}$$

where E_B is the binding energy. This relation is unambigous for a gaseous sample, but for a solid sample it is true as stated only if E_B is the binding energy referred to the spectrometer vacuum level. To obtain the binding energy relative to the sample's vacuum level one must correct for the contact potential between the sample and the spectrometer, which is just the difference between their work functiions,

$$\begin{aligned} K(\text{spect}) &= h\nu - E_B(\text{Fermi}) - \phi(\text{spect}) \\ &= h\nu - E_B(\text{vacuum}) - \phi(\text{spect}) + \phi(\text{sample}) \end{aligned} \tag{2.2}$$

Here ϕ denotes the work function, and K(spect) is the kinetic energy of an electron in the spectrometer. On entering the spectrometer an electron is accelerated by an energy $e[\phi(\text{sample}) - \phi(\text{spect})]$. Although in principle this correction could be made for solid samples, ϕ(sample) is seldom known in practice, and the correction is seldom made. An element of uncertainty is thus introduced into the binding-energy shifts in solids. For this reason detailed comparisons of binding-energy shifts with theory in this article are restricted to data for gaseous samples. This does not imply that shifts in solids cannot be interpreted similarly, but for purposes of evaluating theories of binding-energy shifts, which is our purpose here, gaseous-sample data are clearly preferable. That binding-energy shifts in solids parallel those in gases has been shown explicitly by Gelius et al.,[8] who compared experimental shifts for the same compounds as solids and gases, and implicitly by many workers, who found good correlations between shifts in solids and theoretical parameters for free molecules.

The basic physics of core-level binding-energy shifts can be understood in terms of shielding of the core electrons by electrons in the valence shell. When the charge in the valence shell changes, this shielding changes. A useful, albeit oversimplified, classical analogy is that of a charged conducting hollow sphere. If the charge is Q and the radius R, the potential outside the sphere is

$$\phi(r) = \frac{Q}{r} \qquad r > R \tag{2.3}$$

whereas inside the sphere the potential has a constant value

$$\phi_Q = \frac{Q}{R} \qquad r < R \tag{2.4}$$

Now one component of the energy of a core electron in an atom is the potential energy term due to its interaction with valence electrons. Let us consider the "chemical shift" in binding energy of a core electron between two charge states of the same atom with charges Q and $Q - e$ in the valence shell. That is, we compare the processes for element M

$$\text{I} \quad M^{z}(Q) \rightarrow M^{z+1}(Q, \text{core hole}) + e^{-}$$

and

$$\text{II} \quad M^{z+1}(Q - e) \rightarrow M^{z+2}(Q - e, \text{core hole}) + e^{-}$$

The notation is straightforward. If step I involved the loss of a core electron from ferrous ion, for example, $M^{z}(Q)$ would be $Fe^{2+}(6)$, etc. The binding-energy shift for a core-electron would be in large part given by (minus) the shift in ϕ_Q. Thus

$$\delta E_B \cong -\delta V_Q \cong (-e)(-\delta\phi) \cong (-e)\frac{\delta Q}{R} = \frac{e^2}{R} \tag{2.5}$$

where $V_Q = -e\phi_Q$ is the potential energy of a core electron due to the valence shell and R is assumed to be constant for this estimate.

Three useful inferences can immediately be drawn from this crude model:

1. For any two compounds, the binding-energy shifts of all the core-electron orbitals of a particular atom should be about the same. This follows because ϕ_Q is independent of r, for $r < R$.
2. The sensitivity of E_B (core electrons) to Q varies roughly as the inverse of the valence-shell radius, thus increasing toward the top and right side of the periodic table.
3. For $R = 1$Å, $\delta E_B \cong 14.4$ eV$/|e|$, from (2.5). This gives an order of magnitude estimate for the sensitivity $\delta E_B/\delta Q$.

Fadley et al.[9] made a detailed study of shifts in iodine core levels to confirm these predictions. They also estimated shifts from Hartree-Fock orbital energies in free ions. The agreement of the three approaches (classical, Hartree-Fock, and experimental), shown in Fig. 1, provides evidence that the shifts are qualitatively well understood. For $2s$ through $4d$ orbitals, calculated orbital-energy shifts from atomic $I(5s^2\,5p^5)$ to ionic $I^+(5s^2\,5p^4)$ are essentially constant, as the classical model predicts. Although measurements of binding-energy shifts between these states was not feasible, the total binding-energy shifts from KI to KIO_4 were measured. As Fig. 1 shows, these shifts are essentially the same for all the core orbitals, confirming point 1 above. Point 2 was confirmed by Hartree-Fock calculations on other halogens.[9] The third prediction was also con-

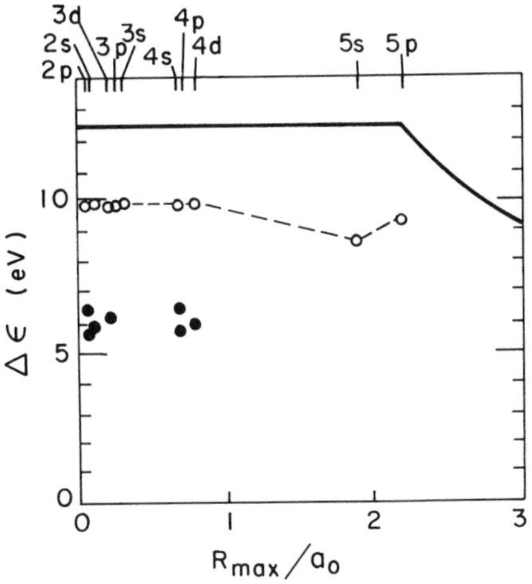

Fig. 1. Core-level binding-energy shifts for electrons in iodine, from Ref. 9. The solid curve is based on the classical model [Eqs. (2.3) and (2.4)], whereas the dashed line connects core-level orbital-energy shifts from I to I+, from Hartree-Fock theory. Filled circles represent experimental shifts from KI to KIO_4. Abscissa is the value of the radial maximum for each orbital, in atomic units.

firmed by the good agreement between the classical and Hartree-Fock results in Fig. 1. Finally the KI–KIO_4 shifts, corresponding to a change of 8 in the oxidation state of iodine, show a change of less than one electron in the iodine valence shell. This implies that an increase of 1 in oxidation state is accompanied by the loss of ~0.1 electron. The ease with which this qualitative conclusion could be drawn is indicative of the directness and power of this method in elucidating chemical structure.

Before discussing methods for calculating the chemical shift in binding energy, which we shall denote as δE, a comment about notation is in order. The complete designation of a chemical shift requires specification of the two compounds as well as the parent atom and orbital. Thus

$$\delta E(\mathrm{C}1s;\ \mathrm{CH_4{-}CF_4}) = E_B(\mathrm{C}1s;\ \mathrm{CF_4}) - E_B(\mathrm{C}1s;\ \mathrm{CH_4}) = 11.0\ \mathrm{eV}$$

indicates that a carbon 1*s* electron is 11.0 eV more tightly bound in CF_4 than in methane. Even this designation is incomplete. This shift has been observed in gaseous sources.[3,10] For solid sources the methane–CF_4 shift is about 1 eV larger.[8] Thus the state of the sample should always be specified. For technical reasons the temperature and pressure should also be given.

A comment on philosophy is also in order. Binding-energy shifts arise from changes in electronic and nuclear charge distributions between one molecular species and another. They are therefore closely related to properties of great chemical interest. Moreover, binding-energy shifts are not only well understood in principle theoretically, but the actual numerical calculation of shifts is straightforward, if tedious. In addition, the origins of these shifts are intuitively obvious: the "back of the envelope" estimates usually give fairly good results, in contrast to the situation for NMR chemical shifts or Mössbauer isomer shifts. Indeed, binding-energy shifts are so well understood that they can be calculated in a number of different ways. Although different approaches may be compared in regard to rigor, accuracy, applicability, etc., it would be impossible to classify them in terms of overall merit because the value of a given method is determined largely by the problem at hand. Instead, the discussion below treats a variety of theoretical approaches to binding energies and shifts approximately in order of decreasing rigor. This ordering is precise only at the beginning, where the different approaches are related by a well-defined series of approximations.

B. Theoretical Description Based on the Calculation of Binding Energies

Chemical shifts in binding energies may be obtained by evaluating the binding energies and taking differences, or they may be estimated directly. The former, more rigorous, approach is discussed first.

1. Methods Involving Both Initial and Final States

a. General Background. The energy relationships among the four states that determine the CH_4–CF_4 chemical shift are indicated in Fig. 2. The total energies of all four states must be calculated in order to predict

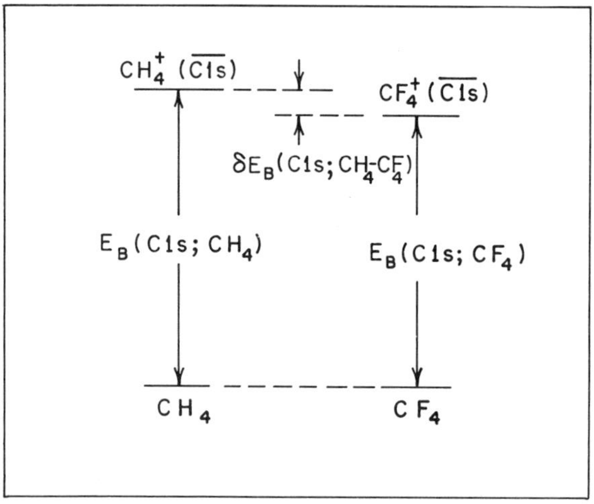

Fig. 2. Relationship of the four states that determine the CH_4–CF_4 shift in the carbon 1*s* binding energy. Bars denote hole states of ions, and equality of initial-state energies indicates that both molecular species are at ground potential.

this shift rigorously. An exact calculation of the total energy of even simple multielectron atomic systems is of course impractical, but self-consistent field (SCF) methods, with corrections, can yield rather accurate total energies. All the binding energies discussed below were calculated by either the Hartree-Fock method or the Dirac-Fock method. The former is based on the (nonrelativistic) Schrödinger equation and the latter on the (relativistic) Dirac equation. Relativistic effects are important for large binding energies, as the electron's velocity in the bound state is no longer negligible compared to the velocity of light. Because the properties of core electrons represent rather unfamiliar territory for most workers, it is useful to have a method for making rough estimates of the magnitude of the relativity correction. The relativistic kinematics of a free particle are described by the relation

$$E^2 = m^2c^4 + p^2c^2 = m^2c^4 + \frac{m^2c^2v^2}{1 - v^2/c^2}$$

Expansion in powers of v/c yields (through order $(v/c)^5$)

$$E = mc^2 + \frac{1}{2}mv^2\left[1 + \frac{3}{4}\left(\frac{v}{c}\right)^2 + \cdots\right]$$

Thus the kinetic energy is increased, and (using the virial theorem) the total energy is decreased, by a fraction $\sim\frac{3}{4}(v/c)^2$ because of relativity. As numerical examples let us consider atomic carbon and argon, with approximate 1s binding energies of 11 a.u. ($\sim$300 eV) and 120 a.u. ($\sim$3200 eV), respectively. The electron's rest mass is 511 keV. Thus.

$$\delta E_B(\text{rel}) \approx \frac{3}{4}\left(\frac{v}{c}\right)^2 E_B \approx \frac{3}{2}\frac{E_B{}^2}{5.11 \times 10^5\ \text{eV}} \approx 0.01 \text{ a.u. for carbon}$$

$$\approx 1.2 \text{ a.u. for argon}$$

In terms of absolute binding energies the relativity correction is therefore fairly unimportant (0.1% or 0.3 eV) for the carbon 1s electron, whereas for the argon 1s case the correction is too large (1% or 30 eV) to ignore. Accurate values of the relativity corrections to the total energies of these two atoms have been given by Veillard and Clementi.[11] They are 0.01381 a.u. (carbon) and 1.76094 a.u. (argon). In both cases most of the correction may be assigned to the two 1s orbitals. Thus our rough estimate of δE_B(rel) is $\sim$50% too large.

The Hartree-Fock Equations for n doubly occupied orbitals have the form (in atomic Units)

$$\left[\left(-\frac{1}{2}\nabla^2 - \sum_k \frac{Z_k}{r_k}\right) + \sum_{j=1}^{n}(2J_j - K_j)\right]\phi_i = \epsilon_i\phi_i$$

Here ϕ_i is a one-electron orbital. The first term in parentheses accounts for an electron's kinetic energy plus its interaction with all nuclei. The Coulomb operator K_j has matrix elements

$$J_{ij} = \langle\phi_i(\mu)|J_j|\phi_i(\mu)\rangle = \langle\phi_i(\mu)|\,\langle\phi_j(\nu)|\frac{1}{r_{\mu\nu}}|\phi_j(\nu)\rangle|\phi_i(\mu)\rangle$$

$$K_{ij} = \langle\phi_i(\mu)|K_j|\phi_i(\mu)\rangle = \langle\phi_i(\mu)|\,\langle\phi_j(\nu)|\frac{1}{r_{\mu\nu}}|\phi_i(\nu)\rangle|\phi_j(\mu)\rangle$$

where μ and ν label electron coordinates. There are n Hartree-Fock equations in ϕ_i, with $i = 1, 2, \ldots, n$. They must be solved iteratively until self-consistency is achieved, yielding a total determinantal wave function

$$\psi = |\phi_1{}^2\phi_2{}^2 \cdots \phi_n{}^2|$$

The individual Hartree-Fock equations are "pseudo-eigenvalue" equations. The left-hand side of the Hartree-Fock equation can be abbreviated to $\mathfrak{F}\phi_i$, where $\mathfrak{F}$ is called the Fock operator. The orbital energies ϵ_i are given by

$$\epsilon_i = \langle\phi_i|\mathfrak{F}|\phi_i\rangle = \epsilon_i{}^0 + \sum_j (2J_{ij} - K_{ij}) \tag{2.6}$$

The one-electron energy $\epsilon_i{}^0$ is the expectation value of the one-electron operator $-\frac{1}{2}\nabla^2 - \sum_n Z_n/r_n$. The total Hartree-Fock energy of a system with n doubly occupied orbitals is

$$E_{HF}^{(2n)} = 2\sum_{i=1}^{n} \epsilon_i{}^0 + \sum_{i,j=1}^{n} (2J_{ij} - K_{ij}) \tag{2.7}$$

To obtain the Hartree-Fock approximation to the adiabatic ionization potential of the kth orbital of this system, another Hartree-Fock energy $E_{HF}^{(2n-k)}$ must be obtained by solving the Hartree-Fock equation with the kth orbital singly occupied. An approximate value $E_{HF}^{(2n-k)'}$ can be estimated, however, simply by striking from (2.7) those terms that can be associated with a single electron in the kth orbital; that is,

$$E_{HF}^{(2n-k)'} = \epsilon_k{}^0 + 2\sum_{i\neq k}^{n} \epsilon_i{}^0 + \sum_{i,j\neq k}^{n} (2J_{ij} - K_{ij}) + \sum_{i\neq k}^{n} (2J_{ij} - K_{ij}) \tag{2.8}$$

This is the final-state energy that the system would have if there were no relaxation of the passive orbitals during ionization, that is, if the ionization were "sudden." The sudden ionization energy of the kth orbital in the Hartree-Fock scheme,

$$E_{HF}^{(2n)} - E_{HF}^{(2n-k)'} = \epsilon_k{}^0 + \sum_{i\neq k}^{n} (2J_{ik} - K_{ik}) + J_{kk} \tag{2.9}$$

is just equal to the orbital energy ϵ_k, as comparison of (2.6) and (2.9) will show. This result was first shown by T. Koopmans.[12] It is known as Koopmans' theorem. Because neglecting final-state relaxation (i.e., taking the integrals in (2.8) to having the same values as in (2.7)) always gives final-state energies that are too high, the sudden approximation overestimates binding energies. By contrast, neglect of relativity clearly tends to give estimates of ionization potentials that are too small. Further discussion of orbital-energy estimates of binding energies is given in Section II.A.2.

The use of the Hartree-Fock method to calculate the energy of a highly excited state (such as a 1*s* "hole" state) is subject to question. For the lowest state of each symmetry type the variational principle guarantees that the Hartree-Fock energy gives an upper bound to the true energy of the system. For a higher state of a given symmetry type this holds only if McDonald's theorem[13] is satisfied, that is, only if the state in question is orthogonal to all lower states of its symmetry type. For most systems a 1*s* hole state is so high in energy that there are many lower-lying states of the same symmetry, and orthogonalizing it to all these states would be a formidable task. Bagus[14] and Verhaegen et al.[15] have pointed out, however, that the 1*s* hole state is unusual in that the 1*s* orbital has little overlap with the valence orbitals. Off-diagonal elements of the interaction matrix between the 1*s* hole state and the lower states will therefore be small in comparison to differences in their energies. Thus the hole state would be changed little by orthogonalizing it to the lower states. In the calculations to date, hole states have simply been treated as if they satisfied McDonald's theorem.

b. Results for Atoms and Ions. Bagus[14] made the first complete Hartree-Fock calculations on the hole states that can be formed by ejecting an electron from the closed-shell configurations F^-, Ne, Na^+, Cl^-, Ar, and K^+. After making a relativity correction, he found that core-electron binding energies calculated from initial- and final-state energies agreed better with experiment than did those estimated from ground-state orbital energies. His results for the *s*-type hole states of neon and argon are set out in Table I. In comparing his results with experiment, Bagus observed that the remaining errors in the binding energies were positive in some cases and negative in others. This was somewhat unexpected, because these errors arise almost entirely from differences between the electron correlation energies of the initial and final states. From a naive point of view the correlation energies might be expected to be pairwise transferable. The initial state would then always have a larger correlation energy than the final state simply because it has more electrons, and the true binding energy would always be larger than the Hartree-Fock value corrected for relativity. In fact the opposite is often true. Bagus pointed out how this can be understood in terms of configuration interaction. The $Ne^{+}(1s^2\,2s\,2p^6)$ state, for example, can interact with configurations made by promoting only one electron to states with principal quantum number $n > 2$ (e.g., $1s^2\,2s^2\,2p^4\,ns$). Thus its total energy is lowered more by electron correlation than is that of neutral neon in its ground state, because the configuration $1s^2\,2s^2\,2p^6$ can interact only with configurations formed by promoting two or more electrons to states with $n > 2$.

TABLE I

The *ns* Binding Energies of Neon and Argon (after Bagus[14])

Final State[a]	E_B(expt)[b]	ΔE_B^{reld}	E_B (sudden)[e]	ΔE_B (sudden)[f]	E_B(adiabatic) $= E_{\text{hole}} - E_0$	ΔE_B(adiabatic)[f]
$Ne(\overline{2s})$	1.7815		1.9303	−0.1488 (−4.049 eV)	1.8123	−0.0308 (−0.838 eV)
$Ne(\overline{1s})$	31.981(4)[c]	0.040	32.812	− 0.831 (−22.6 eV)	31.961	+0.020 (+0.54 eV)
$Ar(\overline{3s})$	1.0745		1.2773	−0.2028 (−5.518 eV)	1.2198	−0.1453 (−3.954 eV)
$Ar(\overline{2s})$	11.992(3)[c]		12.3219	−0.330 (−9.0 eV)	11.938	+0.054 (+1.5 eV)
$Ar(\overline{1s})$	117.83(2)[c]	0.54	119.15	− 1.32 (−35.9 eV)	117.97	−0.13 (−3.5 eV)

[a] Bar denotes hole relative to ground-state neutral atom initial state.
[b] Energies are in Hartrees (1 Hartree = 27.210 eV) unless otherwise indicated.
[c] New experimental value from Ref. 3. Error in last place is given parenthetically.
[d] Relativity correction which has been added to Bagus' theoretical binding energies to give results in columns 4 and 6.
[e] From orbital energies and relativity correction.
[f] Here $\Delta E_B = E_B(\text{expt}) - E_B\,(\text{theory})$.

Even without corrections for correlation, Bagus' hole-state binding energies were accurate to 0.2%.

Rosen and Lindgren[16] carried out relativistic Hartree-Fock-Slater (HFS) calculations on many atoms. They have given results for states in Cu, Kr, I, Eu, Hg, and U. HFS calculations are ordinarily more approximate than

HF calculation because exchange effects are estimated by using the "Slater exchange potential,"

$$V_{ex}(r) = -\left[\frac{81}{32}\frac{\rho(r)}{\pi^2 r^2}\right]^{1/3}$$

However, these workers parameterized this potential as

$$V'_{ex}(r) = -\frac{c}{r}\left[\frac{81 r^n \rho(r)^m}{32 \pi^2}\right]^{1/3}$$

and did variational calculations to optimize c, n, and m. Their total energies obtained with optimized parameters appear to be essentially identical to the Hartree-Fock values. They calculated binding energies in two ways. Their "Method A" was based on "frozen" orbitals and therefore similar to the orbital energy approach (but not identical, as Koopmans' theorem doesn't hold for the HFS approximation). In their "Method B," optimized HFS calculations were carried out on both the initial and the final states, and the binding energies were obtained by difference. The agreement of these binding energies with experiment varied considerably, but Method B tended to be within 1% of experiment and Method A within 2%. In many cases the agreement was much better than these figures. Results for several states with binding energies in the range of interest for ESCA studies are given in Table II.

TABLE II

Relativistic Optimized HFS Binding Energies,[a] after Rosen and Lindgren[16]

State	E_B Method A	E_B Method B	E_B Expt	State	E_B Method A	E_B Method B	E_B Expt
Kr $2p_{1/2}$	64.82	63.75	63.47	Eu $3d_{3/2}$	43.81	42.88	42.80
Kr $2p_{3/2}$	62.82	61.77	61.55	Eu $4s$	14.24	13.79	13.61
Kr $3s$	11.20	10.88	(10.6)	Eu $4p_{1/2}$	11.57	11.15	10.77
Kr $3p_{1/2}$	8.60	8.27	8.18	Eu $4d_{3/2}$	5.99	5.64	5.34
Kr $3p_{3/2}$	8.30	7.97	7.86	Hg $4s$	30.66	30.04	29.58
I $3s$	40.45	39.91	39.55	Hg $4p_{1/2}$	26.12	25.50	25.04
I $3p_{1/2}$	35.27	34.71	34.34	Hg $4d_{3/2}$	14.80	14.27	14.07
I $3d_{3/2}$	24.12	23.52	23.35	Hg $4f_{5/2}$	4.48	4.00	3.93
I $4s$	7.73	7.53	7.00				

[a] All energies are in Hartrees.

The binding-energy calculations of Rosen and Lindgren are extended over the periodic table, and to discuss them in detail would be beyond the scope of this article. In seeking an explanation for the significant residual discrepancies between theory and experiment they invoked the Lamb shift for the inner shells of heavy atoms, and the magnitude of this effect appears to be about correct. For smaller binding-energy cases they discounted correlation effects because they expected correlation to increase the theoretical binding energies, which were already too large. In view of Bagus' discussion this interpretation should be reinvestigated.

Gianturco and Coulson[17] made both orbital-energy estimates (Method A) and hole-state calculations (Method B) of the binding energies of atomic and ionic sulfur. They did both HF and HFS calculations. Their results for neutral sulfur are given in Table III. The binding energies agree well with experiment, especially for the Hartree-Fock Method B case.

TABLE III

Sulfur Binding Energies (in eV) after Gianturco and Coulson[17]

State	Method A HF	Method A HFS	Method B HF	Method B HFS	Expt (X-ray)	Expt (ESCA)
1*s*	2503.6	2496	2474	2484	2476	2477
2*s*	245	239	233	238	231	233
2*p*	181	174	172	173	170	169
3*s*	23	22	22	23	21	—
3*p*	11.5	10	11	10	12	—

These authors also studied the variation of E_B with ionic state for the *K* and *L* shells of sulfur. They found average shifts of $\sim$16 eV per ionization state in each case. This is expected, as Fadley et al.[9] have discussed, because the shielding in both these inner shells is essentially complete. In fact the simple argument that shielding by the *M* shell is less complete for the *L* shell than for the *K* shell correctly predicts the *L*-shell shifts to be slightly smaller than the corresponding *K*-shell shifts. The similarity of magnitudes of the two bodes ill for the sensitivity of shifts in the K_α X-ray emission lines, however, as Gianturco and Coulson noted. In fact the situation for the range of sulfur charge states (not oxidation states) that is chemically realizable, that is, about -1 to $+2$, is much worse than might be inferred from their complete calculation for ionic charge states between -2 and $+6$, because the K_α energies shift little with charge for charge states near zero compared to the shifts in highly ionized states.

Again the physical reason for this is clear: as the charge in the sulfur ion increases through successive loss of the *M* electrons, the radial wave functions of those remaining are drawn in until they can no longer shield the *L* shell as effectively as the *K* shell. For charge states near zero, on the other hand, both *K* and *L* electrons are in more nearly true "core" orbitals, in which their energies are affected equally by the valence shell. The sensitivity advantage of ESCA shifts over K_α shifts is illustrated for this case in Fig. 3.

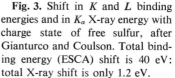

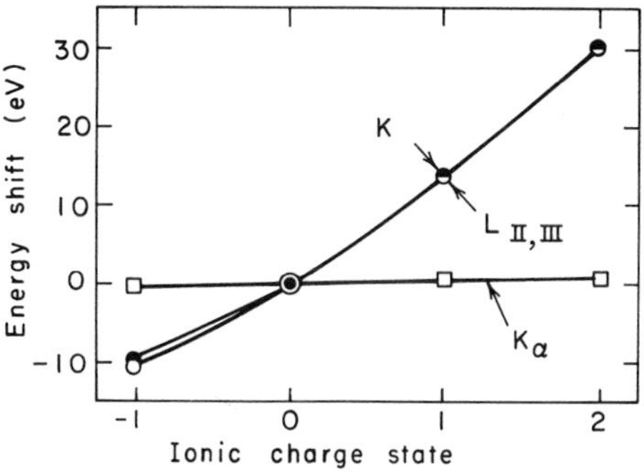

Fig. 3. Shift in *K* and *L* binding energies and in K_α X-ray energy with charge state of free sulfur, after Gianturco and Coulson. Total binding energy (ESCA) shift is 40 eV: total X-ray shift is only 1.2 eV.

Verhaegen et al.[15] described an accurate calculation of the 1*s* and 2*s* binding energies of atomic neon. Their calculation is unique in that they evaluated explicitly the effects of relativity and correlation on the binding energies of the 1*s* and 2*s* orbitals. Table IV displays their results for these two cases. These workers accounted for relativity directly, by solving the Dirac-Fock (or relativistic Hartree-Fock, RHF) equation.

Verhaegen et al. extended core-level binding-energy calculations to a higher level of sophistication by including accurate corrections for correla-

TABLE IV

Binding Energies in Neon (in eV) after Verhaegen et al.[15] and Bagus[14]

Method	$E_B(1s)$ (Ref. 15)	$E_B(1s)$ (Ref. 14)	$E_B(2s)$ (Ref. 15)	$E_B(2s)$ (Ref. 14)
Orbital energy	891.7	891.7	52.5	52.5
HF hole state	868.6	868.6	49.3	49.3
RHF hole state	869.4	869.7	49.4	—
RHF hole state plus correlation	870.8	(871.1)	48.3	(48.2)
Experiment (Ref. 2)	870.2		48.4	

tion energy. It is necessary, in open-shell systems, to distinguish among three kinds of correlation energy which Öksüz and Sinanoğlu[18] have classified as "internal," "polarization plus semi-internal," and "all-external." In internal correlations, two electrons correlate but shift to vacant Hartree-Fock orbitals. One electron in a pair shifts to a vacant Hartree-Fock orbital and one shifts out of the "Hartree-Fock sea" in semi-internal correlation, whereas in the "all-external" case both electrons shift to non-Hartree-Fock orbitals. Clearly only the last kind of correlation is available to closed-shell systems. Verhaegen et al. noted that correlation effects will shift the energy of the atomic $Ne^+(\overline{1s})$ state relative to $Ne(^1S)$ by an amount

$$\Delta E_{\mathrm{corr}} = -[\epsilon(1s,\overline{1s}) + \epsilon(1s,2s) + \epsilon(1s,\overline{2s}) + 3\epsilon(1s,2p) + 3\epsilon(1s,\overline{2p})]$$

where the pair correlation energies ϵ are all negative. This expression is based on the assumption that all the pair correlations are transferable. In fact this is essentially true for the (closed-shell) ground state; whereas the 1*s* hole state also has the possibility of semi-internal correlation. This was assumed to be negligible, however, because of the unusual nature of this highly excited state. Using Nesbet's[19] values for the pair correlation energies in $Ne(^1S)$, obtained by solving the Bethe-Goldstone equations, Verhaegen et al. found $\Delta E_{\mathrm{corr}} = +0.0524$ a.u. for this case. This direction, an increase in E_B because of correlation, is the normal one. For the $Ne^+(\overline{2s})$ case, however, semi-internal correlation is important, lowering the energy by 0.0978 a.u., according to the calculations by Öksüz and Sinanoğlu. This is more than enough to offset the external contribution of 0.0542 a.u. to the energy difference between $Ne^+(\overline{2s})$ and $Ne(^1S)$, and the net contribution of correlation is to decrease the binding energy by 0.0436 a.u. (See also note added in proof, p. 156.)

Table IV deserves careful scrutiny, because neon, with its closed shells, is the ideal test case for calculating core-electron binding energies. Furthermore, the results given for neon represent the first detailed attempt to include electron correlation effects. Bagus' results from Table I are included for comparison. There is a discrepancy of 0.3 eV (0.01 a.u.) in the RHF 1*s* binding energies before the correlation correction. The 0.6 to 0.9 eV discrepancy between theory and experiment is perplexing. The experimental accuracy is 0.1 eV, so the difference must be taken seriously. It is by no means obvious where the error lies, but the assumption that the "unusual" nature of the 1*s* hole state justifies simplifying approximations in the calculation of the state's HF energy and its correlation energy deserves further study. (See note added in proof, p. 156.)

c. Results for Molecules. Schwartz[20] reported the first hole-state calculations in molecules. He used a Gaussian orbital basis set for both excited and ground states, and tested his method of choosing orbitals by comparison of calculations on neon with Bagus' results.[14] Neither relativity nor correlation were explicitly considered. Calculations were made on BH_3, CH_4, NH_3, H_2O, HF, and Ne. At that time experimental 1*s* binding energies were known only for CH_4 and Ne. Since then results for NH_3 and H_2O have become available. The calculated values, shown in Table V, are in excellent agreement with experiment. In fact the agreement is fortuitously good. The relativity correction will increase the binding energy by from 0.1 eV for CH_4 to 1.2 eV for Ne,[21] whereas correlation will raise E_B by about 1.2 eV for each case.[22,23] The values of E_B after these corrections are given in Table V. (See note added in proof, p. 156).

Hillier, Saunders, and Wood[24] took a different approach to calculating $E_B(1s)$ for CH_4, H_2O, and CO. They calculated the ground state using a double zeta basis set of Slater type orbitals, found the Koopmans' theorem E_B, and then accounted for relaxation in the valence orbitals via a CI calculation on the ions using virtual molecular orbitals from the neutral-molecule calculations. It is difficult to evaluate the results of this work, but a few observations can be made. First, the basis set was small, and the total energies were high by up to 0.15 a.u. (4 eV). The orbital energies were in error by only $\sim$0.5 eV, however. The CI calculation presumably accounted for some correlation, but not 1*s*–1*s* correlation, since only valence electrons were considered. Finally, relativity was neglected. The question of whether extraatomic polarization effects differ enough from one molecule to another that hole-state calculations would predict shifts better than would orbital energies is unfortunately left open. The O1*s* shift between CO and H_2O is improved from 4.86 eV using orbital energies to 3.63 eV when relaxation is considered (the experimental shift is 2.94 eV). In contrast the C1*s* shift between CO and CH_4 is worsened (5.46 eV from orbital energies, 7.55 eV with relaxation, and 5.4 eV experimentally).

Brundle, Robin, and Basch[25] have carried out C1*s* hole-state calculations on CH_4, CH_3F, CH_2F_2, and CHF_3. Their values of E_B (C1*s*) are about 4 eV higher than experiment after our (problematical) correction for relativity and correlation. These authors suggested that the correlation energy may be larger in the hole states than in the ground states of these molecules, thus tending to decrease the theoretical binding energies toward the experimental values. This seems unlikely, but even if it were true, it would presumably obviate the basis upon which the hole-state calculations were made to begin with, namely, the "unusual" nature of the 1*s* hole state. The success of this calculation in predicting shifts will be discussed in a later section.

TABLE V

Hole-State 1*s* Binding Energies in Molecules

Molecule	Hole	E_B(expt)[a]	E_B(theory)	E_B(theory, corrected)[l]
BH_3	B1*s*	—	197.5[e]	198.2
CH_4	C1*s*	290.8[10], 290.7[3]	291.0,[e] 290.7,[f], 292.9,[g] 298.0,[h] 283.8[m]	292.3, 292.0, 294.2, 299.3, 285.1
NH_3	N1*s*	405.6[3]	405.7,[e] 408.6,[h] 400.9[m]	407.1, 410.0, 402.3
H_2O	O1*s*	539.7[3]	539.4,[e] 539.7,[f] 540.8,[h] 512.3[m]	541.0, 541.3, 542.4, 512.9
HF	F1*s*	—	693.3[e]	695.1
CO	C1*s*	296.2, 295.9[3]	295.9[f]	297.2
CO	O1*s*	542.6, 542.1[3]	542.1[f]	543.7
CH_3F	C1*s*	293.6, 293.5[c]	296.3[g]	297.6
	F1*s*	692.4[10]		
CH_2F_2	C1*s*	296.4[b]	299.3[g]	300.6
	F1*s*	693.1[b]		
CHF_3	C1*s*	299.1, 298.8[3]	302.7[g]	304.0
	F1*s*	694.1[10]		
CF_4	C1*s*	301.8, 301.8[3]		
	F1*s*	695.0[10]		
C_2H_4	C1*s*	290.7	298.1[h]	299.4
NO	N1*s*, $^3\pi$	410.3[3]	411.17[i]	412.5
	N1*s*, $^1\pi$	411.8[3]	412.52[i]	413.9
	O1*s*, $^3\pi$	543.3[3]	542.1[i]	543.7
	O1*s*, $^1\pi$	544.0[3]	542.5[i]	544.1
O_2	O1*s*, $^4\Sigma_g^-$	543.2, 543.1[10]	542.0[j]	543.6
	O1*s*, $^2\Sigma_g^-$	544.3, 544.2[10]	542.6[j]	544.2
Furan	O1*s*	539.4[d]	547.75[k]	549.3

[a] Values from Refs. 3 and 10 are so labeled. If no reference is given, shifts are from D. W. Davis, J. M. Hollander, D. A. Shirley, and T. D. Thomas, *J. Chem. Phys.*, **52**, 3295 (1970) and hydride reference values from Ref. 10 (C) or 3 (N, O). Accuracy is ±0.1 to ±0.2 eV.

[b] D. W. Davis, D. A. Shirley, and T. D. Thomas, *J. Chem. Phys.*, **56**, 671 (1972).

[c] Ref. 8.

[d] Uppsala group, quoted by P. Siegbahn, *Chem. Phys. Lett.*, **8**, 245 (1971).

[e] Ref. 20.

[f] Ref. 24.

[g] From C. R. Brundle, M. B. Robin, and H. Basch, *J. Chem. Phys.*, **53**, 2196 (1970).

[h] R. Moccia and M. Zandomeneghi, *Chem. Phys. Lett.*, **11**, 221 (1971).

[i] P. S. Bagus and H. F. Schaefer, III, *J. Chem. Phys.*, **55**, 1474 (1971).

[j] P. S. Bagus and H. F. Schaefer, III, *J. Chem. Phys.*, **56**, 224 (1972).

[k] P. Siegbahn, *Chem. Phys. Lett.*, **8**, 245 (1971).

[l] The following corrections for relativity plus correlation have been estimated by the reviewer: B, 1.2 eV; C, 1.3 eV; N, 1.4 eV; O, 1.6 eV; F, 1.8 eV. See Refs. 18, and 21–23, and note added in proof, p. 156.

[m] Ref. 26.

Gianturco and Guidotti[26] have cast doubt on hole-state calculations that simply follow the aufbau criterion of emptying a hole-state orbital in the ground-state basis set. In order to test the effect of basis-set flexibility, these workers employed very large basis sets: 39 Slater-type orbitals for CH_4, 32 for NH_3, and 29 for H_2O. The $1s$ binding energies obtained were low by 7.0 eV, 4.7 eV, and 27.4 eV, respectively. They attributed this discrepancy to the failure of the SCF calculations to provide a true upper bound because of lack of orthogonality to lower states. As especially clear evidence of this effect they noted that although their calculations gave discrepancies of only a few electron volts for CH_4 and NH_3, a larger discrepancy was found for H_2O, for which they used a richer atomic basis set on the heavy atom, and in the ground state of which there are three filled a_1 molecular orbitals.

Moccia and Zandomeneghi[27] have offered a solution to the above problem. They used an approach called the strong orthogonal group function (GF) approximation.[28] They approximated the neutral-molecule wave function by an antisymmetrized product of geminals constructed from sets of orbitals localized around the K shell. The K-hole state was then described by eliminating one electron from the K geminal. They stated that this choice of orbitals has the effect of preventing the exaggerated mixing of states of the same symmetry which tends to spoil hole-state calculations in molecules. Their GF binding energies were too large in all four cases calculated—CH_4, NH_3, H_2O, and C_2H_4 (see Table V). This was also taken as evidence that SCF GF results can be trusted not to collapse in the way that ordinary hole state calculations do. Clearly this whole question needs further study, especially since other workers have not found hole-state calculations to collapse in the way that Gianturco and Guidotti did.

Bagus and Schaefer[29] used a very large basis set of 18 Slater orbitals on each atom to calculate $E_B(1s)$ for transitions to the four hole states $NO^+(\overline{N1s};\, {}^1\pi)$, $NO^+(\overline{N1s};\, {}^3\pi)$, $NO^+(\overline{O1s};\, {}^1\pi)$, and $NO^+(\overline{O1s};\, {}^3\pi)$. Their results agreed very well with experiment for the $\overline{O1s}$ states and were about 2 eV high for the $N1s$ states after estimated corrections for relativity and correlation were applied (Table V). These same authors[30] made hole-state calculations on O_2. With a similar basis set of 18 Slater orbitals on each atom they obtained similar excellent agreement with experiment after corrections were applied: discrepancies of only 0.6–0.7 eV for the $O_2^+(\overline{1s};\, {}^4\Sigma_g^-)$ state and 0.1–0.2 eV for the $O_2^+(\overline{1s};\, {}^2\Sigma_g^-)$ state. An important result of this calculation is that it established the localized nature of the $1s$ hole in these final states. This topic has also been clarified in a lucid discussion on atomic relaxation energies by Snyder.[31] The problem is this: in molecules possessing two or more equivalent atoms, proper symmetry of the total molecular state (${}^3\Sigma_g^-$ in O_2, for example) is often regarded as being achieved by a simple aufbau approach. That is, molecular orbitals such as $1\sigma_g$ and $1\sigma_u$ are successively filled with electrons. It is then natural to imagine that a photoemission event in which an oxygen *K* electron is ejected will result in a final state described by the O_2 molecular orbital designations, with a hole in the σ_g or σ_u shell. In such a "delocalized" hole state each oxygen would have an electron population of $7\frac{1}{2}$. Such a state is unstable, however, relative to localization of the 1*s* hole. Snyder has stated the reason for this result succinctly. After discussing relaxation of the passive orbitals in terms of atomic shielding constants, he noted, ". . . one expects the relaxation energy to be quadratic in the charge of the hole." He argued that distribution of a hole over *n* centers would produce a hole charge of $1/n$ on each center and thus approximately a total atomic relaxation energy $1/n$ times as large as that for a localized hole. For $N_2^+(\overline{1s})$ he showed that an improvement of about 7 eV in the atomic relaxation energy could be expected if the hole were localized. Bagus and Schaefer[30] made hole-state calculations with *g* or *u* symmetry imposed on the 1*s* hole states, finding $E_B(1s) = 554.4$ eV, in poor agreement with the experimental value of 543.1 eV. When the symmetry restriction was relaxed, the Hartree-Fock equations yielded two equivalent solutions at $E_B(1s) = 542$ eV, corresponding to a 1*s* hole on either oxygen atom. They pointed out that a total wave function of the proper Σ_g or Σ_u symmetry can be formed from these two localized hole states. The 12 eV relaxation energy is actually in quite good agreement with Snyder's estimate of 7 eV for a nitrogen atom, because it is clear from the results of Bagus and Schaefer that a great deal of *molecular* relaxation, or polarization of valence electrons toward the localized 1*s* hole, takes place. The gross atomic population that these workers found for the $O_2^+(\overline{1s};\ ^4\Sigma^-)$ state are given in Table VI.

P. Siegbahn[32] made a hole-state calculation on the ion formed by ejecting an oxygen *K* electron from furan. This calculation is of interest because furan is by far the largest and least symmetrical molecule on which hole-

TABLE VI

Gross Atomic Populations for the Localized $O_2^+(\overline{1s};\ ^4\Sigma^-)$ State (Ref. 30)

Shell[a]	Oxygen A	Oxygen B	Shell	Oxygen A	Oxygen B
$1s_A$	1.00	0.00	$\sim 3\sigma_g$	1.03	0.97
$1s_B$	0.00	2.00	$\sim 1\pi_\mu$	3.44	0.56
$\sim 2\sigma_g$	1.13	0.87	$\sim 1\pi_g$	0.26	1.74
$\sim 2\sigma_\mu$	0.92	1.08	Total	7.78	7.22

[a] Molecular orbital designations are approximate.

state calculations have been attempted. The result is a lowering of the binding energy well over halfway from the Koopmans' theorem value toward the experimental value, which may be taken as encouraging, especially because Siegbahn used a smaller basis set for the ion calculation.

Before going on to the less rigorous theories, let us review and criticize the present situation in hole-state calculations. In broad outline core-level binding energies are well understood and can be quite accurately calculated for very small systems. At a slightly finer level of detail there are still several important open questions. In heavy atoms there are large discrepancies between theoretical and experimental binding energies, perhaps arising from quantum electrodynamic (Lamb shift) effects. In neon, the best-studied case to date, for which both relativity and correlation have ostensibly been dealt with rigorously, a residual discrepancy of less than 1 eV remains in $E_B(1s)$. In most of the hole-state binding energy calculations on molecules to date, the results were within 4 eV of experiment after relativity and correlation corrections of questionable applicability had been made. Because the theoretical E_B's tended to be larger than experimental values, an expansion of the basis set for the hole-state calculations would ordinarily be indicated. However, the $1s$ hole states lie above, and are not orthogonal to, other states of the same symmetry, so that these calculations are not protected by a variation principle. Thus basis-set expansion requires care. Although it has been argued that the "unusual" nature of these states obviates the need for orthogonalization, the arguments presented in the molecular case are not very rigorous or quantitative, and more work on this question is needed. On the positive side, it is now clear that core hole states in symmetric molecules are localized. Hole-state calculations predict shifts fairly accurately, and it appears that hole-state calculations can be extended to larger molecules.

2. *Methods Involving the Initial State Only*

a. Connection Between Hole-State and Frozen-Orbital Calculations. As indicated in (2.9), the binding energy can be estimated as simply a one-electron orbital energy. We shall use the notation

$$E_B{}^k(KT) \equiv -\epsilon(k) \tag{2.10}$$

to represent the binding energy of the kth orbital as estimated by a "sudden" approximation in which the passive orbitals are frozen, whether or not Koopmans' theorem is rigorously applicable to the particular case under discussion. When this theorem is applicable, $E_B{}^k(KT)$ is given by (2.9). Before discussing actual results obtained using $E_B(KT)$, it is instructive to relate $E_B(KT)$ to $E_B(\text{hole})$, the theoretical binding energy obtained from hole-state SCF calculations. Of the available discussions of the relationship between $E_B(KT)$ and $E_B(\text{hole})$, three are reviewed below. These three approaches differ in detail, and each affords a unique physical insight.

Hedin and Johansson[33] formulated the correction that must be applied to $E_B{}^k(KT)$ to bring it down to $E_B{}^k(\text{hole})$ in terms of a polarization potential created by the presence of a hole in the kth orbital. Specifically, they wrote the Hartree-Fock Hamiltonians for the ground state and k-orbital hole state in terms of the one-electron operator h and the operators V_i describing two-electron Coulomb plus exchange interaction as

$$H = h + V = h + \sum_i V_i$$

$$H^* = h + V^* = h + V - V_k + V_p \tag{2.11}$$

where an asterisk denotes the hole state. Here V_p is a polarization potential,

$$V_p = \sum_{i \neq k} (V_i{}^* - V_i) \tag{2.12}$$

that describes the change in the Hartree-Fock potential accompanying the removal of an electron from the kth orbital. Hedin and Johansson proved by a straightforward derivation that with the neglect of some small terms the true binding energy and orbital energy of the kth orbital are related by (in our notation)

$$E_B{}^k(\text{hole}) = E_B{}^k(KT) + \tfrac{1}{2}\langle k | V_p | k \rangle \tag{2.13}$$

where $|k\rangle$ is the kth one-electron orbital. These authors suggest that this result can be understood physically by the hypothetical two-step process: (1) adiabatic relaxation of electrons in all other orbitals following the "switching off" of the charge of the electron in the kth orbital, thereby storing energy $\tfrac{1}{2}\langle k | V_p | k \rangle$, followed by (2) ejection of this electron, which would now have a binding energy given by (2.13) (using Koopmans' theorem, now valid because no further relaxation can occur). This result was shown to be very nearly equivalent to Liberman's[34] suggestion that E_B is essentially the arithmetic mean of the orbital energies for the kth orbital in the ground state and hole state; that is,

$$E_B{}^k(\text{hole}) \cong \tfrac{1}{2}[E_B{}^k(KT) + E_B{}^k(KT)^*] \tag{2.14}$$

Using this relation, which he derived using the similar approach of Brenner and Brown,[35] Liberman found $E_B(\text{argon } 1s) = 118.0$ a.u., in excellent agreement with the values in Table I.

TABLE VII
Polarization Potential Corrections to Na^+ Binding Energies
(after Hedin and Johansson[33])

Final State	E_B(expt)[a]	E_B(expt) $- E_B(KT)$[b]	$\frac{1}{2}\langle k \mid V_p \mid k \rangle$	E_B(hole) $- E_B(KT)$
$Na^{2+}(\overline{1s})$	40.000	0.762	0.822	0.828
$Na^{2+}(\overline{2s})$	2.924	0.166	0.102	0.105
$Na^{2+}(\overline{2p})$	1.741	0.056	0.113	0.117

[a] All energies are in a.u. (27.210 eV).
[b] The experimental values have been corrected for relativity and for solid-state effects.

Results from the polarization potential method, applied to Na^+, are given in Table VII. The polarization potential model appears to work very well for core orbitals. From their results on Na, K, Na^+, and K^+, Hedin and Johansson concluded that on formation of a hole in a given shell the relaxation of more tightly bound shells is negligible, and that intrashell relaxation is small in comparison to the relaxation of outer shells. They also found that relaxation effects are only weakly dependent on Z. For "outer core" levels (e.g., the $2s$ level in Na^+) they found that E_B(hole) gave no better agreement with experiment than did $E_B(KT)$. This was attributed to the presence of only intrashell relaxation in these cases.

Manne and Aberg[36] have given an especially clear picture of the relationship between the Koopmans' theorem "state" $\psi_{KT}{}^k(N-1)$ formed (as an abstract concept: it does not exist in nature) by removing an electron from the kth orbital of an N-electron system and the real final states of the system $\psi_i{}^k(N-1)$. Since the latter form a complete set, they could write

$$\psi_{KT}{}^k(N-1) = \sum_{i=0}^{\infty} \langle \psi_i{}^k \mid \psi_{KT}{}^k \rangle \psi_j{}^k(N-1) \tag{2.15}$$

Now $\psi_{KT}{}^k(N-1)$ and $\psi_i{}^k(N-1)$ are, respectively, the initial and final states of the $N-1$ passive orbitals. By assuming the transition moment for photoemission to be energy independent, they showed that the transition moment to each state i is proportional in magnitude to the above overlap integral. Thus the probability of the system going to final state i is proportional to the square of this integral, and the energy sum rule

$$E_B{}^k(KT) = E_B{}^k(\text{hole}) + \sum_{i=1}^{\infty} \mid \langle \psi_i{}^k \mid \psi_{KT}{}^k \rangle \mid^2 (E_i - E^k(\text{hole})) \tag{2.16}$$

follows. The lowest-lying hole state [$i = 0$ in (2.15)] has been separated here from the sum over final state of higher excitation. Such states have been observed as high (binding) energy satellite peaks in photoelectron spectra. By rearranging (2.16) it is evident that $E_B{}^k(KT)$ measures the *average* energy of the spectrum. Manne and Aberg calculated an average binding energy of 886 ± 1 eV from the Ne $1s$ spectrum reported by Krause et al.[37] This is in fair agreement with the Koopmans' theorem value of 892 eV. Finally, these authors pointed out that there is a strong analogy between the phenomena described by (2.16) and the Franck-Condon principle for electronic transitions to vibrational states within a vibrational manifold. We note that this analogy is especially close for molecular photoelectron spectra.

Snyder[31] considered the problem of orbital relaxation in an atom from which an inner electron is ejected and discussed this effect in terms of atomic shielding constants. He gave an equation for the binding energy of an electron in the mth shell, based on atomic shielding constants ideas in the equivalent of a "frozen orbitals" approximation:

$$\begin{aligned} E_B{}^m(\text{“}KT\text{”}) = &-\frac{1}{m^2}(Z - s_m)^2 + \frac{2Z}{m^2}(Z - s_n) \\ &- \frac{2}{m^2}\left(0.85\, N_{m-1} + \sum_{l=1}^{m-2} N_l\right)(Z - s_m) - \frac{2}{m^2}(2N_m - 2) \\ &\times (s_m - s_m{}')(Z - s_m) - 2\sum_{n>m} \frac{N_n}{n^2}(s_n - s_n{}')(Z - s_n) \end{aligned} \qquad (2.17)$$

Here s_n is the shielding constant for the nth shell,[38,39] and the prime denotes the core-ionized state. The five terms in (2.17) denote, respectively, kinetic and potential-energy interactions involving the nucleus, and repulsive interactions between an electron in the n shell and electrons in inner shells, in the m shell, and in outer shells. This equation gave $1s$ binding energies for neon and argon that were within 3–4% of the Koopmans' theorem values from Hartree-Fock calculations, as shown in Table VIII. Also listed for these cases are the derivatives of $E_B(\text{“}KT\text{”})$ with Z, which gives the variation of binding energy for isoelectronic systems such as F^-, Ne, and Na^+. Finally Snyder derived an expression for the relaxation energy in an ion with a $1s$ hole (and possessing electrons up to the $3s$, $3p$ shells)

$$\Delta E^{\text{relax}} = -(1.2 + 2.5\, N_2 + 1.5\, N_3)\ \text{eV} \qquad (2.18)$$

Here N_2 and N_3 are the populations of the $n = 2$ and 3 shells. This equation predicts ΔE^{relax} to be independent of Z. In fact the Z dependence is approxi-

TABLE VIII
Comparison of Energies from Shielding-Constant and Hartree-Fock Calculations[a]

	E_B("KT") (eV)		ΔE_B("KT")/ΔZ (eV)		ΔE^{relax} (eV)	
Final State	Shielding	HF	Shielding	HF	Shielding	HF
$Ne^+(\overline{1s})$	930.3	891.7	209.4	203.0	−21.2	−23.2
$Ar^+(\overline{1s})$	3313.9	3227.4	402.7	397.2	−33.2	−32.2

[a] From Refs. 14 and 31.

mately $Z^{-1/2}$. The predicted magnitudes of ΔE^{relax} for Ne^+ and Ar^+ are within ~5% of the Hartree-Fock values.

As the above discussion indicates, the formal connection between $E_B(KT)$ and E_B(hole) is well understood. For atomic systems the actual magnitudes of ΔE^{relax} can be calculated with good accuracy by alternate approaches, indicating that for these systems the mechanistic details of relaxation in the hole state are known. At a level of sophistication adequate for the discussion of chemical shifts in binding energies, however, atomic relaxation is inadequate, and molecular relaxation, in particular, *differential* molecular relaxation, must be considered. Thus although this section makes a conceptual link between E_B(hole), discussed earlier, and $E_B(KT)$, discussed below, it does not provide a quantitative bridge that would provide a basis for using $E_B(KT)$ in estimating chemical shifts. Such a bridge could take either of two forms: differential molecular relaxation could be shown to be negligible, or the magnitude of molecular relaxation could be estimated for each case. Since in O_2^+ about ~5 eV of relaxation energy can apparently be attributed to extraatomic relaxation (as discussed earlier), it seems probable that the differences of extraatomic relaxation energies from one molecule to another could be a fair fraction of this figure. Thus the failure of $E_B(KT)$ to include this effect could perhaps account for up to 2 of 3 eV of scatter in the comparison of experimental chemical shifts with theoretical shifts by the use of orbital energies.

b. Comparison of Orbital Energy Differences with Experiment. In this section experimental binding-energy shifts in gaseous molecules are compared with orbital energies. Although orbital energies are calculated, there is little reason to compare them with experimental binding energies, because they tend to be high by an amount in excess of the whole range of chemical shifts. The carbon 1*s* orbital in methane is about 305 eV, for example, and the binding energy of this orbital is 290.8 eV. The intercomparison of orbital energies from different calculations on the same compound is more meaningful, but even its value is limited. Although $\epsilon(1s)$ for a particular compound presumably has a unique value in the Hartree-Fock limit, its value is not governed by the variation principle. Thus although one might expect $\epsilon(1s)$ and the total energy E to be correlated far from the HF limit (i.e., for poor choice of basis sets) just on the ground that $2\epsilon_{1s}$ is a reasonably large fraction of $-E$, no such correlation is to be expected near this limit. Hence the goodness of an $\epsilon(1s)$ value cannot be judged by its magnitude, nor can the proximity of $\epsilon(1s)$ to the Hartree-Fock limit necessarily be judged by the value of E alone, without additional information about the particular SCF calculation in question. These conclusions are illustrated by the values of E and $\epsilon(1s)$ for methane given in Table IX. It is difficult to compare results reported by different workers, because their basis sets differ in a variety of ways. Fortunately Gianturco and Guidotti[26] have studied the relationships of $\epsilon(1s)$ to E and to the basis set, by varying the basis set in a systematic way, for the molecular CH_4, NH_3, and H_2O. They found variations in $\epsilon(1s)$ of 1.2, 2.8, and 1.1 eV, respectively, for these three molecules. That $\epsilon(1s)$ and E are not strongly correlated is evident from the entries from Ref. 26 in Table IX. More evidence is given in the original paper, in which appear the results of 6, 5, and 9 SCF calculations, respectively, for these three molecules. It is particularly noteworthy that "double-zeta" basis sets give values of E that are fairly close to those obtained using large basis sets in CH_4, but that the double-zeta $\epsilon(1s)$ results are considerably in error for NH_3 and H_2O. In

TABLE IX

Total and 1*s*-Orbital Energies for CH_4

Basis set[a]	$-E$(a.u.)	$-\epsilon(1s)$ (eV)	Ref.
Extensive GTO	40.1890	305.07	8
GTO	40.1812	304.9	20
GTO 2ζ	40.1303	305.22	24
GTO	40.1823	304.97	25
Large STO set	40.2045	305.15	26
minus C 3*d*'s	40.1866	304.76	26
2ζ	40.1845	305.30	26
Minimal STO	40.1153	305.95	26

[a] STO = Slater type orbital, GTO = Gaussian type orbital. More detailed descriptions of basis sets are (in most cases) given in the references.

fact for these two cases the double-zeta basis sets give worse results for $\epsilon(1s)$ than do minimal basis sets (Table X). In view of the ~1 eV error in $\epsilon(1s)$ that appears to attend the use of double-zeta basis sets, and particularly because this $\epsilon(1s)$ can apparently err in either direction, about 1 eV of scatter can be expected in theoretical chemical shifts based on orbital energies from ab initio calculations of double-zeta quality. Evidently similar scatter can be expected if shifts are estimated as differences between orbital energies from different sources, unless all the values of $\epsilon(1s)$ are obtained from calculations near the Hartree-Fock limit. On the other hand, the results quoted in Table IX can be interpreted as indicating that a careful calculation of $\epsilon(C1s; CH_4)$ with a well-chosen basis set will yield a reproducible value in the range 305.1 ± 0.1 eV.

Basch and Snyder[40] were the first to predict a large number of binding energy shifts, using orbital energies from ab initio (double-zeta quality) calculations on 30 small molecules. Davis et al.[41] measured shifts for some of these molecules, finding good agreement. Several other experimental shifts and orbital energy differences for small molecules are also available.[3,8,25,42–44] We have listed in Table XI, and plotted in Figs. 4*a–c*, those cases for which both experimental and theoretical figures are available. In most cases for which two experimental values are available the agreement is very good. Average experimental values are used in the figures. In all three cases plotted—C, N, and O—straight lines of unit slope have been drawn through the points. Perfect agreement between δE_B(expt) and $\Delta\epsilon$ would correspond to all the points' lying on the lines. In fact in only one

TABLE X
Energies for CH_4, NH_3, and H_2O, after Gianturco and Guidotti[26]

Molecule	Basis set	$-E$ (a.u.)	$-\epsilon(1s)$ (eV)	$\Delta\epsilon(1s)$[a]
CH_4	Extensive	40.2045	305.15	—
CH_4	2ζ	40.1845	305.30	+0.30
CH_4	Mimimal	40.1153	305.95	+0.80
NH_3	Extensive	56.1861	423.52	—
NH_3	2ζ	56.1675	422.34	−0.18
NH_3	Minimal	56.0051	422.65	−0.87
H_2O	Extensive	76.0384	560.05	—
H_2O	2ζ	76.0052	558.73	−1.32
H_2O	Minimal	75.7030	559.53	−0.51

[a] This difference gives an estimate of the expected variation in chemicel shifts that are estimated from orbital energies derived from the smaller basis sets.

TABLE XI

$\delta E_B(1s$, expt) and $\Delta\epsilon(1s)$ for Small Molecules (eV)[a]

		Carbon		Nitrogen		Oxygen		Fluorine	
No.	Molecule	δE_B	$\Delta\epsilon$	δE_B	$\Delta\epsilon$	δE_B	$\Delta\epsilon$	δE_B	$\Delta\epsilon$
1	C_2H_4	−0.1	0.9						
2	HCO_2H	4.99(10)	6.0			0.67(5)	2.0		
						−0.95(5)	0.8		
3	Cyclopropane	−0.23	0.5						
4	C_2H_4O	2.01(5)	2.4			−1.05(5)	0.2		
5	N_2			4.3,[b] 4.35(20)	5.4				
6	C_2H_2	0.4	1.56,[c] 1.4						
7	HCN	2.6(2)	3.00,[c] 2.8	0.55(20)	2.53,[c] 3.0				
8	C_2H_6	−0.2	0.2						
9	CO_2	6.8,[b] 6.84(5)	7.86,[d] 8.3			1.1,[b] 1.44(5)	3.2		
10	CO	5.2,[b] 5.4	5.01,[d] 5.67,[c] 5.5			2.4,[b] 2.94(10)	3.43,[c] 3.3		
11	CH_3OH	1.6,[b] 1.9(2)	2.0			−0.8,[b] 0.80(10)	−0.2		
12	O_2					3.4,[b] 3.84(6)	4.3		
13	N_2O			2.9,[b] 3.17(10)	6.1	1.5,[b] 1.54(10)	2.9		
				6.9,[b] 7.04(5)	9.3				
14	H_2CO	3.3[d]	4.55,[c] 3.90,[d] 4.1						
15	CH_3F	2.8(2)	3.18,[c] 2.89,[d] 3.0,[e], 4.9[f]						
16	CH_2F_2	5.55(5)[g]	5.93,[d] 6.1[e]					0.73(5)[g]	1.2[e]
17	CHF_3	8.1,[b] 8.3(2)	8.81,[d] 9.4,[e] 15[f]						
18	CF_4	11.1,[b] 11.0(2)	12.11,[d] 12.8[e]					2.6(2)[g]	3.6[e]

[a] Reference compounds are hydrides, except for F(1s), which is referred to CH_3F. Unless otherwise annotated, $\Delta\epsilon(1s)$ and $\delta E_B(1s)$ values are from Refs. 40 and 41, respectively.

[b] Ref. 3.

[c] Ref. 42.

[d] Ref. 8.

[e] Ref. 25.

[f] Ref. 43.

[g] Ref. 44.

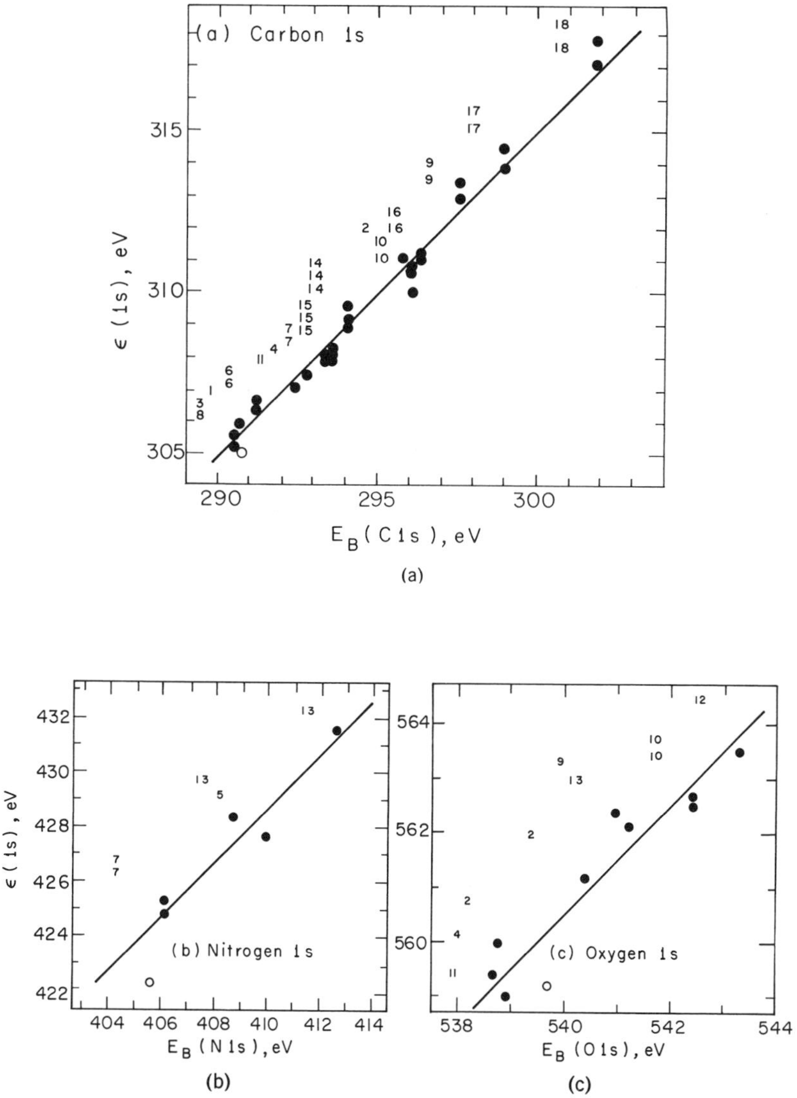

Fig. 4 (*a*) Orbital energies versus binding energies for carbon 1*s* electrons in gaseous carbon-containing molecules. Open circle denotes methane, while other compounds are numbered as in Table XI. Compound numbers stand in the same relative positions as do the points. Multiple entries denote more than one theoretical value. Line has unit slope. (b) Nitrogen data: open circle denotes ammonia. Two points for N_2O (13) denote inequivalent nitrogens. (*c*) Oxygen data: open circle denotes water. Two points for HCO_2H (2) denote inequivalent oxygens.

case (NH_3) is the orbital-energy value off by more than 1.0 eV. Several points are 1.0 eV off the lines, but the average error is only ~0.5 eV. Thus orbital energies appear to provide reasonably accurate values of binding energy shifts, reliable to the ~1 eV level. In view of the foregoing discussion about the expected scatter of ~1 eV in the theoretical values of $\epsilon(1s)$ this is about the best agreement that could be anticipated. Before further improvement can be expected in the agreement between $\Delta\epsilon(1s)$ and δE_B, basis sets of better than double-zeta quality will probably be required. As evidence that a large basis set can give good results, the CH_4–CF_4 shift of 12.11 eV predicted by Gelius et al.[8] by the use of a large basis set agrees well with the experimental value of 11.0 eV, especially if the former is corrected downward to 11.6 eV to allow for the scale factor of 1.05 between $\epsilon(C1s)$ and $E_B(C1s)$. Gelius et al. found a slope of $\Delta\epsilon/\delta E = 1.09$ by fitting a line through the seven points that they calculated with large basis sets.

In summary, chemical shifts predicted from differences in orbital binding energies based on ab initio Hartree-Fock calculations agree with experiment to ~1 eV or better, when basis sets of double-zeta quality are used. Better basis sets will probably improve this agreement, because at the double-zeta level the orbital energies are still not near their optimum values. Finally, the scale for $\Delta\epsilon$ should probably be 1.05 that for δE_B, because orbital energies tend to be about 5% larger than binding energies.

C. Quantum-Mechanical Methods Not Involving Binding-Energy Calculations

The foregoing theoretical approaches are valuable in elucidating the foundations of binding-energy shifts, but ab initio calculations presently constitute a rather cumbersome approach to the actual calculation of shifts for molecules of any size. Fortunately the physical origins of the shifts are understood well enough that they can be calculated directly, using models based on electrostatic potential energy considerations. These models can be subdivided further into those that entail (or require) an accurate evaluation of the local potential and those that do not.

1. Potential Models

It was realized very early that binding-energy shifts arose almost entirely from differences in the electrostatic potential energies of core electrons.[45] However, the first detailed theoretical analyses that demonstrated this result quantitatively were given relatively recently and independently by Basch[46] and Schwartz.[42] The analysis given by Schwartz is summarized below to provide a basis for the potential-model approach.

The orbital energy of a $1s$ electron on nucleus n can be conveniently expressed by rewriting (2.6) as

$$\epsilon_{1s} = \left\langle 1s(1) \middle| -\frac{1}{2}\nabla_1^2 - \frac{Z_n}{r_{1n}} \middle| 1s(1) \right\rangle + J_{1s1s} - \sum_{m \neq n} Z_m \left\langle 1s(1) \middle| \frac{1}{r_{1m}} \middle| 1s(1) \right\rangle + \sum_{i=\text{local}} (2J_{1si} - K_{1si}) + \sum_{i=\text{distant}} (2J_{1si} - K_{1si}) \qquad (2.19)$$

Here it has been assumed that the molecular orbitals have been expressed in terms of a "localized molecular orbital" basis set $\{L_i\}$.[47] The last two sums are then taken over the "local" orbitals L_i that connect atom n and over the "distant" orbitals that do not. Schwartz showed that to a very good approximation $K_{1si} = 0$ for the distant orbitals. By direct calculation he found values of 2×10^{-4} a.u. or less for the three $1s$-distant orbital exchange integrals in CH_3F. He also showed that, to within 10^{-4} a.u., the distant Coulomb integrals J_{1si} are equal in magnitude to $1/Z_n$ times the electrostatic attraction integral between nucleus n and the distant orbital L_i, and that $\langle 1s(1)|Z_m/r_{1m}|1s(1)\rangle = Z_m/R_{nm}$, where R_{nm} is an internuclear distance. Finally the one-electron interaction with nucleus n, $\epsilon_{1s,n} \equiv \langle 1s(1)| -\frac{1}{2}\nabla_1^2 - Z_n/r_{1n}|1s(1)\rangle$, was shown to vary by only a few ten-thousandths a.u. from one molecule to another. In view of these results, shifts in the orbital energy can be related to shifts in the external electrostatic potential evaluated at the nucleus by the approximate expression

$$\Delta(-\epsilon_{1s}) \cong \Delta V_{\text{ext}} + e^2\Delta \sum_{i=\text{local}} \left[2\left\langle L_i(1) \middle| \frac{1}{r_{1n}} \middle| L_i(1) \right\rangle - 2J_{1si} + K_{1si} \right] \qquad (2.20)$$

The last term is just the difference between the actual interaction energy of electrons in the local orbitals with the $1s$ electron and the value that this interaction energy would have if the $1s$ orbital were collapsed to the nucleus and exchange were absent. The results given by Schwartz for CH_4 and CH_3F show that for the C$1s$ orbitals in these molecules the second term in (2.20) amounts to only 0.23 eV, or less than 10% of the measured shift. From calculations on 15 molecules, he found $\Delta(-\epsilon_{1s}) = 1.11\, \Delta V_{\text{ext}}$ on the average. This coefficient exceeds unity as expected (i.e., because the radial extent of the $1s$ orbital makes the Coulomb and exchange integrals in ϵ_{1s} more sensitive to environment than the $1/r$ integrals in V_{ext}). Since orbital energies exceed experimental binding energy shifts by a few percent ΔV might be expected to predict these shifts better than $\Delta(-\epsilon)$ would.

Basch[46] gave a similar derivation, differing mainly in that he allowed $1s$ orbitals to collapse into their nuclei for the purpose of approximating certain Coulomb integrals involving these orbitals as one-electron integrals. With this approximation, he found that the "potential" relation, $\Delta(-\epsilon) \cong \Delta V$, is valid if the quantity

$$\left\langle 1s(1) \left| -\frac{1}{2}\nabla_1^2 - \frac{Z_n}{r_{1n}} \right| 1s(1) \right\rangle + J_{1s1s} - \sum_i K_{1si}$$

does not change appreciably with environment. He established the validity of the potential relation by direct calculation, for the fluorinated methanes, of $\Delta\epsilon$, ΔV, and $\Delta V'$, where the quantity

$$V' = \sum_{m \neq n} \frac{Z_m}{R_{mn}} + 2\sum_i \left\langle i \left| \frac{1}{r_{1n}} \right| i \right\rangle = \sigma_{\text{av}}{}^d(n) \qquad (2.21)$$

is just the diamagnetic shielding coefficient at nucleus n. V' differs from V only in that it contains a term in the sum for the $1s$ orbital. Equation (2.21) establishes a connection between binding-energy shifts and NMR parameters, as the agreement among Basch's values of $\Delta\epsilon$, ΔV, and $\Delta V'$, set out in Table XII, shows.

Comparison of shifts in potential and orbital energies as determined from ab initio calculations are helpful in understanding the origins of shifts, but beyond that their value is limited. Most molecules are too large for ab initio calculations to be feasible, and in those cases for which ab initio calculations can be made, the orbital energies themselves are readily available and may as well be used directly to estimate shifts. The real reason for establishing the relation between $\Delta\epsilon$ and ΔV is that V, but not ϵ, can be reliably estimated for larger molecules by the use of intermediate-level molecular-orbital theories such as the CNDO[48] model. Çelius et al.[8] have studied the potential model using both CNDO and ab initio wave functions. For several small carbon-containing molecules they have done ab initio calculations using large basis sets and have given values for $\Delta\epsilon$,

TABLE XII

Orbital Energy and Potential Shifts for C$1s$ Electrons, after Basch[46]

Method	CH_4	CH_3F	CH_2F_2	CHF_3	CF_4
Experiment	(0.0)	2.8 eV	5.6	8.3	11.0
$\Delta\epsilon(1s)$	(0.0)	3.0	6.1	9.4	12.7
$\Delta V(1s)$	(0.0)	3.0	6.2	9.6	13.1
$\Delta V'(1s)$	(0.0)	3.0	6.2	9.5	13.0

TABLE XIII
Molecular Parameters for Carbon Compounds

Molecule	$\Delta\epsilon_{1s}$ [a] (eV)	Carbon atomic charge		Molecular potential [c] (eV)	
		ab initio [a]	CNDO [b]	ab initio [a]	CNDO [b]
CH_4	0	−0.71	−0.049	9.40	0.65
CH_3F	−2.89	−0.13	0.180	2.73	−1.85
CHF_3	8.81	0.67	0.613	−6.86	−6.67
CF_4	12.11	1.01	0.708	10.99	−8.82
CO	5.01	0.32	0.042	−4.13	−0.53
CO_2	7.86	0.66	0.536	−8.14	−6.64

[a] Ref. 8.
[b] Ref. 49.
[c] Potential energy of 1*s* electron from extraatomic origins.

$q_c(g)$, the gross atomic charge on the carbon atom, and V, the molecular potential arising from the surrounding atoms. The discussion below is based largely on their results, which are set out in Table XIII; although it differs in detail, the conclusions are consistent with those of Gelius et al.

These workers compared experimental shifts δE_B with calculated parameters, finding a good fit to the relation

$$\delta E_B = 18.3\, q_c(g) + V + 3.0 \text{ eV} \tag{2.22}$$

In order to compare $\Delta\epsilon$ with $\Delta q_c(g)$ and ΔV, we have tested for a relation of the form

$$\Delta\epsilon = kq_c(g) + V + b \tag{2.23}$$

by plotting $\Delta\epsilon - V$ against $q_c(g)$, in Fig. 5. These quantities show a very nearly linear relationship, but a slight curvature is also apparent. A line with parameters $k = 18.3$ and $b = 3.0$ fits the points quite well, thereby justifying the linear variation of $\Delta\epsilon$ with $q_c(g)$.

It is useful to examine the relationship between $q_c(g)$ and V as calculated from ab initio wave functions and the comparable quantities from CNDO theory. The latter have been given by Ellison and Larcom [49] and have also been calculated by D. W. Davis.[50] They are also listed in Table XIII. The agreement between either q(ab initio) and q(CNDO) or V(ab initio) and V(CNDO) is poor, but this means little by itself because the two values of q are defined differently. The ab initio gross atomic charges are based on a Mulliken population analysis[51] and thus include "overlap population," whereas the CNDO theory allows for no overlap. As a result the range of atomic charge is more than a factor of two larger in the ab initio theory.

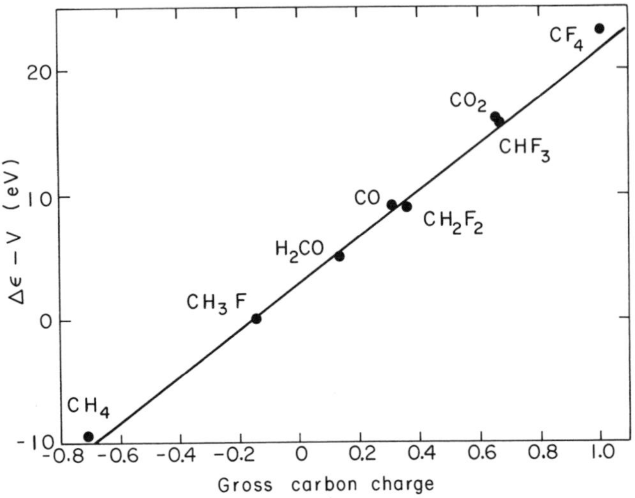

Fig. 5. Plot of energy parameters for C(1*s*) in several compounds, against gross carbon atomic charge, to test linearity. Data were taken from Ref. 8.

This is compensated in part by a smaller value of k in (2.23) and in part by a larger range of the extraatomic potential V. Thus the near agreement of k for the two sets of charges (18.3 vs. 23.5[8]) does not imply that the charges themselves agree that well. However, plots comparing the charges (Fig. 6*a*) or the potentials (Fig. 6*b*) separately show linear relationships for both cases, thereby supporting the validity of relations like (2.23) for either ab initio or CNDO parameters. The CO points, and perhaps the CO_2 points, lie substantially removed from straight lines through the substituted methane results in both Figs. 6*a* and 6*b*. In fact the CNDO model predicts the CO shift poorly. This result is expected. The CNDO theory gives essentially a point-charge treatment of shifts in $E_B(1s)$, and multiple bonds are not well described by point charges.

The above discussion suggests that a potential model based on CNDO theory might predict shifts in good agreement with experiment, with some reservations about multiply bonded systems. Several approaches have been taken to test this possibility. The Uppsala group[3,52] wrote the binding energy shifts as (in our notation)

$$\delta E_i = kq_i + V_i + l \tag{2.24}$$

where $V_i = e^2 \sum_j q_j / R_{ij}$ is the "molecular potential" term. Here a charge q_j is assigned to the jth atom, and it is taken as being located at the nucleus.

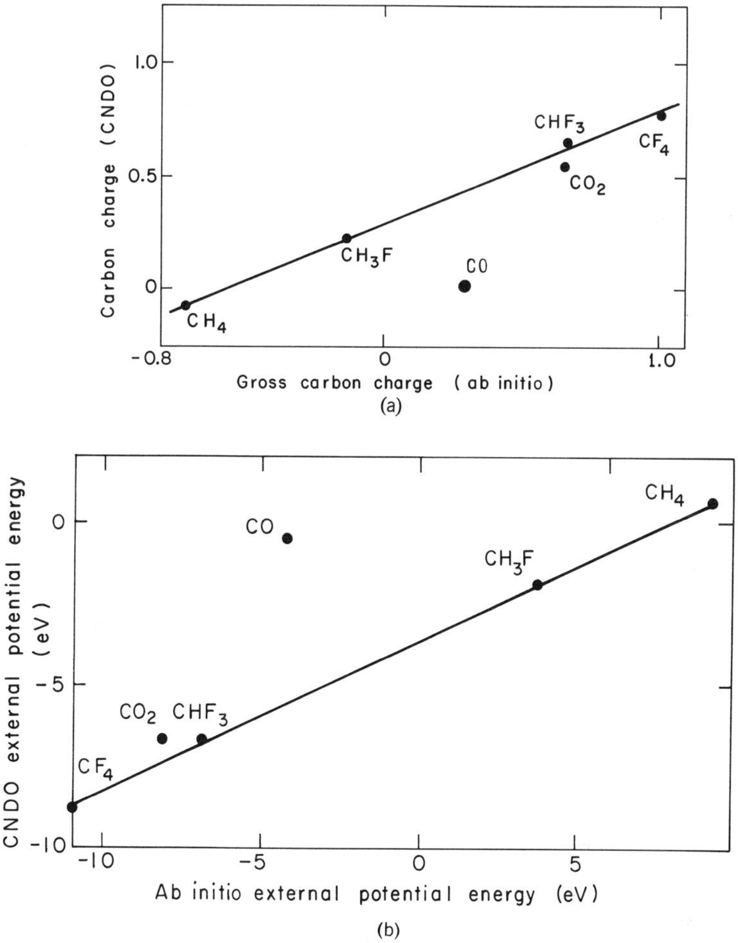

Fig. 6. (*a*) The CNDO carbon charge[49] versus the ab initio gross carbon charge.[8] Multiply bonded CO and CO_2 do not follow trend. (*b*) The Cl*s* external potential energies from CNDO[49] and ab initio[8] calculations. Note that Co and CO_2 are again off the line, but this time above it. Thus the deviations are partially compensated in the sums $kq + V$.

They calculated q_i and V_i for a number of molecules, using CNDO theory, and least-squares fitted experimental shifts δE_i to determine k and l. They made fits for compounds of C, N, O, F, and S. The quality of fit varied from one element to another, and in some cases there was too little variation in q_i to determine k very well. However, the fits tended to be rather good for most compounds in a group, with some points, such as CO in the

C(1*s*) group and N_2 in the N(1*s*) group, falling well off the line. An important result of these fits is that the values of k were quite close to those of the corresponding atomic (1*s*-valence) one-center repulsion integrals. In the format: element symbol, k from fit, k from repulsion integrals, the results were as follows: C, 21.9, 22.0; N, 21.5, 26.4; O, 25.8, 30.7; F, 27.6, 35.1; S, 13.8, 16.5.

Ellison and Larcom[49] have suggested that the above relation could be altered to give separate kq terms for host-atom s and p electrons, by writing

$$\delta E_i = k_s q_{is} + k_p q_{ip} + V \tag{2.25}$$

with $q_i = q_{is} + q_{ip}$ and a reference chosen such that $l = 0$. By carrying out a two-parameter (k_s and k_p) fit, they found that they could correctly predict the C(1*s*) shift in CO as well as the other C(1*s*) shifts reported by Davis et al.[41] They found $k_s = 17.5$ and $k_p = 24.5$ for C(1*s*), and, for several oxygen compounds, $k_s = 18.0$, $k_p = 26.4$. It should be noted, however, that the carbon compounds are fitted better by a two-parameter expression only because CO is among them. If CO is excluded one parameter does essentially as well. Furthermore, a two-parameter fit excluding CO gives values of k_s and k_p for C(1*s*) that are very close to one another. In view of this, of the similarity of the 1*s*–2*s* and 1*s*–2*p* repulsion integrals, and of the deviations shown by CNDO parameters for CO (as indicated in Fig. 6), the value of a two-parameter fit seems questionable.

Davis, Shirley, and Thomas[44,53] have used CNDO theory in a way that differs from either of the above approaches. Without any empirical curve-fitting they simply calculated the expected C(1*s*) and F(1*s*) shifts for a series of fluorinated benzenes and methanes. The results are quite encouraging. Before discussing them, however, a couple of observations should be made, lest the results appear better than they really are. First, rather simple molecules were chosen. Second, comparisons of C(1*s*) shifts are made only within each group (substituted benzenes and methanes). The two scales disagree by 0.9 eV, indicating that the CNDO approach can handle subtle shifts within groups of compounds with similar bonding better than intergroup shifts. Finally a subtlety was introduced into the calculations of V. One can treat electrons in atomic orbitals on neighboring atoms as if they were simply point charges, and evaluate the electrostatic potential they create as $q_{\bar{j}}/R_{ij}$. This is exact for a spherical charge distribution on center j, hence for s orbitals and closed shells. The foregoing estimates of V were made by this "point-charge" method, and Davis et al. also used this approach. However, they also made another estimate, based on the "p–p'" method. In this second calculation the external potential at nucleus i arising from orbitals on center j was evaluated by actual calculation of $1/r$ integrals. These integrals have different values for p_σ and p_π orbitals. Ordinarily in CNDO theory only integrals of the form $\langle p_\sigma j|1/r_{ij}|p_\sigma j\rangle$ or $\langle p_\pi j|1/r_{ij}|p_\pi j\rangle$ would be considered. This is all right if the coordinate axes are chosen normal to the line from i to j. If not, invariance to coordinate transformations requires the retention of off-diagonal elements $\langle p_j|1/r_{ij}|p_j'\rangle$, where p and p' are, for example, p_x and p_y.

The CNDO potential model predicts the fluoromethane shifts very well, as Table XIV shows. The shifts predicted by the p–p' modification agree better with experiment than do any other theoretical estimates. Even the F(1*s*) shifts are predicted well, in contrast to the ab initio results in Table XI.

TABLE XIV

CNDO Shifts in Fluorinated Methanes[44]

	Shifts in E_B(C1*s*)[a]			Shifts in E_B(F1*s*)[b]		
Molecule	Point charge	p–p'	Expt[c]	Point charge	p–p'	Expt
CH_3F	2.58	2.97	2.8(2)	—	—	—
CH_2F_2	4.99	5.58	5.55(5)	1.07	0.82	0.73(5)
CHF_3	7.32	8.54	8.3(2)	2.09	1.60	1.7(2)
CF_4	9.52	11.14	11.0(2)	3.11	2.37	2.6(2)

[a] Shifts in eV, relative to methane.
[b] Relative to CH_3F.
[c] Error in last place is given parenthetically.

For fluorine-substituted benzenes the CNDO potential model also predicts shifts quite well.[53] For this case the p–p' method overestimates the shifts somewhat, whereas the point-charge method gives excellent results, as shown in Fig. 7. None of the 28 shifts deviates by more than 0.4 eV from the experimental value. Apparently this model can predict shifts quite well within a series of related compounds. Its narrow range of applicability is a drawback, however, as is the ambiguity of whether the point-charge or p–p' modification is preferable.

2. *The ACHARGE Approach*

Davis et al.[53] introduced a different approach for interpreting binding energies, called the "atomic charge" analysis, or ACHARGE. In some

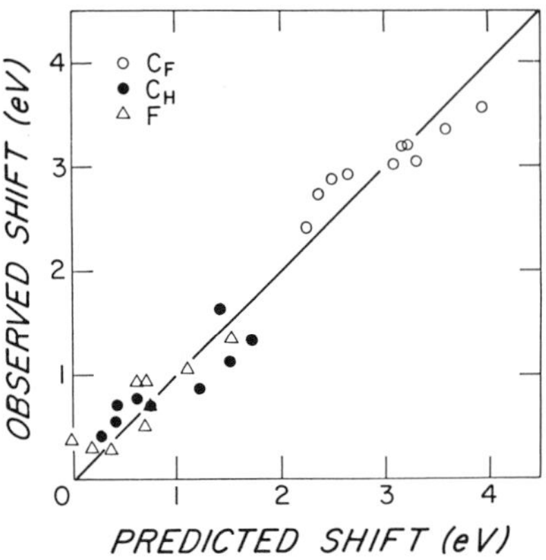

Fig. 7. Observed C(1*s*) and F(1*s*) binding-energy shifts in fluorobenzenes, relative to C_6H_6 (for C1*s*) and C_6H_5F (for F1*s*), plotted against CNDO "point-charge" potential-theory estimates, from Ref. 53. Here C_F and C_H denote energies of peaks assigned to the aggregate of all carbons bonded directly to F or H.

respects ACHARGE is quite similar to the above analyses, but philosophically it is quite different. The idea in ACHARGE is to work backward; that is, to learn chemistry from binding energies rather than using known chemical facts to explain the spectra. The ACHARGE approach is based on assuming point charges to exist on all the atoms in a molecule, measuring a complete set of core-level shifts, and deriving a consistent set of values of the charges from the shifts. ACHARGE is not a molecular-orbital model, but gives rather an *expermental* population analysis. The derived charge values agree very well with CNDO charges, presumably because CNDO is essentially a point-charge theory.

In ACHARGE an equation of the form

$$\delta E_i = k_i q_i + e^2 \sum_{j \neq i}^{n} \frac{q_j}{R_{ij}}$$

is written for each of the n atoms in a molecule. The parameter k_i, which has the same value for all atoms of a given element, is essentially the two-electron repulsion integral for a free atom of that element. If the molecule does not contain hydrogen there exist n equations linear in the charges q_i (for hydrogen-containing molecules some assumption about q_H must be made). If there are equivalent atoms, these equations can be condensed by gathering terms in each q_i and eliminating redundant equations, obtaining finally m linear equations, with $m \leqslant n$. In matrix form these may be written

$$\vec{\delta} = A\vec{q} \tag{2.26}$$

where $\vec{\delta}$ and $\vec{q}$ are vectors whose components are the sets $\{\delta E_i\}$ and $\{q_i\}$, and A is an $m \times m$ matrix. A diagonal element of A has the form

$$A_{ll} = k_l + e^2 \sum_{l'} \left(\frac{1}{R_{ll'}}\right) \tag{2.27}$$

where the sum is taken over all other sites equivalent to the lth site. Only the lth site itself contributes a linear equation to (2.26); equations arising from the sites labeled by l' were redundant and were lost in contracting from n to m equations. The off-diagonal elements have the form

$$A_{lp} = e^2 \sum_{p'} \left(\frac{1}{R_{lp'}}\right) \tag{2.28}$$

with the sum taken over sites that are equivalent among themselves but different from l. The matrix A is usually nonsymmetric.

In studying fluorinated benzenes Davis et al. used $k_c = 22$ eV/$|e|$ for carbon and $k_F = 32.5$ eV/$|e|$ for fluorine. For a given molecule the charges on all hydrogens were assumed equal. An additional equation was obtained by requiring overall charge neutrality. Finally, for each molecule all carbons bound to the same ligand (hydrogen or fluorine) were taken to have the same 1s binding energy, because inequivalent carbons with the same ligand appeared only as unresolved components of the same C(1s) peak in the photoelectron spectrum. Using this model, measured shifts δE(C1s) and δE(F1s), and molecular geometries, Davis et al. deduced atomic charges for several fluorinated benzenes that agreed very well with CNDO charges. Their results for four molecules are given in Table XV, together with the CNDO values. In spite of the approximate nature of the ACHARGE model it yields charges that are consistent with the basic physical and chemical properties of fluorobenzene. For example, withdrawal of electronic charge from the ring by fluorine is manifest as a negative charge on the fluorine atom and polarization of the C–F bond. On a more detailed level the ortho-meta-para alternation in charge, usually invoked to explain the preferential ortho-para orientation of electrophilic substituents, is evident. This alternation is explained classically by the resonance forms

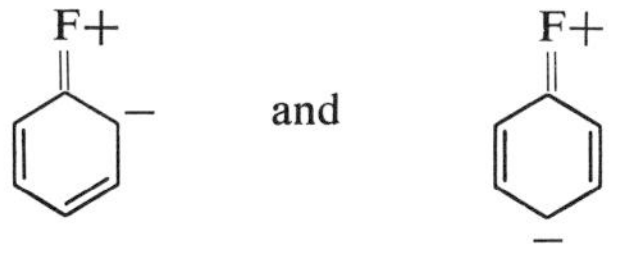

Although Davis et al. presented consistent evidence for ortho-meta-para charge alternation derived from different arguments, the effect is small.

TABLE XV
Atomic Charges in Fluorinated Benzenes (after Ref. 53)

Compound	Atom[a]	q (ACHARGE)	q (CNDO/2)
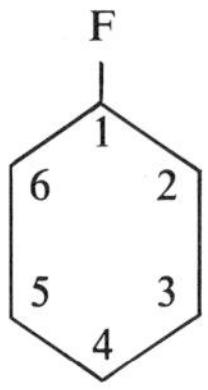	C_1	23[b]	24
	$C_{2,6}$	−4	−5
	$C_{3,5}$	1	3
	C_4	0	−1
	F	−19	−20
	$\overline{H}$	0	0
C_6F_6	C	14	$15._5$
	F	−14	$-15._5$
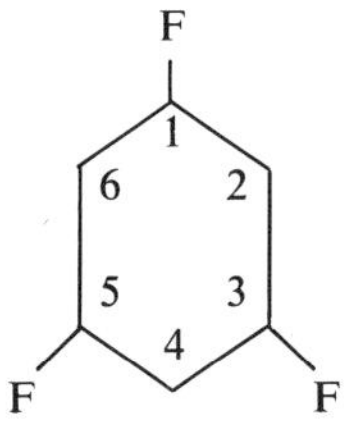	$C_{1,3,5}$	27	$28._5$
	$C_{2,4,6}$	−13	−14
	F	−18	−19
	H	4	4
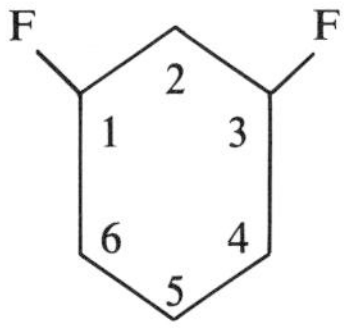	$C_{1,3}$	25	26
	C_2	−9	−12
	$C_{4,6}$	$-4._5$	−7
	C_5	2	5
	F	−18	−20
	$\overline{H}$	$0._5$	2

[a] Here $\overline{H}$ denotes average of all hydrogen charges.
[b] Charges are given in units of 10^{-2} $|e|$.

Larger effects of this nature were found in multiply substituted cases in which the charge shifts caused by two or more fluorines could reinforce one another. In *m*-difluorobenzene, for example, the carbon in position 2 is ortho to two fluorines, and consequently its charge is -0.09 $|e|$, or about twice that of an ortho carbon in fluorobenzene. Carbons 2, 4, and 6 in 1,3,5-trifluorobenzene are each ortho to two fluorines and para to another. Each of these carbons therefore carries the large negative charge of -0.13 $|e|$, in the ACHARGE analysis. Further chemical arguments of this nature can be made on the basis of the atomic charges derived from shifts in other fluorinated benzenes. These arguments are essentially the

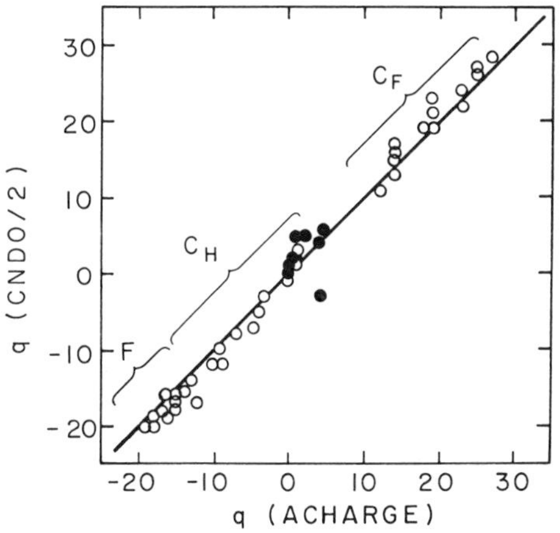

Fig. 8. Comparison of atomic charges in fluorinated benzenes, as derived from CNDO theory and from the ACHARGE analysis.[53] Filled circles are average charges on hydrogens. Open circles represent fluorines, and carbons bonded to hydrogen or fluorine, as labeled.

same that would be made by using CNDO charges, since the two sets of charges agree so well, as shown in Fig. 8.

3. Atomic Charge Correlations

In the early days of ESCA studies, particularly before molecular calculations were widely applied to the estimation of shifts, the shifts were interpreted as arising primarily from the atomic charge on the host atom, without a detailed account being made of contributions from the remainder of the molecule. These interpretations usually took the form of plots of binding energy versus atomic charge. The correlations were usually quite good on a rough scale, but poor on a finer scale.

Figure 9 shows binding energies of C(1s) electrons from a number of small gaseous hydrocarbon molecules, plotted against host-atom CNDO charges. The data were taken from Refs. 3 and 41. The trend is obvious, but individual points scatter by 1–2 eV typically. This is to be expected: neglect of the environment cannot destroy the trend of δE over a large range of charge. The slope of a line "through" the points in (2.8) is only 13 eV/$|e|$. This is in accord with the earlier observation[54] that the molecular potential should decrease this slope by less than a factor of two. The slope without the environment would be given by the one-center, two-electron integral, $k_c \cong 22$ eV/$|e|$ in this case.

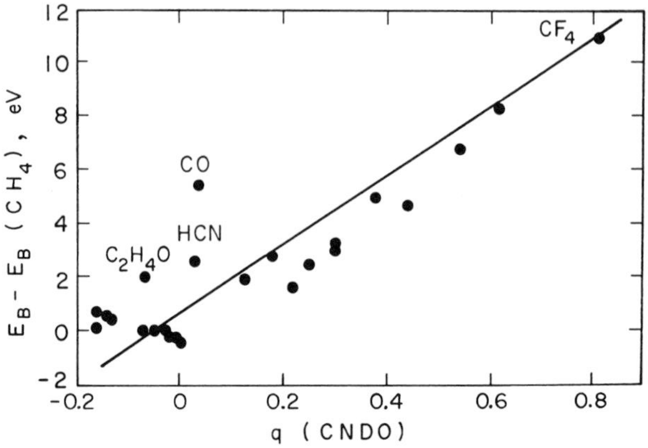

Fig. 9. Experimental C(1*s*) binding-energy shifts for small gaseous molecules plotted against host-atom charge from CNDO theory, after Refs. 3 and 41. The straight line connects the methane and CF_4 points.

Figure 10 shows a comparison of experimental and predicted binding-energy shifts for the same compounds as in Fig. 9. The theoretical values are calculated using the CNDO potential model, as described above. Comparison with Fig. 9 shows that inclusion of the potential makes an important difference. Comparison of Figs. 4*a* and 10 shows that for most cases the CNDO-potential model predictions are nearly as good as ab initio orbital-energy values, but that for the somewhat unusual molecules

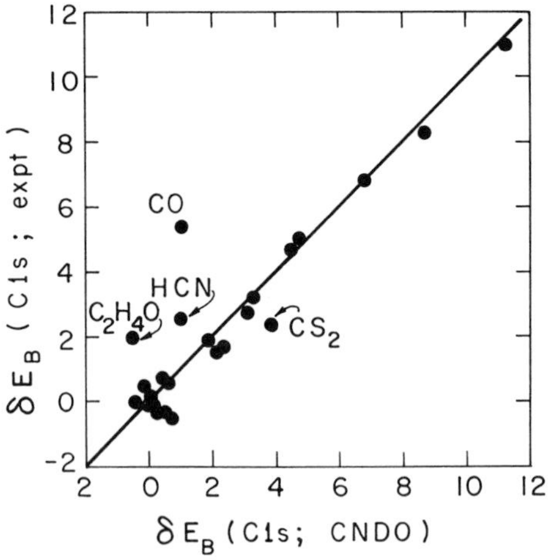

Fig. 10. Experimental C(1*s*) shifts, relative to methane, vs. shifts calculated on CNDO theory using the molecular potential, for the same compounds plotted in Fig. 9. Compare also with ab initio results in Fig. 4*a*.

CO, C_2H_4O, HCN, and CS_2 the ab initio values are distinctly superior (see also Fig. 7 and the related discussion).

If comparisons are restricted to structurally similar compounds such as the fluorinated benzenes, so that the inability of CNDO theory to deal with unusual compounds would not be a factor, the correlation of δE_B with host-atom charge might be expected on the above arguments to break down. That it does is evident in Fig. 11, wherein measured C(1*s*) shifts for these compounds[53] are plotted against CNDO charges. The points are distributed in two groups, composed of carbons bonded to hydrogens and to fluorines. Although the latter group have higher charges and higher binding energies than the former, and the two *groups* would fit rather well onto Fig. 9, the correlation of δE_B with q(CNDO) within each group is essentially nonexistent. The reason for this does not lie in the inadequacies of CNDO theory, for, as Fig. 7 shows, the CNDO potential model deals with these shifts rather well. Instead, the poor correlation in Fig. 11 must arise from neglect of the external potential. From this result and the foregoing discussion it is clear that binding energy–charge correlations are approximately valid over large charge ranges but have little application to subtle shifts.

Extended Hückel molecular orbital (EHMO) theory has been used extensively in discussing binding-energy shifts. The δE_B–atomic charge correlations are of varying quality, but typically they show an overall in-

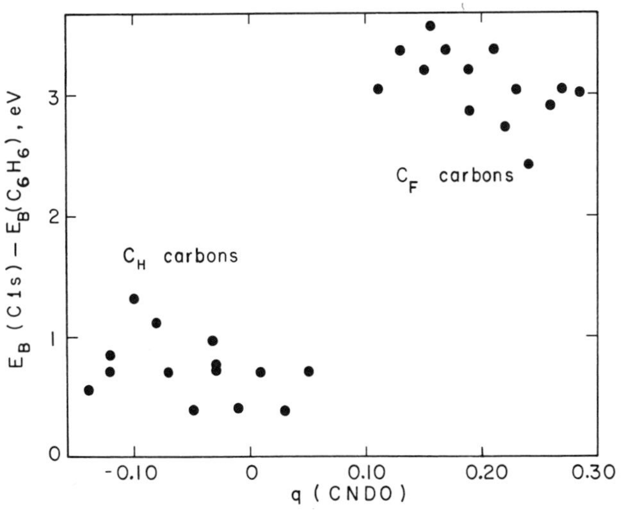

Fig. 11. Carbon 1*s* binding-energy shifts, relative to benzene, for fluorinated benzenes, plotted against CNDO charges. Data were taken from Ref. 53.

crease of E_B with q, with considerable scatter. There is a rather basic deficiency in EHMO theory; it does not account plausibly for electron repulsion. For this reason bond polarities are enormously exaggerated whenever EHMO theory is applied to compounds in which atoms of different electronegativities are bonded together. Atomic charges from EHMO calculations are therefore unrealistic and can be regarded only as empirical parameters. This deficiency showed up early in E_B vs. q correlations as an absurdly small slope, $\Delta E_B/\Delta q$.[54] More recently Schwartz[55] has shown that an improved correlation can be obtained using EHMO theory between observed binding energies and computed average potentials at the host nucleus. The "slope," $-\Delta E/\Delta V$, is still much too small, however. The computed CH_4–CHF_3 potential energy difference is 29.9 eV, for example, whereas the experimental shift is only 8.3 eV.

4. Thermochemical Estimates

Before discussing empirical correlations it is useful to consider a method for estimating core-electron binding-energy shifts that was introduced by Jolly et al.[56–59] This method is based on the similarity of compounds that have isoelectronic valence orbitals and equally charged cores. It has the virtue of using empirical thermochemical data to predict core-level binding-energy shifts, although it could equally well employ total energies calculated by SCF computations on molecular ground states. As this last remark implies, relaxation of passive orbitals is automatically taken into account, as only ground states are finally compared.

Jolly et al. pointed out that the N(1*s*) binding-energy shift from molecular nitrogen to ammonia is given by the reaction

$$NH_3 + N_2^{+*} \rightarrow NH_3^{+*} + N_2 \qquad \Delta E = \delta E_B(N1s;\, N_2\text{–}NH_3) \tag{2.29}$$

where an asterisk denotes a molecule with a nitrogen 1*s* electron missing. There are no thermochemical data available, in most cases, for such highly excited species as NH_3^{+*}. However, OH_3^+, which is isoelectronic in its valence orbitals and in which O has a core (nucleus plus 1*s* shell) of the same charge as N in NH_3^{+*}, is well known. These cores may be exchanged via the reactions

$$\begin{aligned} NH_3^{+*} + O^{6+} &\rightarrow OH_3^+ + N^{6+*} \qquad \Delta E = \delta_1 \\ N_2^{+*} + O^{6+} &\rightarrow NO^+ + N^{6+*} \qquad \Delta E = \delta_2 \end{aligned} \tag{2.30}$$

which can then be added to give

$$NH_3^{+*} + NO^+ \rightarrow OH_3^+ + N_2^{+*} \qquad \Delta E = \delta_1 - \delta_2 \tag{2.31}$$

Now Jolly pointed out that if $\delta_1 - \delta_2 \cong 0$, that is, if the energy of exchanging the O^{6+} and N^{6+*} cores is essentially insensitive to the chemical environment, addition of (2.29) and (2.31) yields a reaction

$$NH_3 + NO^+ \rightarrow OH_3^+ + N_2 \qquad \Delta E = \delta E_B(N1s;\ N_2\text{–}NH_3) \qquad (2.32)$$

with an energy that can be calculated from the energies of formation of the four species involved. But this reaction energy is just the N1s binding-energy shift from N_2 to NH_3, which is thereby predicted. From similar equations core-level shifts can be predicted from thermochemical data for compounds of other elements. For example the methane–CF_4 shift can be derived using the reaction

$$CH_4 + NF_4^+ \rightarrow NH_4^+ + CF_4 \qquad (2.33)$$

Estimated and measured shifts for several gaseous carbon compounds[57,59] are shown in Fig. 12.

This thermochemical approach is very valuable because it gives good results. Clearly it can be expanded to employ SCF total energies or energies of formation, as well as thermochemical data. Because of the potential usefulness of the method, it seems worthwhile to study its theoretical basis a bit further. The core-exchange step represented by (2.30) seems particu-

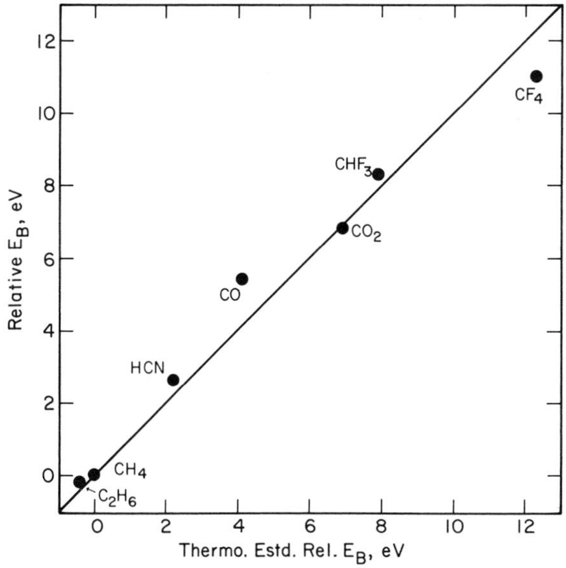

Fig. 12. Experimental and thermochemically estimated C1s binding energy shifts in gaseous carbon compounds, after Jolly et al.[57,59]

larly in need of justification. In the following abbreviated discussion[60] this step is justified in the Hartree-Fock formalism.

Consider the core-exhange "half-reaction"

$$ {}_7\mathrm{N}1s^2 + (({}_6\mathrm{C}1s)) \rightarrow (({}_7\mathrm{N}1s^2)) + {}_6\mathrm{C}1s \qquad (2.34)$$

wherein a 5^+ core consisting of a nitrogen nucleus plus a filled $1s$ shell replaces a 5+ core consisting of a carbon nucleus plus a half-filled $1s$ shell. The double parentheses denote the molecular environment, which has identical electronic configurations on the two sides, although the radial wave functions may vary slightly. The nuclear positions are assumed identical. The total energy of the nitrogen compound may be written in Hartree-Fock notation as

$$\begin{aligned} E_N(\mathrm{cpd}) &= 2\,\epsilon_0{}^N(1s) + J_N(1s1s) + 2\sum_{i\neq 1s}^{n}[2J_N(1s\,i) - K_N(1s\,i)] \\ &\quad - 2\sum_{m\neq N} Z_m \left\langle \mathrm{N}1s(1)\left|\frac{1}{r_{1m}}\right|\mathrm{N}1s(1)\right\rangle - 2\sum_{i\neq 1s}^{n} Z_N \left\langle \phi_i(1)\left|\frac{1}{r_{1N}}\right|\phi_i(1)\right\rangle \\ &\quad + \sum_{m\neq N}\frac{Z_m Z_N}{R_{mN}} + \cdots \end{aligned} \qquad (2.35)$$

where J and K are Coulomb and exchange integrals, ϕ_i is a molecular orbital, and R_{mN} is the internuclear distance from the host N nucleus to any other. The system is assumed to possess n doubly occupied orbitals. Because interactions between any two particles or orbitals outside the N $1s^2$ core should be only negligibly affected by core exchange, only those terms that directly involve the $1s$ core particles (N nucleus and $1s$ electrons) are written explicitly in (2.35). Expressions for the energies of the other three entities in (2.34) can be arrayed conveniently as

$$\begin{aligned} \Delta E &= \left.\begin{aligned} &2\,\epsilon_0{}^N(1s) - 2\,\epsilon_0{}^{N'}(1s) - \epsilon_0{}^C(1s) + \epsilon_0{}^{C'}(1s) + J_N(1s1s) \\ &\qquad - J_{N'}(1s1s) \end{aligned}\right\} \quad (\mathrm{I}) \\ &\quad \left.\begin{aligned} &+ \sum_i^n \Big\{ 2[2J_N(1s\,i) - K_N(1s\,i)] - [2J_C(1s\,i) - K_C(1s\,i)] \\ &- 2Z_N \left\langle \phi_i(1)\left|\frac{1}{r_{1N}}\right|\phi_i(1)\right\rangle + 2Z_C \left\langle \phi_i(1)\left|\frac{1}{r_{1C}}\right|\phi_i(1)\right\rangle \Big\} \end{aligned}\right\} \quad (\mathrm{II}) \\ &\quad \left.\begin{aligned} &+ \sum_{m\neq N} Z_m \left[\frac{Z_N - Z_C}{R_{mN}} - 2\left\langle \mathrm{N}1s(1)\left|\frac{1}{r_{1m}}\right|\mathrm{N}1s(1)\right\rangle \right. \\ &\qquad \left. + \left\langle \mathrm{C}1s(1)\left|\frac{1}{r_{1m}}\right|\mathrm{C}1s(1)\right\rangle \right] \end{aligned}\right\} \quad (\mathrm{III}) \end{aligned} \qquad (2.36)$$

Here primes denote cores. Term III should in principle vanish identically. In fact the calculations and arguments presented by Schwartz[42] in his justification of potential models can be used to show that both I and III are negligibly small. Term II is not so simple. If the orbitals i are expressed in terms of a "localized" orbital basis set,[47] and the sum over i is split into sums over local and distant orbitals, the former can be shown, by Schwartz's calculations, to be negligibly small. The sum over local orbitals is nonzero, however. The attraction of the $_7$N $1s^2$ core is systematically greater for these orbitals than that of the $_6$C $1s$ core. This is a result both of incomplete shielding by the $1s$ electrons and of the contributions of the exchange integrals (the two effects reinforce one another). Each local orbital contributes a term of the order of -0.1 a.u. to ΔE. This term is similar in nature to the terms under the sum in (2.20). It is different in detail, however, and somewhat larger. Still the same arguments[42] should apply to show that term II in (2.36) varies with environment an amount similar to the variation of the sum in (2.20). Thus the thermochemical model is justified to about the same level of approximation as the potential model.

Jolly et al. have used the thermochemical model to estimate heats of formation from core levels shifts and thus to predict the possible stabilities of hypothetical compounds. From the $1s$ binding energy of the middle nitrogen in sodium azide, together with the known heats of formation of $NaN_3(s)$ and $Na^+(g)$, and an estimated sublimation energy for NON, Jolly and Hendrickson[56] used the hypothetical reaction

$$NaN_3(s) + NON(g) + O^{6+}(g) \rightarrow 2\ NON(g) + Na^+(g) + N^{6+*}(g) + e^-(g)$$

to predict $\Delta H^0 = -100$ kcal/mole for the isomerization

$$NON(g) \rightarrow NNO(g)$$

Using similar reasoning Jolly[59] was able to predict bond energies of essentially zero for the hypothetical molecules ArO_3 and ArO_4. The thermochemical model can be applied to molecules that are too large for accurate Hartree-Fock calculations at present. For example, Hollander and Jolly[57,59] made very good estimates of Xe($3d_{5/2}$) shifts in xenon fluorides, as shown in Fig. 13. The success of these predictions establishes the validity of the thermochemical approach for core levels other than $1s$ levels.

D. Correlations of Binding-Energy Shifts with Other Properties

There is of course no sharp distinction between prediction and correlation except for approaches that are completely rigorous in the first case or completely without theoretical justification in the latter. Thus most of the

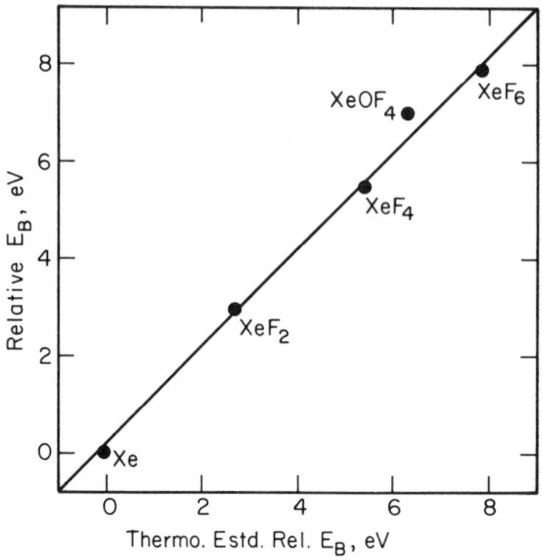

Fig. 13. Experimental and thermochemically estimated xenon core level binding-energy shifts, after Refs. 57 and 59.

correlations discussed below could be turned around and used to predict shifts, and they are all theoretically understood to a greater or lesser degree. The common thread that links these methods is their ability in each case to illuminate some aspect of atomic or molecular structure by connecting two quantities—binding-energy shift and another property—whose relationship might not be obvious. The correlations discussed below represent but a miniscule sample of the very wide range of possibilities. In fact the statement, "Each chemist can correlate binding-energy shifts with his favorite property," is essentially true. Because of its direct connection to the molecular charge distribution, the shift in core-level binding energy is related to practically every parameter of chemical interest.

1. Correlation with Other Binding-Energy Shifts

Perhaps the most obvious quantity with which the binding-energy shifts of a given core orbital may be compared is the shift of a different core orbital in the same atom. The physical insight afforded by a construct such as the conducting-sphere model leads to the expectation that all "deep" core levels in a given atom should show equal shifts upon a given change in environment. Fadley et al.[9] found that this is true for iodine, as discussed in Section II.A and shown in Fig. 1. Similar behavior has been observed for core levels in other atoms: if two core levels are described by wave functions whose radial extents are significantly smaller than that of the valence shell, these core levels will show very similar binding-energy shifts.

The above conclusions were valid—and even useful— in the early days of ESCA. Indeed an early disappointment of the method was the realization that the magnitudes of binding-energy shifts depended only on the principal quantum number, and not on the orbital angular momentum, of the valence electrons. Now, however, the field has moved to a higher level of sophistication, both experimentally and theoretically, and in favored cases some sensitivity to details of orbital composition can be obtained. In a recent very careful study of shifts in the binding energy of the iodine $3d_{5/2}$ orbital in alkyl iodides and HI, Hashmall et al.[61] found a strong correlation with iodine $5p_{1/2}$ binding energies from UV photoemission studies,[62] as shown in Fig. 14. There are two significant features about this figure. First, the slope of the line through the alkyl iodide points is 1.22 ± 0.05, or significantly greater than unity. Thus binding-energy shifts are greater for the more corelike $3d_{5/2}$ orbitals than for the outer $5p_{1/2}$ orbitals. This result was actually anticipated in Fig. 1, wherein the orbital energies for core levels in ionic iodine were found to shift with charge state by essentially the same amount for $1s$ through $4d$ orbitals and somewhat less for the $n = 5$ orbitals (in two cases the $5p$ shifts were anomalously high because of the small basis sets). A more reliable estimate of this effect can be obtained directly from the Coulomb and exchange integrals that involve

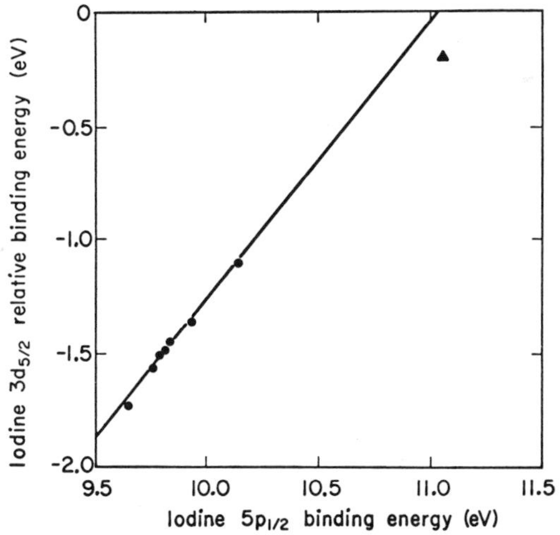

Fig. 14. Correlation of iodine $3d_{5/2}$ and $5p_{1/2}$ binding energies in alkyl iodides (circles) and HI (triangle) after Hashmall et al.[16]

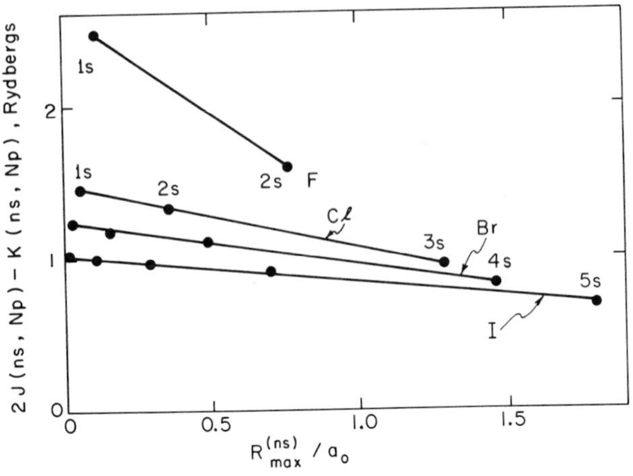

Fig. 15. Sensitivity of s-electron binding energies to valence-electron population, for s electrons in halogens. Note decrease from next-to-last to outer shell.

valence and core electrons. An estimate of this type is shown in Fig. 15, in which the function [see (2.6)].

$$2J(ns,Np) - K(ns,Np) = F^0(ns,Np) - (1/6)G^1(ns,Np)$$

was plotted for each s orbital, $n = 1, \ldots, N$, and for each halogen as a free atom. Here N is the principal quantum number of the valence shell, and F^0 and G^1 are Slater integrals. The values of F^0 and G^1 were given by Mann.[63] The above function is essentially equal to the "slope" $k = \partial E_B/\partial q$ that appears in the potential models (2.24). The abscissa in Fig. 15 is the radial maximum, R_{max}, of the ns shell. In each case there is a large decrease in k from the penultimate to the outermost shell. Thus the slope observed in Fig. 14 is expected, and it can be at least partially attributed to a variation of the I $5p_\sigma$ orbital population, in the alkyl iodides, with induction through the C–I bond.[61]

The other interesting feature of Fig. 14 is the deviation of the HI point from the straight line through the aklyl iodide data. Hashmall et al. attributed this to hyperconjugation. The lone-pair $p_{1/2}$ orbitals are relatively large and can be destabilized by interaction with σ orbitals on the alkyl groups (not in the C–I bond itself). This effect is negligible, however, for the $3d_{5/2}$ orbitals. It is absent, of course, in HI. Thus the horizontal displacement of 0.14 eV of the HI point is a measure of the hyperconjugative destabilization energy of the $5p_{1/2}$ orbitals, and the role of the core-level shifts is to calibrate the inductive effect. This case provides an example of how detailed bonding information can be extracted from core-level shifts. Further applications of this type can be expected.

2. Correlations with Diamagnetic Shielding Constants

There is in general no direct relation between core-level binding-energy shifts and NMR frequencies, but Basch[46] showed that δE_B should be closely related to the diamagnetic shielding constant $\sigma_{\mathrm{Av}}{}^d$, which is given by the relation

$$\sigma_{\mathrm{Av}}{}^d(\nu) = 2\left(\frac{e^2}{3mc^2}\right)\sum_i \left\langle \phi_i(1) \left| \frac{1}{r_{1\nu}} \right| \phi_i(1) \right\rangle \tag{2.37}$$

This expression gives the shielding constant at nucleus ν. The sum is taken over doubly occupied one-electron orbitals ϕ_i. The potential energy of a core $1s$ orbital k on nucleus ν due to other electrons and nuclei is similar:

$$V_\nu = 2e^2 \sum_{i \neq k} \left\langle \phi_i(1) \left| \frac{1}{r_{1\nu}} \right| \phi_i(1) \right\rangle - e^2 \sum_{n \neq \nu} \frac{Z_n}{R_{\nu n}} \tag{2.38}$$

Although the sums over ϕ_i are different in the two cases, *shifts* in $\sigma_{\mathrm{Av}}{}^d(\nu)$ and V_ν are comparable because the $i = k$ term varies negligibly with environment. One can therefore define a potential energy

$$\begin{aligned} V_\nu' &= 2e^2 \sum_i \left\langle \phi_i(1) \left| \frac{1}{r_{1\nu}} \right| \phi_i(1) \right\rangle - e^2 \sum_{n \neq \nu} \frac{Z_n}{R_{\nu n}} \\ &= 3mc^2\, \sigma_{\mathrm{Av}}{}^d(\nu) - e^2 \sum_{n \neq \nu} \frac{Z_n}{R_{\nu n}} \end{aligned} \tag{2.39}$$

so that to a very good approximation

$$-\delta E_B = \Delta V_\nu = \Delta V_\nu' = 3mc^2 \Delta\sigma_{\mathrm{Av}}{}^d(\nu) - e^2\, \Delta\left[\sum_{n \neq \nu} \frac{Z_n}{R_{\nu n}}\right] \tag{2.40}$$

Here the first equality follows from the potential model, the second from the constancy of the $i = k$ term in (2.39), and the third by definition. Basch demonstrated the accuracy of (2.40) by direct calculation for the fluorinated methanes. Thus a link has been established between ESCA shifts and NMR shifts.

Several workers have observed correlations of δE_B with ^{13}C chemical shifts,[64] but (2.40) has as yet been little used. The problem is that a measured NMR shift δ is sensitive to paramagnetic shielding as well as to $\sigma_{\mathrm{Av}}{}^d$. Thus a smooth variation of δE_B with δ can be expected only in restricted groups of compounds for which σ^p varies smoothly with σ^d. However, (2.40) can be used, together with measured values of δE_B, to test theoretical estimates of $\sigma_{Av}{}^d$. For example, Flygare and Goodisman[65] proposed the following approximate equation:

$$\sigma_{Av}{}^d(\nu) \cong \sigma_{Av}{}^d(\nu, \text{free atom}) + \left(\frac{e^2}{3mc^2}\right) \sum_{n \neq \nu} \frac{Z_n}{R_{n\nu}} \tag{2.41}$$

They found that this relation gave excellent predictions of $\sigma_{Av}{}^d(\nu)$ in a number of molecules. At first this might be surprising, because (2.41) can be interpreted as representing a model in which the molecule is taken as a collection of atoms, each with a spherically symmetrical electronic charge distribution and zero net charge. As Flygare and Goodisman pointed out, however, the first term in (2.41) is relatively large, and the second term actually does give a reasonably good representation of the effects on σ^d of electrons outside the host atom. Thus (2.41) should always be approximately correct, and good enough to give a fair estimate of σ^d. As a means of estimating core-level binding-energy shifts, however, the assumptions behind (2.41) represent too low an order of approximation. After combining (2.39) and (2.41) and taking the shift between two compounds, we have

$$-\delta E_B \cong \Delta V_\nu' = 3mc^2 \Delta\sigma_{Av}{}^d \text{ (free atom)} \equiv 0 \tag{2.42}$$

Thus in this approximation all binding-energy shifts would be zero! Basch thus demonstrated that the existence of such shifts is possible only because of inaccuracy in (2.41).

This approach may be turned around, and measured core-level shifts can be used to assess the accuracy of (2.41) for a given case. Substituting in values for physical constants, we have

$$\Delta\sigma^d(\nu) \text{ (ppm)} = -0.65\ \Delta E_B(\nu) \text{ (eV)} \tag{2.43}$$

as the range over which $\sigma_{Av}{}^d(\nu)$ can deviate from estimates based on the Flygare-Goodisman estimate (2.41). In (2.43) ΔE_B is the maximum range of binding energy shifts for core levels of atom ν. In carbon, for example, ΔE_B is 11 eV, so the Flygare-Goodisman estimates for carbon could never be in error by more than ~7–8 ppm.

The next obvious step is to use measured binding-energy shifts to check proposed values of $\sigma_{Av}{}^d$. For example Ditchfield, Miller, and Pople[66] have calculated σ^d values for ^{13}C in a number of carbon-containing molecules. For methane and methyl fluoride they gave values of $\sigma^d(CH_4) = 296.2$ ppm and $\sigma^d(CH_3F) = 320.2$ ppm, or a difference of 24.0 ppm. The binding-energy shift of 2.8 eV would give a shift of −1.8 ppm, whereas the $\sum Z/R$ term in (2.40) would add about 56 ppm to σ^d. Thus the difference between the values of σ^d for these two molecules must in fact be about 50 ppm, or twice the difference proposed by Ditchfield et al.[66] These authors gave

$\sigma^d(CH_2F_2) = 376.5$ ppm, or 80.3 ppm above $\sigma^d(CH_4)$. From $\delta E_B = 5.6$ eV (Table V) and (2.40) a difference of ~110 ppm is obtained. Thus the σ^d values of Ditchfield et al. seem reliable to ~30 ppm, or 10%. This is reasonable , since their σ^d values were calculated for the center of mass.

3. *Correlations with "Pauling Charges" and Electronegativity*

Pauling[67] suggested that the fractional ionic character of a bond between atoms A and B can be estimated from the expression

$$I = 1 - \exp[-0.25(X_A - X_B)^2] \tag{2.44}$$

where X_A and X_B are electronegativities. Using the values $X_H = 2.1$, $X_C = 2.5$, $X_{Br} = 2.8$, $X_{Cl} = 3.0$, and $X_F = 4.0$, the percent ionic characters for the carbon–ligand bonds are as follows: C–H, 4; C–Br, 2; C–Cl, 6; C–F, 43. From these bond ionicities charges can be calculated for all the atoms in halogenated methanes. These will be referred to as "Pauling charges," q_P. Both Thomas[10] and Siegbahn et al.[3] found linear correlations between $\delta E_B(C1s)$ and $q_P(C)$ for halogenated methanes, provided that only a single halogen (F, Cl, or Br) was considered. That is, $\delta E_B(C1s; CH_{4-n}X_n)$ varies linearly with $q_P(C)$ as n is varied from 0 to 4. Values of q_P calculated from (2.44) are given in Table XVI, together with measured C(1s) shifts.[3,10,44] The slopes of the δE_B vs. q_P correlations differed by a factor of about two among the different halogens.[3,10] Before accepting this as evidence for the inadequacy of (2.44), we should plot δE_B against $kq_P + V$ rather than just q_P, to take the molecular potential into account, as discussed in Section II.C. Such a plot is shown in Fig. 16. For this plot k_C was taken as 22 eV/$|e|$ and V was estimated on a point-charge model as

$$V = e^2 \sum_{\text{ligands}} \frac{q_P(\text{ligand})}{R(\text{carbon–ligand})} \tag{2.45}$$

TABLE XVI

Pauling Charges and C1s Shifts in Halogenated methanes[3,10,44]

Compound	X = Br		X = Cl		X = F	
	q_P(C)	δE_B(C1s)	q_P(C)	δE_B(C1s)	q_P(C)	δE_B(C1s)
CH_4	−0.16	(0)	−0.16	(0)	−0.16	(0)
CH_3X	−0.10	1.0 eV	−0.06	1.6 eV	0.31	2.8
CH_2X_2	−0.04	2.2	0.04	3.1	0.78	5.6
CHX_3	0.02	3.0	0.14	4.3	1.25	8.3
CX_4	0.08	4.0	0.24	5.5	1.72	11.0

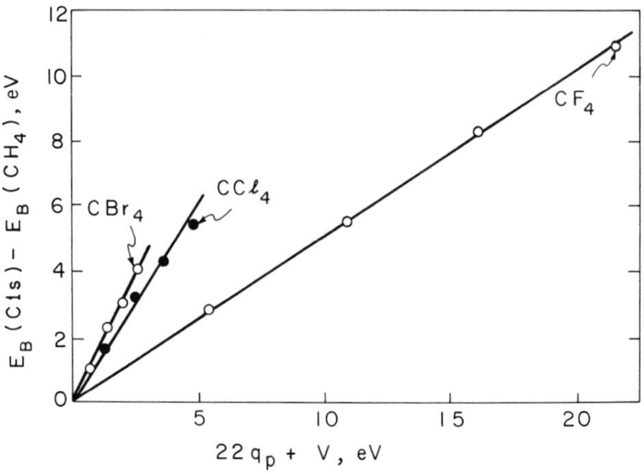

Fig. 16. Binding energies of C1*s* electrons in halogenated methanes, plotted against a potential function, deduced from parameters in Table XVI and in text.

with q_P estimated from (2.44) and R taken as 1.1 Å for C–H, 1.4 Å for C–F, 1.8 Å for C–Cl, and 2.0 Å for C–Br throughout. The factor of two variation in slopes is still present, and the answer to this discrepancy must be sought elsewhere. Fortunately Fig. 16 tells us where to look. Because δE_B is plotted against potential energy rather than an empirical parameter, the points in Fig. 16 should all lie on a straight line of unit slope passing through the origin. In assessing why they don't, one is obliged to question q_P, because both $k_C = 22$ and V are on theoretically firm ground. Siegbahn et al.[3] indicated that increasing the electronegativity of Br from 2.8 to 3.3 would yield [through (2.44)] values of q_P for the brominated methanes that would bring them into agreement with the fluorinated methanes. Thomas[10] preferred to abandon (2.44) and to use a relation such as that proposed by Gordy[68] for relating ionic character to electronegativity,

$$I = \frac{X_A - X_B}{2} \tag{2.46}$$

This would give

$$q_G(\mathrm{C}) = \sum_i \frac{X_i - X_C}{2} \tag{2.47}$$

as the carbon charge, with the sum taken over the four ligands for each molecule. Finally, in the "charge correlation" approximation, the C1*s* binding-energy shift for a halogenated methane, relative to methane, should be given by[10]

$$E_B(\mathrm{C}1s) - E_B(\mathrm{CH_4}) = (\text{const}) \sum_i (X_i - X_H) \qquad (2.48)$$

A plot testing this relation is shown in Fig. 17. Four points have been added to the plot given by Thomas.

In reviewing the above results it should be noted that the excellent empirical correlation of binding energy with electronegativity, shown in Fig. 17, does not support the validity of q_P as calculated from (2.44) or q_G from (2.47). In both cases the range of charges on carbon in the fluorinated methanes is too large. Figure 16 illustrates this for q_P. For q_G the range is even larger, in fact unreasonably large, as Thomas [10] observed. Since neither of the proposed relationships between ionicity and electronegativity gives charges that are consistent with binding-energy shifts, it is of some interest to derive charges that are consistent from a point-charge model and ascertain their relationship to electronegativity. Writing for a halomethane CX_4

$$E_B(\mathrm{C}1s;\, \mathrm{CX_4}) = kq_{\mathrm{C}}(\mathrm{CX_4}) + \frac{4e^2}{R_{\mathrm{CH}}}\, q_{\mathrm{X}}(\mathrm{CX_4}) = \left(k - \frac{e^2}{R_{\mathrm{CX}}}\right) q_{\mathrm{C}}(\mathrm{CX_4}) \quad (2.49)$$

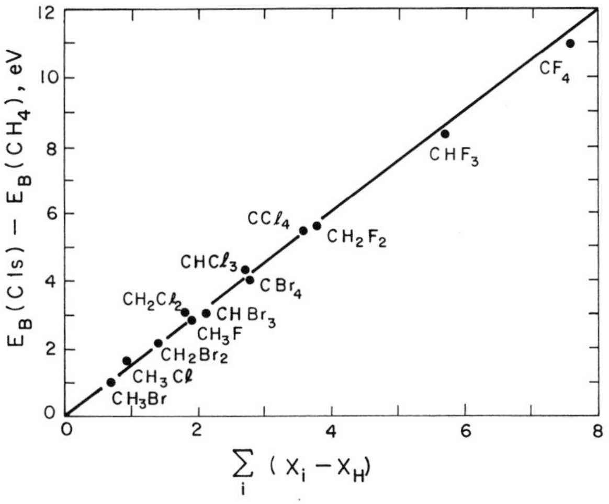

Fig. 17. Binding energies of C1*s* electrons in halogenated methanes, vs. total ligand electronegativity difference, after Thomas.[10] Four points have been added, from Refs. 3 and 44.

and similarly

$$E_B(\mathrm{C}1s;\ \mathrm{CH_4}) = \left(k - \frac{e^2}{R_{\mathrm{CH}}}\right) q_{\mathrm{C}}(\mathrm{CH_4}) \tag{2.50}$$

for methane, it is clear that even with k, R_{CX}, and R_{CH} known, the binding-energy shift can give only Δq_{C}. Unique values of q_{C} are obtained for the compounds CH_4, CBr_4, CCl_4, and CF_4, however, if the constraint is imposed that the ionic character of an AB bond be an even function of $(X_{\mathrm{A}} - X_{\mathrm{B}})$. Using $k_{\mathrm{C}} = 22\ \mathrm{eV}/|e|$ and the bond distances given above, we find $q_{\mathrm{C}}(\mathrm{CH_4}) = -0.20$, $q_{\mathrm{C}}(\mathrm{CBr_4}) = 0.15$, $q_{\mathrm{C}}(\mathrm{CCl_4}) = 0.26$, and $q_{\mathrm{C}}(\mathrm{CF_4}) = 0.79$ as the set of charges that will satisfy these criteria. These charges can be predicted by the linear equation

$$I = 0.129\,(X_{\mathrm{A}} - X_{\mathrm{B}}) \tag{2.51}$$

for the carbon–ligand bonds. C1s shifts relative to methane are then predicted by the potential model

$$\begin{aligned} \Delta E_B &= k\Delta q_{\mathrm{C}} + \Delta V \\ &= (22)(0.129) \sum_{i=1}^{4} [X_X(i) - X_{\mathrm{H}}) \\ &\quad - e^2(0.129)\left[\sum_{i=1}^{4}\left(\frac{X_X(i) - X_{\mathrm{C}}}{R_{\mathrm{CX}}}\right) - \frac{4(X_{\mathrm{H}} - X_{\mathrm{C}})}{R_{\mathrm{CH}}}\right] \end{aligned} \tag{2.52}$$

Shifts based on this equation are plotted, together with experimental shifts, in Fig. 18. This approach combines the advantages of preserving the excellent agreement found by Thomas (Fig. 17) and of giving both a quantitative relationship between ΔE_B and a reasonable set of charges on the carbon atom. Comparison with Fig. 6*a* supports this last point. The charge $q_{\mathrm{C}}(\mathrm{CF_4}) = 0.77$ predicted by (2.51) lies between the value 0.76 of CNDO theory and the ab initio value 1.01, which may be exaggerated by the overlap terms. Either (2.44), or (2.47) gives a carbon charge in CF_4 that is much too large (1.72 or 3.0).

Equation (2.51) can hardly be regarded as the final answer to ionicities in halomethanes. Some specific problems remain. For example, the F(1s) shift between CH_3F and CF_4 is 2.6[10] eV, but this model predicts 5 eV. This shift is sensitive to q_{F}, which (2.51) would predict to be the same for these two molecules, whereas some saturation must take place in electron transfer from C to F. Still the charges predicted by (2.51) provide a good starting point for further improvements and extension to more complex molecules.

4. Correlations with "Group Shifts"

The foregoing discussion leads naturally to the concept of "group shifts," wherein the various groups bonded directly to the host atom cause additive shifts in core-level binding energies. Thus for the halomethanes the C(1*s*) shifts relative to CH_4 can be written

$$\Delta E = \sum_{\text{group}} (\Delta E_{\text{group}} - \Delta E_{\text{H}}) \tag{2.53}$$

In fact this equation can be obtained by rearranging terms in (2.52). The coordinates along the abscissa of Fig. 18 can be reproduced with the values $\Delta E_{\text{X}} - \Delta E_{\text{H}} = 1.033$ eV, 1.362 eV, and 2.725 eV, respectively, for X = Br, Cl, and F. Gelius et al.[8] have exploited the concept of group shifts to correlate C1*s* shifts in a number of carbon compounds in the solid state, but excluding ionic compounds. They found a very good correlation for compounds involving a total of 34 different functional groups. For the above three cases their least-squares procedures gave values of 0.88 eV, 1.55 eV, and 2.78 eV (the shifts in solids are slightly different from those in gases).

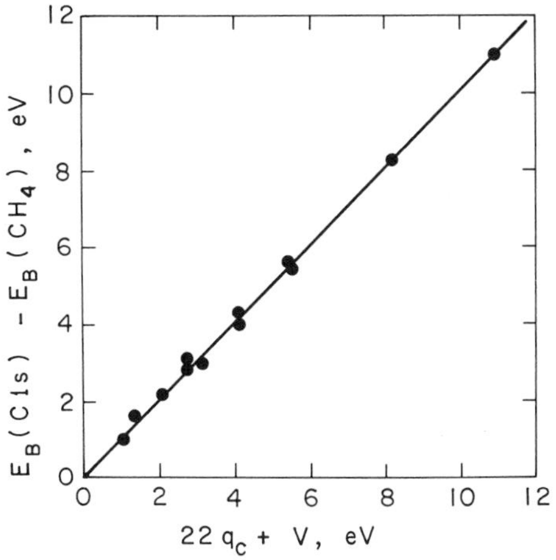

Fig. 18. Binding-energy shifts in halomethanes versus predictions of a potential model based on (2.51) and (2.52).

Another very impressive example of the use of group shifts has been given by Hedman et al.[69] These workers correlated the phosphorus 2*p* shifts in a large number of phosphorus compounds with excellent results. Figure 19 shows about half of their data. The success of the group shift approach for carbon and phosphorus shifts indicates that this empirical

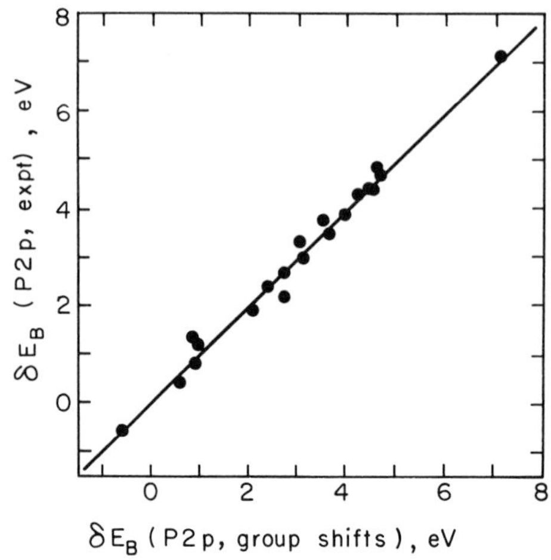

Fig. 19. Experimental $P2p$ core-level shifts versus group shifts, after Hedman et al.[69] These workers showed 23 additional points between 2 and 5 eV.

procedure may ultimately prove the best way for predicting core-level shifts, especially for relatively large molecules and in cases for which large amounts of core-level shift data are already available.

III. VALENCE-SHELL STRUCTURE STUDIES

A. Introduction

The lower binding energies of valence-shell orbitals makes them accessible to lower-energy photons in the ultraviolet (UV) region. Valence orbitals may also be studied by various methods other than photoelectron spectroscopy. Thus, although XPS can make certain unique contributions to valence-shell studies, it is only one of several complementary techniques, and a relatively new one at that. Furthermore, XPS is at present a relatively low-resolution technique. In this section the main objective will be to point out the ways in which XPS can contribute to the elucidation of valence-shell structure. The approach that will be used is to cite specific examples of contemporary valence-shell studies, without making an exhaustive coverage of the literature. Effective use of XPS for valence-shell studies is only beginning, but these examples show that the method holds considerable promise. Applications to the valence shells of metals, molecules, and salts are discussed separately.

B. Valence Bands in Metals

Many metallic properties are attributable to the itinerant electrons in the valence shells. The valence orbitals form bands, and electrons fill the bands essentially up to the Fermi energy E_F. As fermions, electrons fill bands according to the Fermi-Dirac distribution function

$$f = \frac{1}{e^{(E-E_F)/kT} + 1}$$

For $(E_F - E) \gg kT$, f is essentially unity, whereas for $(E - E_F) \gg kT$, f is essentially zero. Thus at $T = 0$ f is 1 below E_F and 0 above. At room temperature $kT = 0.026$ eV, so on the 0.1–1 eV scale of X-ray photoemission the function f is still quite sharp. The number of states available varies with energy according to a function $N(E)$, which is termed the density of states. The *occupied* density of states is then

$$\rho(E) = f(E)N(E)$$

Often no clear distinction is made between $\rho(E)$ and $N(E)$, as it is usually clear from context which is meant. A distinction is usually made between valence bands (below E_F) and conduction bands (above E_F). We shall use this nomenclature.

The nature of Fermi statistics has two consequences for studying valence bands. First, transport properties of metals and many of their other properties can be understood in terms of electronic states very near E_F, and those states farther down in the "Fermi sea" can be ignored. For this reason nearly all the research done on metals to date has in fact studied $N(E_F)$. This approach has obvious merits, particularly in predicting one transport property from another, but even a very detailed knowledge of $N(E_F)$ is totally inadequate for understanding band structure in a fundamental way.

The second consequence of Fermi statistics is that $N(E)$ can never really be studied directly, because the act of studying $N(E)$ disrupts it. This also follows from Fermi statistics. For $(E_F - E) \gg kT$ there are no vacant states nearby in energy, and an electron must be removed entirely, to above E_F at least, in order to be observed. When a hole is thus created, relaxation toward this hole will change $N(E)$. This relaxation can be (and apparently is) small, but it may set a limit on the subtlety of information about $N(E)$ that can be obtained from photoemission.

X-ray and UV photoemission should be compared, because the superior resolution of the latter method would seem to obviate the need for the former. There are some rather strong arguments in favor of X-rays. The greater mean depth from which electrons can be ejected by X-rays implies that this method comes closer to studying bulk properties. In situ monitoring of the surface is also feasible by this method.[70] Finally with X-rays the final-state energy of the ejected electron is so high that this state can be treated as a continuum state which is essentially unaffected by the crystal potential and therefore structureless. Hence the X-ray photoemission spectrum is a relatively accurate representation of the valence band density of states. By contrast UV spectra are strongly modulated by final state structure in the conduction band, as illustrated in Fig. 20.

Another source of modulation in the XPS spectrum is the variation of the radial wave function of the initial *d*-band state with $E - E_F$. This is important because the transition matrix element $\langle 5d|\vec{r}\,|\text{free electron}\rangle$ would vary with energy, and the spectrum would not resemble $\rho(E)$ closely. The extent of modulation is difficult to estimate, but the rather close resemblance between $\rho(E)$ and XPS spectra suggests that it is not very great.

Fadley and Shirley showed the utility of XPS in early valence-band studies of Fe, Co, Ni, Cu, and Pt.[70] At that time even the general features of $\rho(E)$ were experimentally still in doubt for the 3*d* bands. To achieve clean surfaces in the relatively poor vacuum then available in electron spectrometers, the samples were heated and gaseous hydrogen was passed over them continuously during the experiments. This work was later extended to the 4*d* and 5*d* group analogues of Fe, Co, Ni, and Cu.[71] The same cases were also studied by Baer et al.[72] Their results were in good agreement. Although the early work on these metals was of low resolution, the $\rho(E)$ results gave

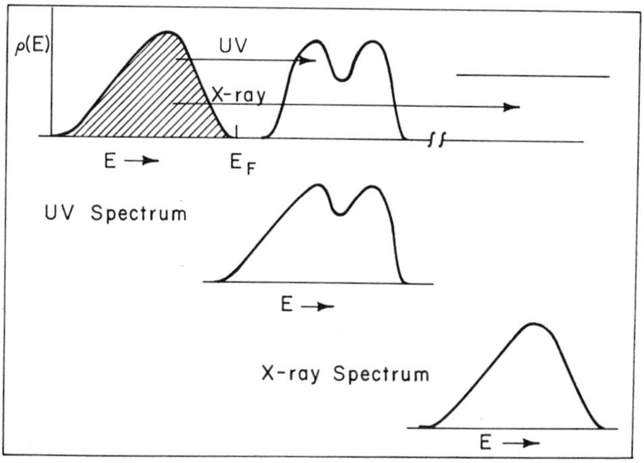

Fig. 20. Illustration of the relationship of UV and X-ray photoemission spectra to band structure. For X-rays the ejected electron's final-state density of states varies slowly with energy, and the spectrum closely resembles the valence band $\rho(E)$. The UV spectrum is affected by conduction band structure.

d band widths and positions. In addition some systematic variations in $\rho(E)$ were observed.[71]

Hagström and co-workers[73–76] studied rare earth metals under better vacuum conditions. They found peaks that could be attributed to the $4f$ electrons. This confirms the expectation[77] that higher orbital angular momentum electron bands should be prominent in XPS spectra. Thus the $4f$ peak in Eu is prominent in the XPS spectrum, while low-energy UV photoemission spectra of Eu and Ba (which differs from Eu in having no $4f$ electrons) are similar.[78]

These workers found narrow $4f$ bands in those rare earths with filled shells (Yb, Lu) and those with half-filled shells (Eu, Gd). Single peaks were found in Eu and Gd, consistent with the $4f^7\ ^8S$ structure, whereas both Yb and Lu showed double peaks, which were assigned to the $4f_{5/2}{}^6$, $4f_{7/2}{}^8$ doublet. In the rest of the rare-earth metals very complex structure was observed. This was attributed to the rather complicated multiplet structure that is possible in all but the simplest cases (i.e., filled or half-filled shells). The rare earth metals are of special interest because the $4f$ shell provides both well-defined localized magnetic moments and (presumably) also conduction electrons. Comparison of XPS spectra of ionic salts and metals should lead ultimately to an understanding of the valence-band structures of these elements.

An illustration of the power of XPS for solving valence-band problems is given by its application to the AuX_2-type intermetallic compounds $AuAl_2$ and $AuGa_2$, by Chan and Shirley.[79] For some time a "$AuGa_2$ dilemma" had existed, making the explanation of certain Knight shifts elusive.[80] Switendick and Narath[81] resolved this enigma by a band-structure calculation that located the $5d$ bands of gold about 7 eV below the Fermi energy. This was contrary to the then-common belief that the $5d$ bands in these compounds lay close to E_F and were responsible for their interesting optical properties. This "d-band dilemma" was settled by the measurements of Chan and Shirley, which showed the d bands centered about 6 eV below E_F, thus confirming the band-structure results.

Recent improvements in resolution, signal-to-background ratio, and particularly vacuum quality, exemplified by the Hewlett-Packard ESCA spectrometer, with a monochromatic X-ray source, promise to yield much better valence-band information. A comparison[82] of the gold valence-band spectrum with theoretical density-of-states results (shown in Fig. 21)[83–89] gives the first example of the power of the newer, second-generation spectrometers. This comparison establishes the necessity of relativistic band-structure calculations for gold. It also appears to favor calculations with full (rather than fractional) Slater exchange. Finally, the good agreement

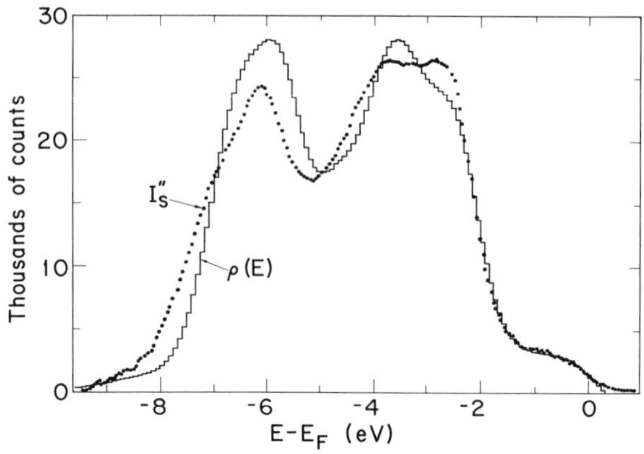

Fig. 21. Comparison of the high-resolution valence-band spectrum of gold (points) with density-of-states function from Ref. 85 (broadened).

of the spectral shape with both high-energy UV spectra[90,91] and theory shows that matrix-element modulation does not distort the spectrum appreciably and that at the He(II) resonance energy (40.8 eV) the spectral shape already resembles the XPS spectrum closely. A detailed comparison of the XPS spectrum with theory is given in Table XVII.

C. Valence Orbitals in Gases: Cross Sections

This important topic will be treated briefly because there are available both comprehensive discussions of experimental spectra[3] and recent reviews of the implications of cross-section studies.[92,93] The reader is referred to these sources, and references cited therein, for detailed discussion. The comments below are confined to a few major points, especially in cross-section studies.

For ten years molecular photoelectron spectroscopy of valence orbitals has been identified mainly with UV excitation, primarily with He(I) or He(II) resonance lines, at 21.2 and 40.8 eV, respectively.[94] The resulting spectra show resolution in the 10^{-2} eV range. Final-state vibrational structure can be observed, and very detailed interpretations can be made. Siegbahn and co-workers[3] have shown that molecular-orbital spectra can also be studied by X-ray photoemission. They have studied a number of small molecules, identifying most or all of the valence-shell molecular orbitals in each case. Typical results are given in Table XVIII.

The low inherent energy resolution of the X-ray protoemission spectra obviates direct competition with UV spectra in making energy assignments

TABLE XVII
Experimental XPS Parameters for Gold Valence Bands and Broadened Density-of-States Parameters

Reference	ΔE_B[a] (eV)	d-band FWHM (eV)	$E_F - E_d$[b] (eV)
82 (expt)	—	5.24	2.04
86	0.79	5.25	1.89
87	0.54	5.54	1.56
89	0.78	4.90	2.21
85	0.85	5.07	2.17
83	0.92	5.67	2.34

[a] FWHM of Poisson broadening function by which theoretical band-structure histograms were multiplied.
[b] Energy difference from Fermi level to a point halfway up the higher-energy d-band peak.

TABLE XVIII
Some Molecular Orbital Binding Energies (after Ref. 3)

H_2O		H_2S		CF_4	
Orbital	E_B (eV)	Orbital	E_B (eV)	Orbital	E_B (eV)
$1b_1$	12.6	$1b_1$	10.3	$3t_2$	16.1
$2a_1$	14.7	$2a_1$	13.2	$1t_1$	17.4
$1b_2$	18.4	$1b_2$	15.1	$1e$	18.5
$1a_1$	32.2	$1a_1$	22	$2t_2$	22.2
				$2a_1$	25.1
				$1t_2$	40.3
				$1a_1$	43.8

for molecular orbitals. However, the X-ray method has considerable value as a complementary technique that can be used to clarify certain assignments. In addition it has great potential as a method for assigning atomic-orbital parentage to molecular orbitals. These advantages are derived from the variation of photoemission cross section with energy and angular momentum.

Price et al.[92] considered the energy variation of photoemission cross section for $2s$ and $2p$ electrons. They presented straightforward overlap arguments that show how the cross-section ratio $\sigma(2s)/\sigma(2p)$ should increase, for valence orbitals, from UV photoemission to X-ray photoemission energies. The consequent effect on molecular orbital photoemission

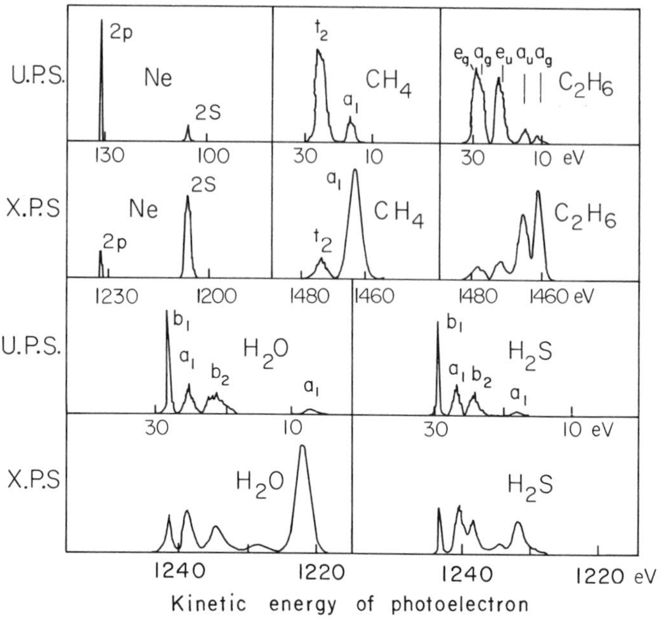

Fig. 22. Comparison of ultraviolet and X-ray photoelectron spectra, from Ref. 92, for neon and four small molecules, showing the increase with energy of $\sigma(s)/\sigma(p)$.

spectra can be dramatic, as indicated in Fig. 22. Here Price et al. have compared UV and X-ray photoemission spectra of the valence orbitals of several small molecules. In both H_2S and H_2O the molecular orbitals b_1, a_1, and b_2, which are derived from p atomic orbitals, retain their relative intensities for the two photon energies, but the a_1 orbital (with s character) shows a relative increase in intensity from low to high photon energy. The potential value of this approach in assigning atomic s or p character to molecular orbitals is obvious. Price et al. also indicated how subtler phase information can be derived from cross-section studies.

Gelius et al.[93,95,96] have made quantitative predictions of XPS spectra from valence orbitals of several small molecules. They gave an argument for the constancy of the photoemission cross section σ of an atomic orbital from one molecule to another. The de Broglie wavelength of a photoelectron ejected from a molecular orbital by $MgK\alpha_{12}$ X-rays is 0.35 Å. Therefore only the innermost regions of the atomic orbitals, where the orbital amplitude varies appreciably over 0.35 Å, can contribute significantly to σ. Hence σ should be nearly independent of the shape of the interatomic portion of the molecular orbital. By assuming that the cross section of the jth molecular orbital could be expressed as a sum over atomic orbitals,

$$\sigma_j^{MO} = \sum_A \sigma_{Aj}$$

and expanding the molecular orbitals in terms of atomic orbitals,

$$\phi_j = \sum_{A\lambda} C_{A\lambda j}\phi_{A\lambda}$$

where λ labels atomic orbital symmetry, Gelius[93] derived the relation

$$\sigma_{Aj} = \sum_{\lambda} P_{Aj\lambda}\sigma_{Ak}$$

Here $P_{A\lambda j}$ is the gross atomic population on atom A of the atomic orbital $A\lambda$ in molecular orbital j. Gelius et al. worked with relative, rather than absolute, cross sections which they determined by careful studies of the rare gases. Spectra were then fitted using gross atomic populations from ab initio calculations. The results for CF_4 are shown in Fig. 23. The cross-section ratios $\sigma(F2s)/\sigma(F2p) = 10$ and $\sigma(F2s)/\sigma(C2s) = 2.0$ were used by Gelius for the theoretical curve. The fit is generally excellent, with the extra intensity in the $3a_1$ region perhaps arising from two-electron effects. Although this was one of the best fits, good results were obtained for a number of molecules. This approach consequently appears to have great

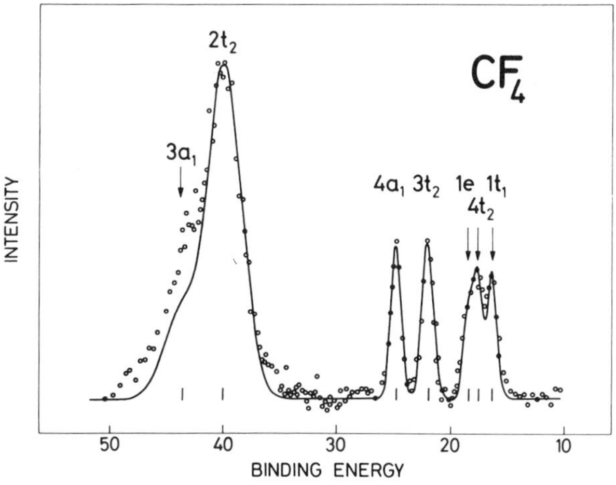

Fig. 23. Experimental photoemission spectrum of CF_4 molecular orbitals, using MgK radiation, and theoretical curve (after Gelius[93]).

potential in elucidating molecular-orbital structure in terms of atomic-orbital composition.

D. Valence Orbitals in Inorganic Anions

Prins and Novakov[97] studied molecular-orbital spectra of perchlorate and sulfate anions in anhydrous salts of lithium and other metals. They observed six peaks and assigned them to seven molecular orbitals. Their results for $LiClO_4$ and Li_2SO_4 are given in Table XIX. Qualitative assignments of peak intensities as strong, medium, or weak have been made by the reviewer. Prins and Novakov observed that theoretical descriptions of the molecular orbital structure of these isoelectronic anions tended to yield three groups of orbitals. The lowest-energy group consists of two levels—a_1 and t_2—formed from the ligand oxygen $2s$ orbitals. The high intensities of the two highest-binding-energy peaks, the relative intensities, and the constancy of their intensities from one salt to another all support this assignment. The next two peaks have been assigned to a_1 and t_2 orbitals derived from the central atom $3s$ and $3p$ orbitals plus the oxygen $2s$ and $2p\sigma$ orbitals. The intensities of these lines in ClO_4^- were about equal to those of the $3s$ and $3p$ lines of Cl^-, thus supporting this assignment. The least-bound group of three orbitals—e, t_2, and t_1—is formed from oxygen $2p\pi$ orbitals, and the low intensity of these peaks is a consequence of the low photoemission cross section of the oxygen $2p$ orbitals. Again the power of intensity-ratio arguments in making spectral assignments is clearly illustrated in the work of Prins and Novakov.

IV. MULTIPLET SPLITTING

A. Introduction

When substances with unpaired electrons in their valence orbitals are studied, their core-level peaks may be split by exchange interaction. This effect has been termed multiplet splitting to distinguish it from other effects that can give rise to extra peaks (e.g., Auger peaks, "shake-off" peaks, "shake-up" peaks, and multiple valence states). In order to identify multiplet splitting it is necessary to eliminate these other effects convincingly. Long experience in the reviewer's laboratory has shown that this can be an extremely tricky problem. Since 1966 a very large number of extra peaks have been identified, but not reported, either because confirmatory experiments showed them to be irreproducible or because they were found to be of different or ambiguous origin. The main difficulty is that the surfaces of most oxides and salts will decompose or at least acquire structural and/or chemical characteristics different from the bulk when placed

TABLE XIX
Valence-Orbital Binding Energies in $LiClO_4$ and Li_2SO_4, after Prins and Novakov (Ref. 97)

Orbital	$E_B(LiClO_4)$ (eV)	$E_B(Li_2SO_4)$ (eV)	Intensity
$t_1(02p\pi)$	6.3	5.8	Weak
e, $t_2(02p\pi)$	9.0	7.7	Weak
$t_2(3p)$	13.4	11.4	Medium
$a_1(3s)$	16.5	14.3	Medium
$t_2(02s)$	27.0	25.3	Strong
$a_1(02s)$	34.4	(29.0)	Medium

in a good vacuum, let alone the vacua that prevail in most photoelectron spectrometers.

Multiplet splitting may be conveniently categorized by reference to a diagram such as that shown for an atomic $n = 2$ shell in Fig. 24. This one-electron diagram is conceptually imprecise in that it refers to the initial state (and to atomic levels), but it gives a qualitative idea of the types of splitting that are possible. Electrostatic splitting is splitting that arises through the angular dependence of Coulombic interactions between electrons bound in different orbitals. Both Coulomb and exchange integrals contribute to this effect. The absolute value of m_j is indicated in Fig. 24 to

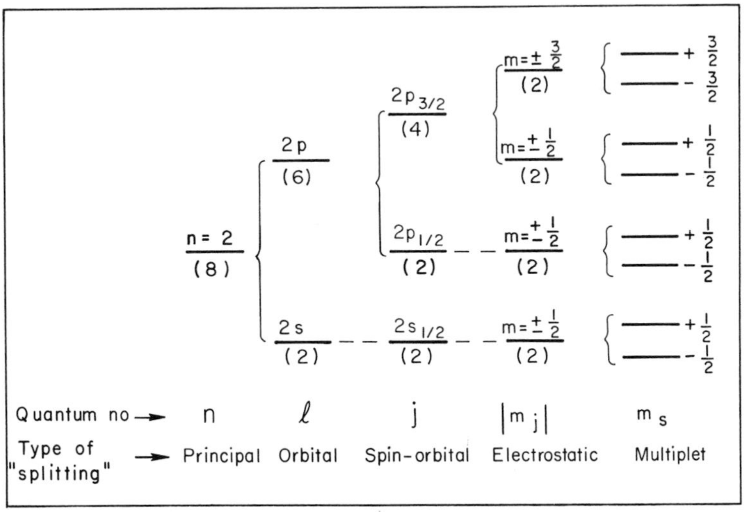

Fig. 24. Classification of core-level splittings, in the one-electron approximation.

emphasize that the electric "field" cannot lift the twofold degeneracy associated with the sign of m_j. Electrostatic splitting was discussed in more detail by Hollander and Shirley.[54]

In multiplet splitting the final spin degeneracy of a core level is lifted through interaction with an unpaired spin in the valence shell. This interaction is mainly attributable to exchange, as discussed below, although correlation effects and (to a very slight extent, through differences between radial wave functions for spin-up and spin-down core orbitals) even Coulomb integrals can make finite contributions. Leaving the oversimplified one-electron diagram shown in Fig. 24 and considering the final states that are accessible in a photoemission process, we can write for the simple case of atomic lithium

$$\mathrm{Li}(1s^2 2s, {}^2S) \rightarrow \mathrm{Li}^+(1s2s, {}^1S \text{ or } {}^3S) + e^- \tag{4.1}$$

The final states can be described by products of symmetric space and antisymmetric spin functions, or vice versa. The energies of these two states may be estimated by adding to E_0, the sum of the one-electron energies (which is the same for the ${}^1S$ and ${}^3S$ states), the electron repulsion term, given by the expectation value of e^2/r_{12}. This leads to the Coulomb and exchange integrals

$$H_c = \int 1s(1)\, 2s(2) \left(\frac{e^2}{r_{12}}\right) 1s(1)\, 2s(2)\, dv_1\, dv_2$$

$$H_x = \int 1s(1)\, 2s(2) \left(\frac{e^2}{r_{12}}\right) 1s(2)\, 2s(1)\, dv_1\, dv_2$$

The resultant energies are

$$E({}^1S) = E_0 + H_c + H_x$$

$$E({}^3S) = E_0 + H_c - H_x$$

and the final-state splitting is given by

$$\Delta E = E({}^1S) - E({}^3S) = 2H_x \tag{4.2}$$

The relative intensities of the multiplet peaks is given by the statistical (multiplet) ratio,

$$\frac{I({}^3S)}{I({}^1S)} = 3 \tag{4.3}$$

The generalization of this discussion to an arbitrary case is complicated, but we can generalize to the case of any spin S in the valence shell, provided that photoemission only from a core level of s character is considered. Thus we are interested in the process

$$M^Z(\cdots ns^2\cdots,{}^{2S+1}X) \rightarrow M^{Z+1}(\cdots ns\cdots,{}^{2S}X \text{ or } {}^{2S+2}X) \qquad (4.4)$$

The generalization of (4.2) is

$$\Delta E = E({}^{2S}X) - E({}^{2S+2}X) = (2S + 1)H_x \qquad (4.5)$$

where H_x is the exchange integral between an ns orbital and a valence-shell orbital. The intensity ratio is

$$\frac{I({}^{2S+2}X)}{I({}^{2S}X)} = \frac{S + 1}{S} \qquad (4.6)$$

from the multiplicities.

B. Multiplet Splitting in Atoms

Fadley and Shirley first suggested multiplet splitting in X-ray photoemission, and they reported an unsuccessful search for splitting in the $3p$ photopeak of metallic iron.[70] Later they attempted to study high-spin atomic systems[98] in order to clarify the reason for this negative result. With the technique and apparatus then available only atomic europium could be studied, as a vapor at 600°C. Poor counting statistics dictated that only the intense $4d_{3/2} - 4d_{5/2}$ doublet could be used. This doublet was significantly perturbed, presumably because of interaction with the valence configuration $4f^7$; ${}^8S$. Careful least-squares curve fitting yielded a value of 2.44 ± 0.15 for the intensity ratio $I(4d_{5/2})/I(4d_{3/2})$, in contrast to an expected unperturbed ratio of 3/2. Auxiliary experiments on gaseous Xe, which has no $4f$ electrons, gave 1.47 for this ratio, and with gaseous Yb (with a filled $4f$ shell) the ratio was 1.49. Thus a multiplet effect is clearly present. A quantitative interpretation would require a rather large configuration interaction calculation because of the large angular momenta involved. Further work on atomic gases would be desirable as a means of testing atomic structure calculations. Advances currently under way in spectrometer design should permit studies of atomic systems that are theoretically more tractable.

C. Multiplet Splitting in Molecules

Hedman et al.[99] first reported splitting in core levels of molecular O_2 and NO. In oxygen they found two lines, of relative intensity 1:2, and spaced by 1.1 eV, with the higher-intensity line having the lower binding energy. The electronic ground state of O_2 is ${}^3\Sigma_g^-$. Ejection of a $1s$ electron therefore leads to states of ${}^2\Sigma^-$ and ${}^4\Sigma^-$ character, with the remaining $1s$

electron on the oxygen atom from which an electron was ejected coupling antiparallel or parallel to the valence-orbital spin $S = 1$. For NO the splitting was 1.5 eV in the N1s line, with an intensity ratio of 1:3 for the two components. The O1s line was broadened to 1.2 eV, as compared to 0.9 eV FWHM for O1s in O_2, and a splitting of 0.7 eV was derived. theoretical estimates of the splittings were in reasonably good agreement with these results[100,101] (Table XX).

TABLE XX

Binding Energies of 1s Electrons in NO and O_2 (in eV)

Case[a]	Binding energy	Measured splitting, ΔE	ΔE from final-state calculations	ΔE from frozen-orbital estimates	ΔE(expt), Hedman et al.[99]
$\underline{N}O(^1\Pi)$	411.5(5)[b]	1.412(16)[c]	1.35[e]	1.23[e] 1.26[g] 0.88[h]	1.5
$\underline{N}O(^3\Pi)$	410.1(5)				
$N\underline{O}(^1\Pi)$	543.6(5)	0.530(21)[c]	0.48[e]	0.73[e] 0.77[g] 0.68[h]	0.7
$N\underline{O}(^3\Pi)$	543.1(5)				
$O_2(^2\Sigma)$	544.2(5)	1.115(9)[d]	0.61[f]	1.68[f] 1.20[h]	1.1
$O_2(^4\Sigma)$	543.1(5)				

[a] The atom losing a 1s electron is underlined. Assumed final-state symmetry is denoted parenthetically.

[b] Standard deviation in the last digit is given parenthetically.

[c] Ref. 102.

[d] Ref. 103.

[e] Ref. 29.

[f] Ref. 30.

[g] Ref. 104.

[h] Ref. 100.

Recently theoretical estimates of the core-level splitting based on hole-state calculations by Bagus and Schaefer[29,30] have become available. Davis and Shirley[102,103] remeasured the splittings in both O_2 and NO, taking care to obtain good statistical accuracy and making extensive least-squares fits of their spectra. These spectra are shown as part of Fig. 25, and the derived splittings are given in Table XX. Also given are theoretical estimates by Bagus and Schaefer and by Schwartz.[104]

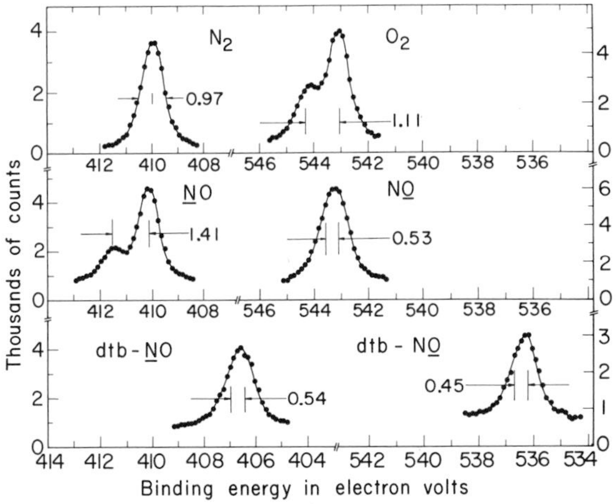

Fig. 25. Nitrogen and oxygen 1*s* peaks in (from top): diamagnetic N_2 and paramagnetic O_2 (showing multiplet splitting); paramagnetic NO, showing that the spin is mostly on the N atom; paramagnetic di-*tert*-butyl nitroxide, showing spin migration to alkyl group and, from decrease in binding energies, charge migration to NO group. From Refs. 3,99, 102, and 103.

The results are intriguing. For NO the hole state calculations and the more precise experimental results show very good agreement. In O_2, however, the most approximate theoretical estimate of splitting actually agrees better with experiment than the hole-state calculation. This is probably fortuitous, because the latter show a very substantial transfer of electronic charge toward the hole state[30] (see Section II.B.1.c). Such an effect is totally absent in frozen-orbital calculations that involve the initial state alone.

The NO results of Hedman et al. showed that the unpaired spin density resides mainly on the N atom in NO, as expected from molecular-orbital calculations. Davis and Shirley[102] studied the N1*s* and O1*s* lines from di-*tert*-butyl nitroxide. They found a "splitting" of 0.448 ± 0.026 eV in the oxygen 1*s* line, only slightly smaller than the NO result. For the nitrogen line, however, the splitting was reduced from 1.412 ± 0.016 eV to only 0.530 ± 0.021 eV. Since to first approximation the splitting goes as $(2S + 1)H_x$ [Eq. (4.5)], these authors noted that an atom i upon which a fraction f_i of the unpaired spin resides will show a splitting of approximately

$$\Delta E^i \approx f_i H_x^{\,i}(2S + 1) \tag{4.7}$$

They therefore interpreted the di-*tert*-butyl nitroxide results as indicating that the $p\pi$ antibonding orbital of NO (in which the unpaired spin resides) expands from nitrogen over the alkyl groups in the larger molecule, whereas the oxygen atom retains most of its population in this orbital. At the same time the decreased N1*s* binding energy (406.5 eV in dtb-NO vs. 410.5 eV in NO) and O1*s* binding energy (536.3 eV in dtb-NO vs. 543.2 eV in NO) indicate considerable electron transfer from the alkyl groups to the NO group. Thus core level binding energies can provide useful information about spin and charge migration in free radicals. In comparison with ESR studies, ESCA is much less sensitive but also less ambiguous.

D. Multiplet Splitting in Salts

Fadley et al.[105] first observed multiplet splitting in transition-metal ions in salts. Fadley and Shirley[98] discussed this work in more detail. In Mn^{2+} or Fe^{3+} the outer electrons have the configuration d^5 and form a ^6S ground state with five unpaired spins coupled to $S = 5/2$. The exchange integral H_x between a $3d$ electron and a core nl electron depends on n but is nearly independent of l. The factor $(2S + 1)H_x(3d;nl)$ would give splittings of $\sim$12 eV for $n = 3$ orbitals and $\sim$3 eV for $n = 2$ orbitals. Line widths and the expectation that both correlation effects and spin migration to anions would reduce the splitting from this figure dictated that early experiments should concentrate on the $n = 3$ orbitals. The a priori obvious choice was the more intense $3p$ peak, but it showed a complicated spectrum, with no clearly defined, simple splitting. The reason for this result is straightforward. After ejection of a $3p$ electron the remaining open shell configuration $3p^5[3d^5(^6S)]$ can couple to form a ^7P final state in only one way (since the spin configuration is "stretched," with all six spins parallel). The complementary ^5P state can be formed in three ways, however, from d^5 terms of ^6S, ^4P, and ^4D. Thus the less intense ^5P "peak" intensity is in fact distributed among the three eigenstates formed from these levels. These eigenstates are spread over 20 eV, and their intensities are low enough to obviate the immediate advantages of studying the $3p$ peaks on the basis of total intensity. The $3s$ peaks are simpler, however: there are two final states, ^5S and ^7S, split by $6H_x(3d,3s)$. The splitting of these peaks was observed in several materials,[98,105] with results given in Table XXI.

In interpreting these results several points were made. First, the splittings in MnF_2 and FeF_3 were smaller by half than estimates based on free atom spin-unrestricted Hartree-Fock, on restricted Hartree-Fock, or on multiplet hole theory estimates.[105] Agreement was good with estimates based on unrestricted Hartree-Fock calculations on MnF_6^{4-} clusters. however, suggesting that spin migration to ligands is important. The slightly

TABLE XXI

Multiplet Splitting in 3*s* Peaks (after Fadley and Shirley[98])

Atom	Compound	Electron configuration	3*s*(1) − 3*s*(2) separation (eV)	3*s*(1):3*s*(2) intensity ratio
Mn	MnF_2	$3d^5\ {}^6S$	6.5	2.0:1.0
	MnO	$3d^5\ {}^6S$	5.7	1.9:1.0
	MnO_2	$3d^3\ {}^4F$	4.6	2.3:1.0
Fe	FeF_3	$3d^5\ {}^6S$	7.0	1.5:1.0
	Fe	$(3d^64s^2)$	(4.4)	(2.6:1.0)
	$K_4Fe(CN)_6$	$(3d^6)$	. . .	>10:1
	$Na_4Fe(CN)_6$	$(3d^6)$	. . .	>10:1
Ni	Ni	$(3d^84s^2)$	(4.2)	(7.0:1.0)
Cu	Cu	$(3d^{10}4s^1)$	. . .	>20:1

smaller splitting in MnO_2 than in MnF_2 may be a consequence of the fact that Mn^{4+} has only three 3*d* electrons. The single 3*s* lines in $K_4Fe(CN)_6$ and $Na_4Fe(CN)_6$ may be attributed to covalent bonding in these compounds.

At first sight, multiplet-splitting studies in transition metals appears to be a powerful diagnostic tool for elucidating spin distributions, and this may yet prove to be true. There is, however, a very severe technical problem. Many compounds are not stable in a vacuum at room temperature. Oxides can lose oxygen, hydrates can lose water, and halides can undergo reactions of the type

$$MnCl_2 + O_2 \rightarrow MnO_2 + Cl_2 \qquad K_{eq} = 10^5$$

by reacting with residul oxygen. Only fluorides appear to possess adequate thermodynamic stability for such studies at room temperature.[98] In spite of these limitations a number of studies of multiplet splitting in the 3*d* group have been reported.[106–108] Recently Werthein et al.[109] have reported a very definitive multiplet splitting of over 8 eV in the 4*s* peak of Gd in GdF_3. In this case seven unpaired *f* electrons in the ^8S state couple to the 4*s* electrons to produce ^7S and ^9S components.

E. Multiplet Splitting in Metals

The effects of multiplet structure on valence-band spectra, as reported by Hagstrom et al.,[73–76] was discussed in Section III.B. Fadley and Shirley[98] showed that core-level splittings can serve as a diagnostic tool for detecting localized magnetic moments. Their results for the 3*d* ferromagnets are

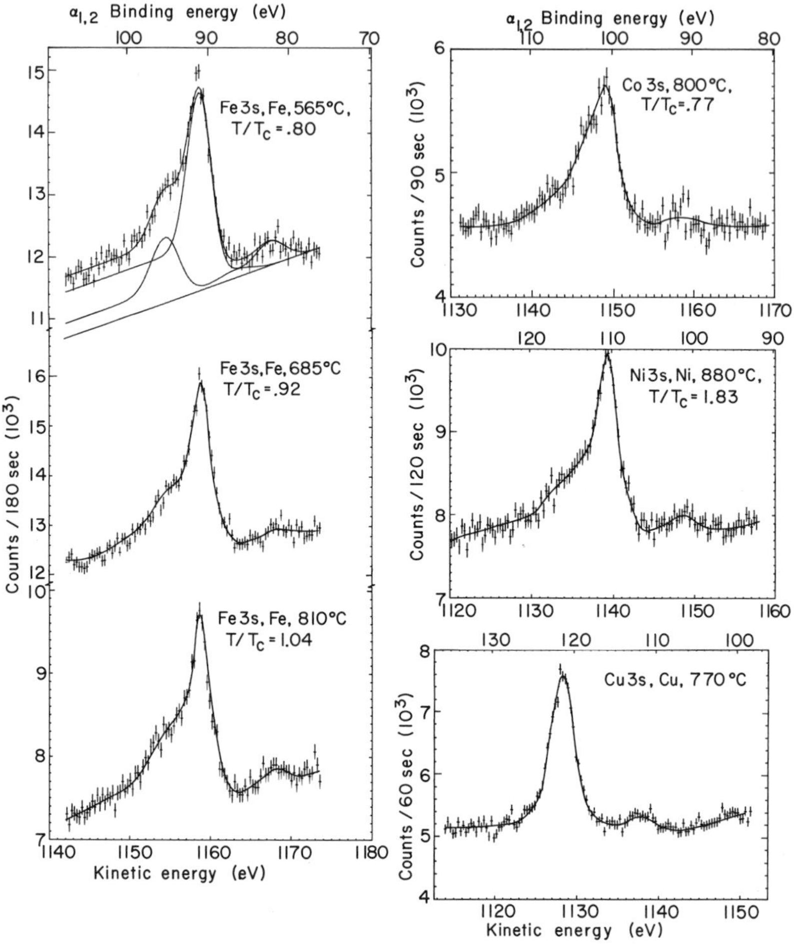

Fig. 26. Multiplet structure of the $3p$ line in $3d$ ferromagnets above and below the Curie point, and the unsplit line of copper.

shown in Fig. 26. The unique power of photoemission for this kind of work lies in its speed. Because the photoemission process takes only $\sim 10^{-16}$ sec, multiplet splitting can detect localized moments that are relaxing too fast to be observed by any other process, or in iron above its Curie point (Fig. 26). Thus ESCA appears to have significant potential for studies in magnetism.

NOTE ADDED IN PROOF: An important paper by C. M. Moser, R. K. Nesbet, and G. Verhaegen [*Chem. Phys. Lett.*, **12**, 230 (1971)] appeared after this article was written.

These authors found for the Ne(1*s*) case that ΔE (corr) = -0.6 eV, not -1.4 eV, and E_B (1*s*) = 870.0 eV. Thus hole-state calculations may give better binding energies than previous work would indicate.

References

1. S. B. M. Hagström, C. Nordling, and K. Siegbahn, *Z. Physik*, **178**, 433 (1964).
2. K. Siegbahn, C. Nordling, A. Fahlman, R. Nordberg, K. Hamrin, J. Hedman, G. Johansson, T. Bergmark, S.-E. Karlsson, I. Lindgren, and B. J. Lindberg, *ESCA—Atomic, Molecular and Solid State Structure by Means of Electron Spectroscopy*, Nova Acta Regiae Soc. Sci. Upsaliensis Ser. IV, Vol. 20, 1967.
3. K. Siegbahn, C. Nordling, G. Johansson, J. Hedman, P. F. Hedén, K. Hamrin, U. Gelius, T. Bergmark, L. O. Werme, R. Manne, and Y. Baer, *ESCA Applied to Free Molecules*, North-Holland, Amsterdam, 1969.
4. D. A. Shirley, Ed., *Electron Spectroscopy*, North-Holland, Amsterdam, 1972.
5. T. A. Carlson, Ref. 4, p. 53.
6. J. M. Hollander and D. A. Shirley, *Ann. Rev. Nucl. Sci.*, **20**, 435 (1970).
7. G. D. Mateescu, Ref. 4, p. 661.
8. U. Gelius, P. F. Hedén, J. Hedman, B. J. Lindberg, R. Manne, R. Nordberg, C. Nordling, and K. Siegbahn, *Phys. Scr.*, **2**, 70 (1970).
9. C. S. Fadley, S. B. M. Hagström, M. P. Klein, and D. A. Shirley, *J. Chem. Phys.*, **48**, 3779 (1968).
10. T. D. Thomas, *J. Am. Chem. Soc.*, **92**, 4184 (1970).
11. A. Veillard and E. Clementi, *J. Chem. Phys.*, **49**, 2415 (1969).
12. T. Koopmans, *Physica*, **1**, 104 (1933).
13. J. K. L. McDonald, *Phys. Rev.*, **43**, 830 (1933).
14. P. S. Bagus, *Phys. Rev.*, **139**, A619 (1965).
15. G. Verhaegen, J. J. Berger, J. P. Desclaux, and C. M. Moser, *Chem. Phys. Lett.*, **9**, 479 (1971).
16. A. Rosén and I. Lindgren, *Phys. Rev.*, **176**, 114 (1968).
17. F. A. Gianturco and C. A. Coulson, *Mol. Phys.*, **14**, 223 (1968).
18. I. Öksüz and O. Sinanoğlu, *Phys. Rev.*, **181**, 42 (1969).
19. R. K. Nesbet, *Phys. Rev.*, **175**, 2 (1968).
20. M. E. Schwartz, *Chem. Phys. Lett.*, **5**, 50 (1970).
21. C. L. Pekeris, *Phys. Rev.*, **112**, 1649 (1958).
22. Schwartz suggested 1.2 eV for correlation following Clementi (Ref. 23). The figure 1.2 comes from this plus the constancy of the $1s^2$ correlation energy with Z (Ref. 18).
23. E. Clementi, *IBM J. Res. Develop.*, **9**, 2 (1965).
24. I. H. Hillier, V. R. Saunders, and M. H. Wood, *Chem. Phys. Lett.*, **7**, 323 (1970).
25. C. R. Brundle, M. B. Robin and H. Basch, *J. Chem. Phys.*, **53**, 2196 (1970).
26. F. A. Gianturco and C. Guidotti, *Chem. Phys. Lett.*, **9**, 539 (1971).
27. R. Moccia and M. Zandomeneghi, *Chem. Phys. Lett.*, **11**, 221 (1971).
28. P. F. Franchini and C. Vergani, *Theoret. Chim. Acta*, **13**, 46 (1969); P. F. Franchini, R. Moccia, and M. Zandomeneghi, *Int. J. Quantum Chem.*, **4**, 487 (1970).
29. P. S. Bagus and H. F. Schaefer, III, *J. Chem. Phys.*, **55**, 1474 (1971).
30. P. S. Bagus and H. F. Schaefer, III, *J. Chem. Phys.*, **56**, 224 (1972).
31. L. C. Snyder, *J. Chem. Phys.*, *J. Chem Phys.*, **55**, 95 (1971).
32. P. Siegbahn, *Chem. Phys. Lett.*, **8**, 245 (1971).
33. L. Hedin and G. Johansson, *J. Phys. B*, **2**, 1336 (1969).

34. D. Liberman, *Bull. Am. Phys. Soc. Ser. II*, **9**, 731 (1964).
35. S. Brenner and G. E. Brown, *Proc. Roy. Soc. (London) Ser. A*, **218**, 422 (1953).
36. R. Manne and T. Aberg, *Chem. Phys. Lett.*, **7**, 282 (1970).
37. M. O. Krause, T. A. Carlson, and R. D. Dismukes, *Phys. Rev.*, **170**, 37 (1968).
38. C. Zener, *Phys. Rev.*, **36**, 51 (1930).
39. J. C. Slater, *Phys. Rev.*, **36**, 57 (1930).
40. H. Basch and L. C. Snyder, *Chem. Phys. Lett.*, **3**, 333 (1969).
41. D. W. Davis, J. M. Hollander, D. A. Shirley, and T. D. Thomas, *J. Chem. Phys.*, **52**, 3295 (1970).
42. M. E. Schwartz, *Chem. Phys. Lett.*, **6**, 631 (1970).
43. T. K. Ha and L. C. Allen, *Int. J. Quantum Chem.*, Symposium, 199 (1967).
44. D. W. Davis, D. A. Shirley, and T. D. Thomas, *J. Chem. Phys.*, **56**, 671 (1972).
45. Ref. 2; also C. S. Fadley, S. B. M. Hagstrom, J. M. Hollander, D. A. Shirley, and M. P. Klein, *Lawrence Radiation Laboratory Report UCRL*-**17299**, January, 1967, p. 233 (unpublished).
46. H. Basch, *Chem. Phys. Lett.*, **5**, 337 (1970).
47. C. Edmiston and R. Ruedenberg, *Rev. Mod. Phys.*, **35**, 457 (1963).
48. J. A. Pople, D. P. Santry, and G. P. Segal, *J. Chem. Phys.*, **43**, S129 (1965).
49. F. O. Ellison and L. L. Larcom, *Chem. Phys. Lett.*, **10**, 580 (1971).
50. D. W. Davis, Lawrence Berkeley Laboratory, private communication, December, 1971.
51. R. S. Mulliken, *J. Chem. Phsy.*, **23**, 1833 (1955).
52. U. Gelius, B. Roos, and P. Siegbahn, *Chem. Phys. Lett.*, **4**, 471 (1970).
53. D. W. Davis, D. A. Shirley, and T. D. Thomas, *J. Chem. Phys.*, **56**, 671 (1972).
54. J. M. Hollander and D. A. Shirley, *Ann. Rev. Nucl. Sci.*, **20**, 435 (1970).
55. M. E. Schwartz, *Chem. Phys. Lett.*, **7**, 78 (1971).
56. W. L. Jolly and D. N. Hendrickson, *J. Am. Chem. Soc.*, **92**, 1863 (1970).
57. J. M. Hollander and W. L. Jolly, *Acc. Chem. Res.*, **3**, 193 (1970).
58. P. Finn, R. K. Pearson, J. M. Hollander, and W. L. Jolly, *Inorg. Chem.*, **10**, 378 (1971).
59. W. L. Jolly, in *Electron Spectroscopy*, D. A. Shirley, Ed., North-Holland, Amsterdam, 1972.
60. D. A. Shirlěy, *Chem. Phys. Lett.* **15**, 325 (1972).
61. J. A. Hashmall, B. E. Mills, D. A. Shirley, and A. Streitwieser, Jr., *J. Am. Chem. Soc.*, **94**, 4445 (1972).
62. F. Brogli, J. A. Hashmall, and E. Heilbronner, *Helv. Chim. Acta*, in press.
63. J. B. Mann, "Atomic Structure Calculations I. Hartree Fock Energy Results for the Elements Hydrogen to Lawrencium," *LA*-**3690**, *TID* **4500**.
64. See, for example, R. E. Block, *J. Mag. Res.*, **5**, 155 (1971).
65. W. H. Flygare and Jerry Goodisman, *J. Chem. Phys.*, **49**, 3122 (1968).
66. R. Ditchfield, D. P. Miller, and J. A. Pople, *Chem. Phys. Lett.*, **6**, 573 (1970).
67. L. Pauling, *The Nature of the Chemical Bond*, 3rd ed., Cornell University Press, Ithaca, New York, 1960.
68. W. Gordy, *Discussions Faraday Soc.*, **19**, 14 (1955).
69. J. Hedman, M. Klasson, C. Nordling, and B. J. Lindberg, *Univ. Uppsala Inst. Phys. Rept. UUIP*-**744** (1971).
70. C. S. Fadley and D. A. Shirley, *Phys. Rev. Lett.*, **21**, 980 (1968).
71. C. S. Fadley and D. A. Shirley, *J. Res. Natl. Bur. Std. A*, **74**, 543 (1970).
72. Y. Baer, P.-F. Hedén, J. Hedman, M. Klasson, C. Nordling, and K. Siegbahn, *Phys. Scr.*, **1**, 55 (1970).

73. G. Brodén, S. B. M. Hagström, P.-O. Hedén, and C. Norris, *Proc. 3rd IMR Symp. Natl. Bur. Std. Spec. Publ.*, **323** (1970).
74. P.-O. Hedén, H. Löfgren, and S. B. M. Hagström, *Phys. Rev. Lett.*, **26**, 432 (1971).
75. G. Brodén, S. B. M. Hagström, and C. Norris, *Phys. Rev. Lett.*, **24**, 1173 (1971).
76. S. B. M. Hagström, Ref. 4, p. 515.
77. F. Combet Farnoux, *J. Phys. Radium*, **30**, 521 (1969).
78. J. G. Endriz and W. E. Spicer, *Phys. Rev. B*, **2**, 1466 (1970).
79. Dorothy P.-Y. Chan and D. A. Shirley, *Natl. Bur. Std., Spec. Publ.* **323**, 791 L. H. Bennett, ed., 1971).
80. V. Jaccarino, M. Weber, J. W. Wernick, and A. Menth, *Phys. Rev. Lett.*, **21**, 1811 (1968).
81. A. C. Switendick and Albert Narath, *Phys. Rev. Lett.*, **22**, 1423 (1969).
82. D. A. Shirley, *Phys. Rev.*, **B5**, 4709 (1972).
83. N. V. Smith and M. U. Traum, "Spin Orbit Coupling Effects in the Ultraviolet and X-ray Photoemission Spectra of Metallic Gold," Ref. 4, p. 541.
84. C. B. Sommers and H. Amar, *Phys. Rev.*, **188**, 1117 (1969).
85. N. E. Christensen and B. O. Seraphin, *Phys. Rev. B*, **4**, 3321 (1971).
86. J. W. D. Connolly and K. H. Johnson, *MIT Solid State Mol. Theory Group Rept. No.* **72**, 19 (1970) (unpublished); and private communication.
87. M. G. Ramchandani, *J. Phys. C, Solid State Phys.*, **3**, S1 (1970).
88. M. G. Ramchandani, *J. Phys. F, Metal Phys.*, **1**, 169 (1971).
89. S. Kupratakuln, *J. Phys. C, Solid State Phys.*, **3**, S109 (1970).
90. D. E. Eastman and J. K. Cashion, *Phys. Rev. Lett.*, **24**, 310 (1970).
91. D. E. Eastman, *Phys. Rev. Lett.*, **26**, 1108 (1971).
92. W. C. Price, A. W. Potts, and D. G. Streets, Ref. 4, p. 187.
93. U. Gelius, Ref. 4, p. 311.
94. D. W. Turner, C. Baker, A. D. Baker, and C. R. Brundle, *Molecular Photoelectron Spectroscopy*, Wiley-Interscience, New York, 1970.
95. U. Gelius, C. J. Allan, G. Johansson, H. Siegbahn, D. A. Allison, and K. Siegbahn, *Phys. Scr.*, **3**, 237 (1971).
96. U. Gelius, C. J. Allen, D. A. Allison, H. Siegbahn, and K. Siegbahn, *Chem. Phys. Lett.*, **11**, 224 (1971).
97. R. Prins and T. Novakov, *Chem. Phys. Lett.*, **9**, 593 (1971).
98. C. S. Fadley and D. A. Shirley, *Phys. Rev. A*, **2**, 1109 (1970).
99. J. Hedman, P.-F. Hedén, C. Nordling, and K. Siegbahn, *Phys. Lett.*, **29A**, 178 (1969).
100. Ref. 3, pp. 59–61. Integrals from Ref. 101 were used in these estimates.
101. H. Brion, C. Moser, and M. Yamazaki, *J. Chem. Phys.*, **30**, 673 (1959).
102. D. W. Davis and D. A. Shirley, *J. Chem. Phys.*, **56**, 669 (1972).
103. D. W. Davis and D. A. Shirley, unpublished data, 1971.
104. M. E. Schwartz, *Theoret. Chim. Acta*, **19**, 396 (1970).
105. C. S. Fadley, D. A. Shirley, A. J. Freeman, P. S. Bagus, and J. V. Mallow, *Phys. Rev. Lett.*, **23**, 1397 (1969).
106. T. Novakov, private communication, March, 1970. Reported at Uppsala Conference on Electron Spectroscopy, September, 1970.
107. G. K. Wertheim, reported at Uppsala Conference on Electron Spectroscopy, September, 1970.
108. J. C. Carver, T. A. Carlson, L. C. Cain, and G. K. Schweitzer, Ref. 4, p. 803.
109. G. K. Wertheim, R. L. Cohen, A. Rosencwaig, and H. J. Guggenehim, Ref 4, p. 813.

AB INITIO CALCULATIONS ON SMALL MOLECULES

J. C. BROWNE

Departments of Computer Science and Physics, The University of Texas, Austin, Texas

F. A. MATSEN

Departments of Chemistry and Physics, The University of Texas, Austin, Texas

CONTENTS

I. INTRODUCTION

It is the purpose of this review to relate the predictions and concepts of coarse structure quantum chemistry to some of the more significant experiments of atomic and molecular physics. We will touch only briefly upon the methods and conceptual foundations of quantum chemistry and then only for the purpose of providing some insight into the problems of carrying out research and the spirit with which it is undertaken.

Ab initio coarse structure quantum chemistry is, in principle, capable of predicting molecular geometry, bond energies, force constants, dipole moments, chemical reactivity, etc. It can have and has had important experimental implications. It can predict* quantities which are not directly obtainable from experiments, and which can be used to calibrate or elucidate a particular experiment. It can even render certain experiments unnecessary.

We present in this paper the results of calculations on several three- and four-electron diatomic molecular species made over the last two decades. In particular we discuss the helium molecule (He_2) and its molecule ion (He_2^+) and the lithium hydride molecule (LiH) and its molecule ion (LiH^+). During this period considerable collaboration occurred between *ab initioist* and experimentalist. The two groups both compete with and complement each other. In order to reflect this profitable interplay, we will discuss separately certain important properties of the several systems in a historical format.

II. THE HAMILTONIAN AND COMPUTATIONAL ORGANIZATION

The Hamiltonian (in the absence of external fields) for the computation of virtually all observable energy states of diatomic molecules is the Breit-Pauli Hamiltonian.[1-3]

$$\mathrm{H} = \frac{1}{2m_a}\vec{P}_a^{\,2} + \frac{1}{2m_a}\vec{P}_b^{\,2} + \frac{1}{2}\sum_{i=1}^{N}\vec{P}_i^{\,2} - \sum_{i=1}^{N}\left[\frac{Z_a}{r_{ai}} + \frac{Z_b}{r_{bi}}\right] + \sum_{i=1}^{N}\sum_{j=1,\, j<i}^{N} r_{ij}^{-1} + \frac{Z_a Z_b}{R_{ab}} + \alpha^2 H(\alpha) \qquad (2.1)$$

where all variables are expressed from a space-fixed reference system and atomic units are used. Here, i indexes the electrons, m_a m_b are nuclear masses, Z_a and Z_b are nuclear charges, R_{ab} is the internuclear distance, and $H(\alpha)$ contains first-order relativistic and magnetic terms. This Hamil-

* We use the word "predict" to refer to observations made in the past as well as the future in spite of the prefix "pre." We regard a theory as a predictive device only. We do not employ the words "interpretation," "explanation," or "understanding" since each appears to be simply a synonym for a theory which makes predictions about past observations. (See F. A. Matsen, "An Epistemology for Science," *Wash. State Univ. Res. Bull.* (June, 1969).

tonian is accurate to second order in the fine structure constant, α. The energies associated with $H(\alpha)$ are, for the light diatomic molecules, much smaller (by approximately five orders of magnitude) than the Coulomb and kinetic energy terms, so that for most purposes these terms can be neglected.

We now introduce relative coordinates (located at the CMN) for the internuclear distance, and express the resulting Hamiltonian in terms of a reference frame rotating with the internuclear axis. On removal of the center of mass motion we obtain for the Hamiltonian

$$H = -\frac{1}{2\mu}\nabla_R^2 - \frac{1}{2}\sum_{i=1}^{N}\nabla_i^2 + V - \frac{1}{2\mu_0}\sum_{i=1}^{N}\sum_{j=1}^{N}\vec{\nabla}_i\cdot\vec{\nabla}_j \tag{2.2}$$

where $V = V(\vec{r},R)$ contains the Coulomb potential terms, $\mu = (m_a\, m_a)/(m_a + m_b)$, and $\mu_0 = m_a + m_b$. If we specify that the electronic angular momentum is quantized along the internuclear axis and the spin is quantized in the fixed reference frame (Hund's case b coupling[5]), then the most general solution of (2.2) can be written in the form

$$\Psi = \sum_k \psi_k(\vec{r},R)\frac{1}{R}X_{k,\nu,K}(R)D_{\Lambda M}{}^K(\Phi,\Theta,0) \tag{2.3}$$

where Θ and Φ are the angles of R, k indexes the set of "quantum numbers" which characterize the electronic state, ψ_k, ν indexes the energy states of the translation (vibration) motion of the nuclei, K is the total (electronic plus nuclear) angular momentum, Λ is the component of electronic angular momentum in the direction of R, and M is the component of K along the z axis of the space-fixed coordinate system. (Browne[4] has discussed the construction of molecular wave functions which are eigenfunctions of the symmetry operators for the molecule). If we neglect the last term on the right-hand side of (2.2), which is approximately four orders of magnitude less than the remaining electronic terms, and also the effect of ∇_R^2 on the ψ_k, then the solutions corresponding to a given set of symmetry operators are decoupled and ψ_k is determined by the Hamiltonian

$$H_{el}\psi_k = E_k\psi_k \tag{2.4}$$

$$H_{el} = \frac{1}{2}\sum_{i=1}^{N}\nabla_i^2 + \sum_{i=1}^{N}\sum_{c=a,b}\frac{Z_c}{r_{ci}} + \sum_{\substack{i,j\\ i>j}} r_{ij}^{-1} + \frac{Z_aZ_b}{R} \tag{2.5}$$

This Hamiltonian (2.5) is the nonrelativistic, spin-free clamped nuclei Hamiltonian. We shall refer to (2.5) as the "*coarse structure*" Hamiltonian, to distinguish it from fine and hyperfine structure Hamiltonians which retain terms neglected in the derivation of (2.5) from (2.1). The corresponding equation for the determination of $X_{k,\nu,K}(R)$ is

$$\left[(2\mu)^{-1}\left(\frac{\partial^2}{\partial R^2}\right) + (2\mu R^2)^{-1}[K(K+1) - \Lambda^2] + E_k(R)\right] \cdot X_{k,\nu,K}(R) = E_{k,\nu,K}X_{k,\nu,K}(R) \qquad (2.6)$$

The effects of ∇_R^2 on ψ_k determine the coupling of the set of fixed nuclei (adiabatic) states by nuclear motion. Clearly if the velocity of the nuclei is comparable to electronic velocities the matrix elements of $\langle\psi_k|\nabla_R^2|\psi_k\rangle$ will become large. In such circumstances different factorizations[6] of the Hamiltonian are appropriate. These tend to the "diabatic" molecular states.[7] The retention of the diagonal matrix elements of ∇_R^2 in the determination of ψ_k and E_k (and the use of the corrected E_k in (2.6)) defines the adiabatic states. The approximation which neglects the ∇_R^2 in the determination of ψ_k but corrects E_k by the addition of $\langle\psi_k|\nabla_R^2|\psi_k\rangle$ may be labeled a "first-order adiabatic approximation." The prediction of molecular energies and properties based on the use of (2.4) and (2.5) to compute wave functions and energies we define as *coarse structure quantum chemistry.*

Ab initio coarse structure quantum chemistry computations are generally based on the variation principle which gives a lower bound to the function:

$$E(\phi) = \frac{\int \phi^* H_{el}\phi \, d\vec{r}}{\int \phi^*\phi \, d\vec{r}} \geq E(\text{exact}) \qquad (2.7)$$

where $\phi = \psi_k$ and E(exact) refers to the exact eigenvalue of H_{el}. There exist two equivalent symmetry formulations of the coarse structure problem: (*a*) the spin formulation in which one employs antisymmetric wave functions which are eigenfunctions of S^2; and (*b*) the spin-free[8] formulation in which one employs spin-free functions which are symmetry-adapted to an appropriate representation of the symmetric group

Virtually all the results we describe here were obtained with a straightforward configuration interaction (CI) expansion of the electronic molecular wave function

$$\psi_k(\vec{r},R) = \sum_i c_i\phi_i \qquad (2.8)$$

where the ϕ_i are configurations symmetry-adapted to the electronic state of interest via (*a*) or (*b*). Application of this procedure, which is sufficiently flexible and general to represent a wide class of molecular states and which is capable of almost arbitrary accuracy for sufficient expenditure of effort has yielded most of the accurate results yet obtained via ab initio computation of electronic molecular wave functions and energies.

The Hartree-Fock or self-consistent-field[9] (SCF) wave functions and energies are based upon a further approximation usually developed from (2.5). The SCF Hamiltonian is determined by a complete variation in (2.7) of a wave function constructed from a single-particle (orbital) basis. The resulting Hamiltonian thus depends on the wave function used in the variational expression. The SCF Hamiltonian determines *orbitals* rather than total wave functions. The energies obtained by the SCF method can be shown to be the lowest obtainable with a given orthogonal single-particle basis set for a specific wave function form. The SCF Hamiltonian for a single-determinant double-occupied representation of a molecular wave function is[9]

$$\left[-\tfrac{1}{2}\nabla_1^2 - \sum_{c=a,b} \frac{Z_c}{r_{ci}}\right] u_i(1) + \sum_j \left[\int u_j^2(2)\, r_{12}^{-1} u_j(2)\, d\vec{r}_2 - \delta(s_i,s_j)\right.$$
$$\left.\int u_j^*(2)\, r_{12}^{-1} u_i(2)\, d\vec{r}_2\right] \cdot u_j(1) = -\sum_j \lambda_{ij}\delta(s_i,s_j) u_j(1)$$

where s_k is the spin coordinate associated with the orbital u_k and λ_{ij} are LaGrangian multipliers to ensure the orthogonality of the u_k.

The deficiencies in the application of the SCF procedure to the determination of molecular wave functions and energies arise principally from the use of a *single determinant* or even a *single* configuration as the representation for the molecular wave function for the determination of the SCF Hamiltonian, orbitals ,and energies. Regarding this point, Wahl and Das,[10] leading proponents of extended SCF procedures, state with respect to the shortcomings of the single configuration SCF method of computing molecular wave functions. "There is no flexibility built into the wave functions which can be expected to adequately describe molecular formation and dissociation." Equally seriously a single determinant can represent only a restricted class of possible electronic molecular symmetry states. Despite these severe limitations the analytic expansion method of Roothaan[11] and others[12] for obtaining SCF molecular wave functions based on single determinant representations of the wave function dominated the field of small molecule computations for nearly a decade.[13]

The development of self-consistent-field methods based on multiconfiguration representations of the molecular wave functions (MCSCF methods) has been carried out by a number of workers.[9] The optimized-valence-configuration (OVC-MCSCF) method developed by Wahl, Gilbert, Das,[14] and their co-workers has been applied with very considerable success to molecules of intermediate (4–30 electrons) size. True

MCSCF methods should not be confused with the use of the orbitals determined from a single-determinant SCF computation in a configuration interaction expansion [see (2.7)].

In the actual computation, by whatever method, the variation function, ϕ, is expanded in some basis set, the expansion coefficients being chosen so as to have proper permutation and point-group symmetry and to minimize $E(\phi)$. The basis set chosen is that basis set which yields the lowest $E(\phi)$ subject to the skill, the patience, and the computer access of the *ab initioist*. The basis sets are of two general types: (*a*) products of one-electron functions (orbitals), and (*b*) functions with interelectronic coordinates. For the second type, a smaller basis is required, but the integrals are more difficult to evaluate. The representation of the Hamiltonian in a given basis set is diagonalized in the conventional way. These basis sets contain so-called nonlinear parameters (usually orbital exponents) which are "optimized" with respect to a particular root of the secular equation.

In the fourth and fifth decades of this century extensive ab initio coarse structure calculations were carried out on the two-electron molecules, H_2 and HeH^+, employing interelectronic coordinates. Predictions made on the basis of these calculations were found to be accurate to the number of significant figures in the experimental data. An excellent review of recent computations on one- and two-electron systems has been given by Kolos.[15] Hirschfelder and Meath[16] also discuss the hydrogen molecule in detail.

Accurate computations on polyelectronic molecules are considerably more difficult for the following reasons:

1. With an interelectronic coordinate function basis set, the difficulty of integrating over the interelectronic coordinate basis set is greatly increased. Recent advances in solving this problem are sketched in Appendix A.
2. With an orbital product basis set the size of the basis set must be greatly increased.
3. With an orbital product basis set the number of integrals to evaluate increases with N^4.
4. With an orbital product basis set the number of linearly independent symmetry-adapted sets from each orbital product is greatly increased.

As a consequence of these difficulties, it is absolutely necessary for the *ab initioist* to have available: (*a*) large quantities of computer time on a high-speed computer with good word length and storage; (*b*) rapid integration procedures; (*c*) sophisticated data management into which is built the symmetry-adaptation procedure; and (*d*) accurate diagonalization procedures for large matrices. Computational research programs organized along the lines outlined above have been established by F. E. Harris and H. H. Michels of the Unviersity of Utah and United Aircraft Research Laboratories; E. R. Davidson and co-workers at The University of Washington; H. Shull and S. Hagstrom at Indiana University; A. C. Wahl and colleagues at Argonne National Laboratory; R. K. Nesbet of IBM Research, San Jose; E. Clemente, A. D. McLean and co-workers of IBM Research, San Jose; and the present authors at The University of Texas at Austin. More detailed discussions of typical computational procedures are given by Browne and Matsen[17] and Harris and Michels[18] and reviewed by Browne.[4]

In the past ten years there has been enormous activity in the field of quantum chemistry aimed at developing new methods of calculation. There is exciting promise of greatly increased scope for the application of fundamental theory in the coming years. Appendix A sketches briefly the origins of some of this work and lists references for further coverage.

We turn now to the consideration of the three- and four-electron molecules which have been principal subjects of attempts at accurate ab initio computation and which have been of special interest to the authors.

III. THE HELIUM MOLECULE AND MOLECULE ION

A. The Dissociation Energy of He_2^+

The helium ion, He_2^+, although well known as a constituent of helium discharges and plasmas, has never been observed spectroscopically. Early estimates of its dissociation energy ranged from 1.4 to 3.2 eV with no reason for preferring one value over the other. This uncertainty in the dissociation energy hampered the interpretation of the spectrum of the helium molecule (see below). In 1963, Reagan, Browne, and Matsen[19] made what was the first really accurate ab initio calculation on a polyelectronic system and established rigorously that $D_e(He_2^+) \geq 2.2$ eV with an approximate upper bound of 2.5 eV. In 1967 Gupta and Matsen[20] made an improved calculation and applied the semitheoretical adjustment procedure of Klein, Greenawalt, and Matsen[21] which utilizes the experimentally determined united atom and separated atom energies. The Gupta and Matsen potential curve is shown in Fig. 1. These calculations yielded a $D_e(He_2^+)$ of 2.30 eV. In 1970 Ginter and Battino[22] by extrapolating potential curves of the Rydberg states of the helium molecule gave an experimental value of $D_e(He_2^+) = 2.33 \pm 0.02$ eV. The quality of the computer potential curve is reflected in Fig. 2 (taken from Ginter and Battino) where the difference between the extrapolated experimental curve and the Gupta-

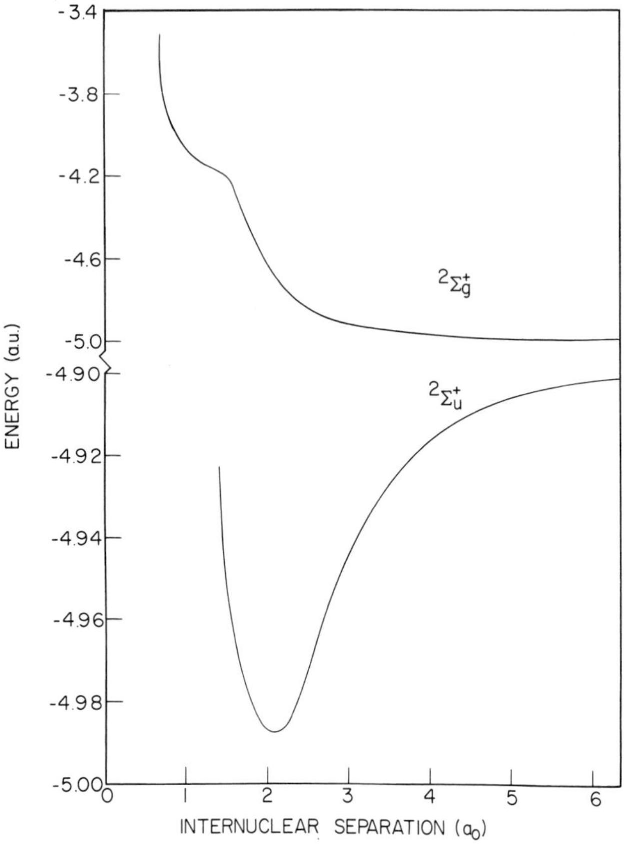

Fig. 1. The $^2\Sigma_u^+$ $(1\sigma_g^2\ 2\sigma_u)$ and $^2\Sigma_g^+$ $(1\sigma_g\ 2\sigma_u^2)$ potential curves of He_2^+ as computed by Gupta and Matsen.[20]

Matsen curve is less than 400 cm^{-1} over a range of internuclear separations $0.9a_0 < R < 1.5a_0$.

More recently, however, Ginter and Brown[23] utilizing analyses of the "600 Å" band of the helium molecule by Sando[24] and Sando and Dalgarno[25] together with their own data conclude that $D_e(^2\Sigma_u^+, He_2^+)$ must exceed 2.40 eV. A definitive calculation by Liu[26] gives a lower bound of 2.454 eV for D_e and a best estimate of 2.469 + 0.006 eV. This is in excellent agreement with the $D_e(^1\Sigma_u^+, He_2) = 2.50$ eV given by Sando since $D_e(^1\Sigma_u^+, He_2)$ and $D_e(^2\Sigma_u^+, He_2^+)$ are known to differ by less than 100 cm^{-1}. The excellent agreement in Fig. 2 between computation and experiment still holds as to shape but the curves must be shifted downward by approximately 0.15 eV.

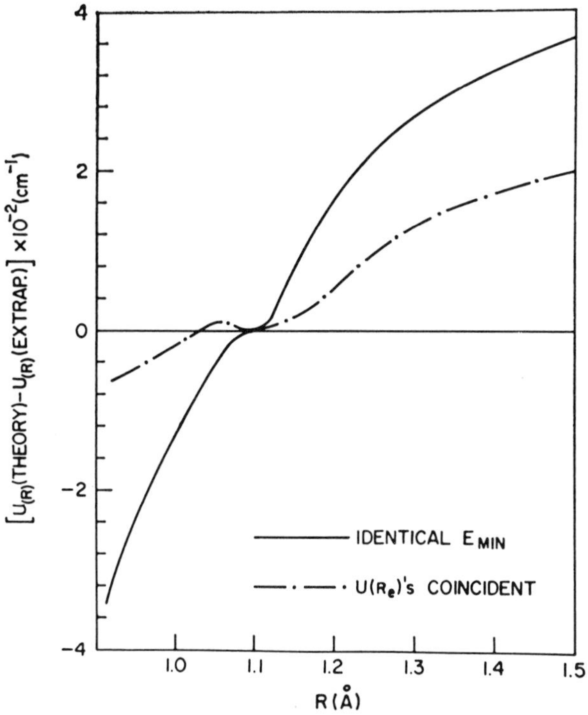

Fig. 2. Comparison of the potential $X^2\Sigma_u^+(He_2^+)$ curves constructed from molecular constants extrapolated from He_2 (see Ginter and Battino[11]) with the theoretical values of Gupta and Matsen.[9] The solid line corresponds to requiring the same minimum energy for the two potential curves, and the broken line corresponds to an additional shift which brings the minima into coincidence (see Ref. 22). This figure is taken from Ref. 22.

B. The Appearance Potential of He_2^+

The helium molecule ion He_2^+ is observed in a mass spectrometer at an appearance potential of 23.1 eV. The formation of He_2^+ is thought to proceed via the Hornbeck-Molnar process.

$$He^* + He \rightarrow He_2^+ + e$$

Therefore the appearance potential for He_2^+ should be given by the relation

$$\text{A.P.}(He_2^+) \geq \text{I.P.}(He) - D_0(He_2^+)$$

Thus a farily precise knowledge of $D_0(He_2^+)$ is necessary before one can determine which of the He* states can participate in the Hornbeck-Molnar process. As noted by Franklin and Matsen[27] in 1964, the $D_e(He_2^+)$ calculated by RBM gives A.P.(He_2^+) = 22.4 eV. This effectively rules out He($2p\ ^1P$) and enables an assignment of He($3p\ ^1P$) as the atomic state responsible for the first appearance of He_2^+ in a mass spectrometer.

C. Elastic Scattering of He^+ by He

In 1965 Marchi and Smith[28] made a theoretical analysis of the elastic differential cross sections of Lorents and Aberth[29] for the scattering of He^+ by He. This analysis requires a knowledge of the lowest $^2\Sigma_u^+$ and $^2\Sigma_g^+$ states of the He_2^+ molecule. Marchi and Smith used the RBM curve for the $^2\Sigma_u^+$ state but were forced to use an approximate $^2\Sigma_g^+$ state curve derived from Phillipson's[30] $^1\Sigma_g^+$ He_2 curve. Lorents, Marchi, and Smith[31] in 1965 observed aberrations in the elastic cross sections which they attributed to zero-order avoided crossings in the lowest $^2\Sigma_g^+$ curves of He_2^+. These zero-order avoided crossings were also observed in some modest calculations by Browne[32] in 1966. Both of these situations called for accurate calculations to be made for the lowest $^2\Sigma_g^+$ state and Gupta and Matsen,[20] responding to a suggestion of F. T. Smith, undertook calculations for the $^2\Sigma_g^+$ state of equivalent or superior accuracy to the RBM $^2\Sigma_u^+$ calculations (see Fig. 1). The calculations, published in 1967, agreed well with the experiments as to location of the avoided crossings. Also in 1967 Mueller and Olson[33] inverted the differential cross-section data of Lorents and Aberth[29] to obtain estimated potential curves for both the $^2\Sigma_u^+$ and the $^2\Sigma_g^+$ states. These curves were in excellent agreement with the RBM $^2\Sigma_u^+$ curve. In 1969, F. J. Smith and M. Kennedy[34] applied a new technique to invert Lorents and Aberth's data and obtain new estimates for both the $^2\Sigma_u^+$ and $^2\Sigma_g^+$ states of He_2^+. These authors conclude that for internuclear separations greater than 0.7 Å (this separation corresponds to about a 20 eV repulsive interaction on the $^2\Sigma_g^+$ curve) the ab initio calculated fixed nuclei potential curves precisely reflect the potentials obtained by inverting the cross-section data. The cross sections computed with the derived potential curves are compared with those computed from the potential curves of Gupta and Matsen in Fig. 3. The potential curves themselves are compared in Fig. 4. Below R = 0.7 Å, a separation reached by laboratory scattering energies $\geq$ 20 eV, the $^2\Sigma_g^+$ curve must be replaced by non-adiabatic curves. This observation leads to a discussion that is beyond the scope of this review. (See Smith[5] and Browne[4] for discussions of this subject.)

D. The Repulsive Potential Curve Between A Pair of Ground State Helium Atoms

The approximate shape of the He_2 potential curve has been known for a long time from equation of state and transport properties data. It has

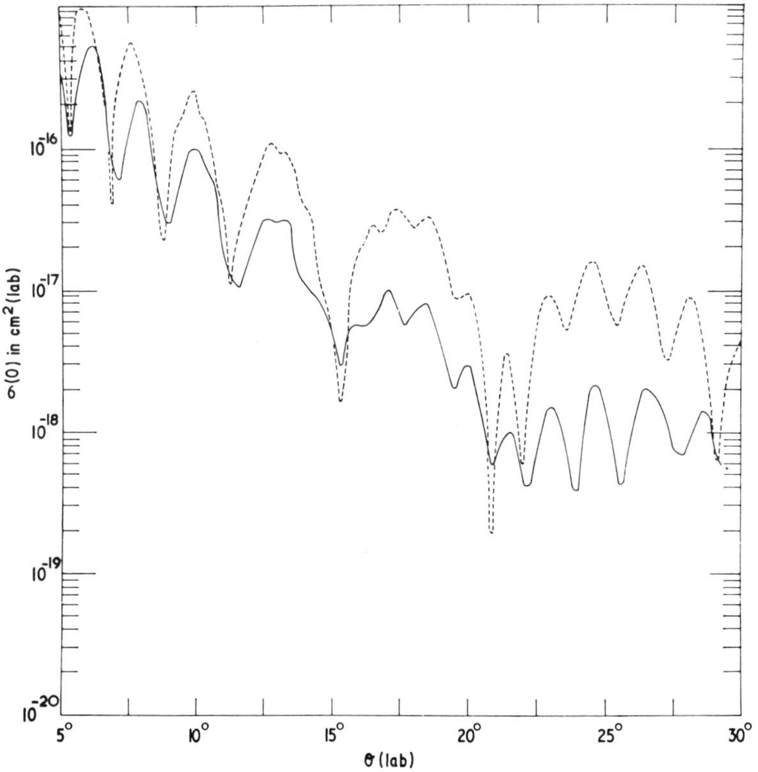

Fig. 3. Experimental differential cross sections at 600 eV incident energy for scattering of He^+ by He (solid line) compared to the differential cross sections computed from the potential shown in Fig. 4. This figure is taken from Ref. 34.

an exceedingly shallow (van der Waals) minimum and a strongly repulsive part.

In 1954 Amdur and Harkness[35] made a direct determination of the repulsive curve by high-velocity helium–helium scattering experiments. In 1956 T. Wu[36] pointed out that the extrapolation of the Amdur-Harkness curve to the united atom limit yielded an energy considerably below the energy of the united atom, beryllium. In 1962 Phillipson[30] made a heavy *ab initio* calculation of the repulsive curve which lay more than 10 eV higher than the Amdur-Harkness potential over the region where both were available. In 1963 Thorson[37] analyzed the high-energy scattering problem and concluded that nonadiabatic effects could not account for the discrepancy. The year 1967 was a year of intense activity. Highly accurate calculations were made by Matsumoto, Bender, and Davidson,[38] Barnett,[39]

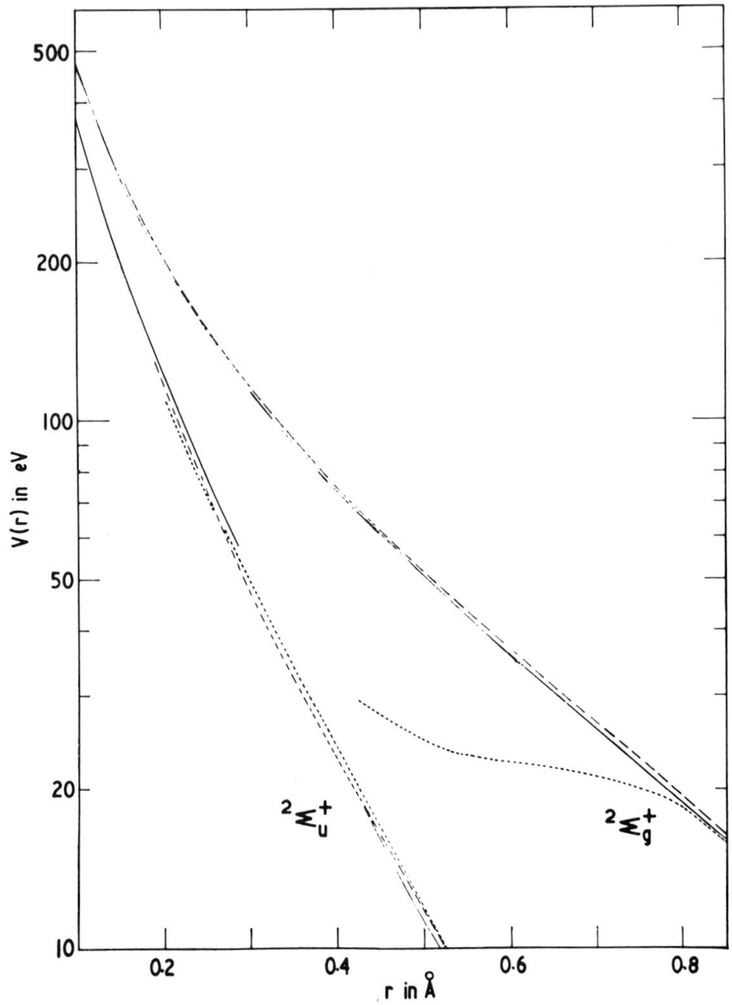

Fig. 4. A comparison of potential curves for He^+ and He. The solid lines are the potentials of Marchi and Smith,[28] the dashed lines the potentials of Kennedy and Smith,[34] and the dotted lines the potentials of Gupta and Matsen.[20]

and Klein, Rodriguez, Browne, and Matsen.[40] These calculations were all in agreement with each other and lay about 2 eV lower than the Phillipson calculations. These calculations have a probable error of not more than 0.3 eV. In the same year Jordan and Amdur[41] published the results of new experiments with much higher beam resolution. Their curve lies from 2–3 eV below the ab initio curve. (See Fig. 5). The remaining discrepancy

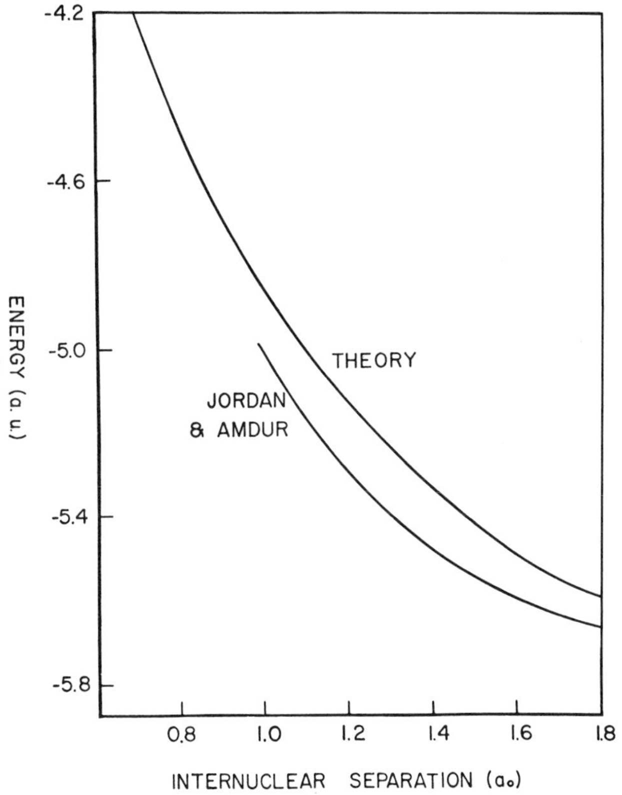

Fig. 5. Theoretical and experimentally derived potential curves for $X^1\Sigma_g^+$ He_2.

is to be attributed to nonadiabatic corrections or to experimental error or both.

E. The Long-Range Attractive Potential Between Two Ground State He Atoms

The experimental observations on the nature of the long-range attractive potential come mostly from bulk properties. Bruck and McGee[42] have reviewed the determination of the form of the long-range potential from analysis of these measurements. They estimate a van der Waals well depth on the order of 10–12°K. The most recent theoretical calculations, Bertoncini and Wahl[43] and Schaefer, McLaughlin, Harris, and Alder,[44] calculate well depths in the vicinity of 11 and 12°K. These calculations represent a breakthrough in the direct application of ab initio calculations to the determination of long-range forces. Previous efforts to obtain long-range forces by direct calculation have been stymied by the difficulty of making calculations sufficiently consistent and sufficiently accurate so that the subtraction of separated atom energies from the energies of the interacting systems did not cause errors orders of magnitude greater than the very small van der Waals energies of interaction. The Schaefer, McLaughlin, Harris, and Alder calculation was a direct application of relatively new pair correlation computational formalism (see Appendix A). The Bertoncini and Wahl calculation was a very careful extended Hartree-Fock calculation.

F. Excited States of the Helium Molecule

The helium molecular spectrum is an extremely rich spectrum and has been intensively studied for a long time. However, the detailed characterization of the helium molecular states has proved very difficult since the ground state is unstable and the molecular ion has not been studied spectroscopically. As a consequence, interpretation of collision processes has also been very difficult. Thus ab initio calculations on the excited molecular states of helium have been of great assistance to spectroscopists and the atomic collision experimentalists. We confine ourselves to a discussion of the lowest set of excited states, all of which have maxima (barriers) in their potential cuves.

1. Nonobligatory Maxima in the $^1\Sigma_u^+$, $^3\Sigma_u^+$ *Potential Curves of* He_2

The existence of these maxima, although suggested as early as 1935 by Nickerson[45] for the $A^1\Sigma_u^+$ state to account for a continuum in the $A^1\Sigma_u^+ \leftrightarrow X^1\Sigma_g^+$ spectrum, was largely discounted until the pioneering ab initio calculations of Buckingham and Dalgarno[46] in 1952 which yielded maxima in both the $^1\Sigma_u^+$ and $^3\Sigma_u^+$ curves.

In 1957 Phelps and Molnar[47] pointed out that rate coefficients derived from helium discharge processes were consistent with a maximum in the $^3\Sigma_u^+$ potential curve maximum. Later diffusion cross-section measurements of Phelps[48] (1966) appeared to support this suggestion. Since the experiments only inferred the existence of the maxima and did not assign precise values, additional and successively more accurate calculations of the $^3\Sigma_u^+$ potential curves were made by Brigman, Brient, and Matsen[49] (1961) and Poshusta and Matsen[50] (1963). The Poshusta and Matsen calculation yielded a barrier height of 0.145 eV. Matsen and Scott[51] review the calculations on $^3\Sigma_u^+$ prior to 1965 in detail. In 1969 Kolker and Michels[52] reported new calculations for the $a^3\Sigma_u^+$ (and $b^3\Sigma_g^+$) potential curves. These curves, the most accurate yet reported for these states, having an error at the separated atom limit of only 0.06 eV, yield a barrier height of 0.11 eV for the $A^3\Sigma_u^+$ state. The chronological development of the ab initio $^3\Sigma_u^+$ potential is shown in Fig. 6. The Kolker and Michel calculation is not sufficiently distinct from the 1965 calculation to be resolved on the coarse scale of the graph. The measurement of the temperature dependence of excitation transfer rate coefficient.

$$\mathrm{He}(1s2s\ ^3S) + \mathrm{He}(1s^2\ ^1S) \rightarrow \mathrm{He}(1s^2\ ^1S) + \mathrm{He}(1s2s\ ^3S)$$

by Colegrove, Shearer, and Walters[53] in 1964 provided the most direct experimental evidence for the existence of the maxima in the $^3\Sigma_u^+$ curve and gave an estimate of the height as 0.06 eV. Richards and Muschlitz[54] (1964) and Rothe, Neynaber, and Trujillo[55] (1965) measured the total elastic cross section for scattering of He($1s2s\ ^3S$) by He($1s^2\ ^1S$). These measurements, however, did not contribute significantly to the estimate of

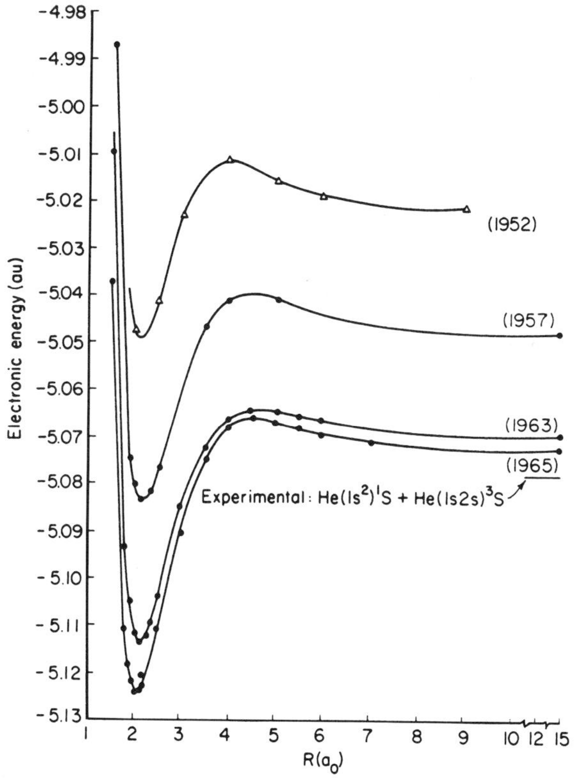

Fig. 6. Historical development of ab initio calculated potential curves for the $A^3\Sigma_u^+$ state of He_2.

the height of the barrier. Ludlum, Larson, and Coffrey[56] (1967) suggested a barrier height of 0.067 eV from rate coefficient measurements of the three-body process.

$$\mathrm{He}(^3S) + 2\ \mathrm{He}(^1S) \rightarrow \mathrm{He}_2(^3\Sigma_u^+) + \mathrm{He}$$

Fitzsimmons, Lane, and Walters[57] (1968) computed diffusion cross sections using the computed ab initio potential curves and found that these have too strongly repulsive long-range tails to give good agreement with their new and previous measurements. This implied that the computed barrier heights might be too large. In 1963, Tanaka and Yoshino,[58] by analysis of the $A^1\Sigma_u^+ \rightarrow X^1\Sigma_g^+$ emission continuum, established that the barrier height of the $A^1\Sigma_u^+$ potential curve did not exceed 0.2 eV.

Browne[59] (1965) with a fairly modest wave function calculated a barrier height of 0.19 eV for the $A^1\Sigma_u^+$ state. Allison, Browne, and Dalgarno[60] (1966) using a somewhat more elaborate wave function obtained a barrier height of 0.083 eV for this state. Scott, Browne, Greenawalt, and Matsen[61] (1966) included closed-shell correlation in a 17 configuration wave function, obtained a barrier height of 0.153 eV, and were able to put a rigorous upper bound of 0.364 eV on the barrier height. The latter calculations focuses on the wave function in the neighborhood of the barrier and not on the equilibrium separation. The predicted binding energy of 1.7 eV would therefore be expected to be low. Tanaka and Yoshino[62] (1969), continuing their work on the $^1\Sigma_u^+ \leftrightarrow {}^1\Sigma_g^+$ spectrum, suggest a minimum barrier height of 0.059 eV for the $^1\Sigma_u^+$ state. Ginter and Battino[22] suggest that the barrier heights should be less than 0.07 eV.

Smith and Chow[63] have established a renumbering of the vibrational assignments suggested by Tanaka and Yoshino[62] for $A^1\Sigma_u^+$ and demonstrated that the barrier height should be 0.05 + 0.01 eV. Sando and Dalgarno[25] have developed a potential curve for $^1\Sigma_u^+$ by analyzing the absorption data of Tanaka and Yoshino. They fix the $^1\Sigma_g^+$ potential curve and use the transition moments calculated by Allison, Browne, and Dalgarno. They then vary the $^1\Sigma_u^+$ potential curve until the densitometer tracing of Tanaka and Yoshino are reproduced. They obtain an excellent reproduction of the Tanaka and Yoshino data with a barrier of 0.05 eV. Smith and Chow and Sando and Dalgarno find somewhat different assignments for the vibrational levels of the Tanaka-Yoshino data. The two assignments, although they yield similar values for the barrier height, of course give different values for $D_e(^1\Sigma_u^+, \mathrm{He}_2)$. Sando[24] reconstructs the emission data of Tanaka and Toshino still using the transition moments of Allison, Browne, and Dalgarno and finds a barrier height of 0.05 eV and a D_e of 2.50 eV. Liu,[26] using his highly accurate computed value for

$D_e(^2\Sigma_u^+, He_2^+)$ together with the vibrational assignment of Sando, estimates a barrier height of 0.08 + 0.02 eV. Ginter and Brown[23] review the work on vibrational assignment including the Liu computation and suggest a barrier height of 0.07 eV. Thus the combination of computation and experiment has yielded a value of the barrier height for the $^1\Sigma_u^+$ state of He_2 with a much higher confidence interval than could be obtained from either separately.

2. Obligatory Maxima in the Potential Curves of the Helium Molecule

R. S. Mulliken[64] suggested in 1964 that many states of He_2, for example, $c^3\Sigma_g^+$, $C^1\Sigma_g^+$, $F^1\pi_u$, $f^3\pi_u$, should have substantial maxima in their potential curves as a result of zero-order avoided crossings. (The interested reader should consult Mulliken[65] for a detailed discussion of the origins of these maxima.) Figures 7 and 8 summarize the structure of these curves. Ginter's[66] (1965) spectroscopic investigations also suggested the occurrence of maxima in the potential curves of the low-lying excited states of the

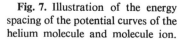

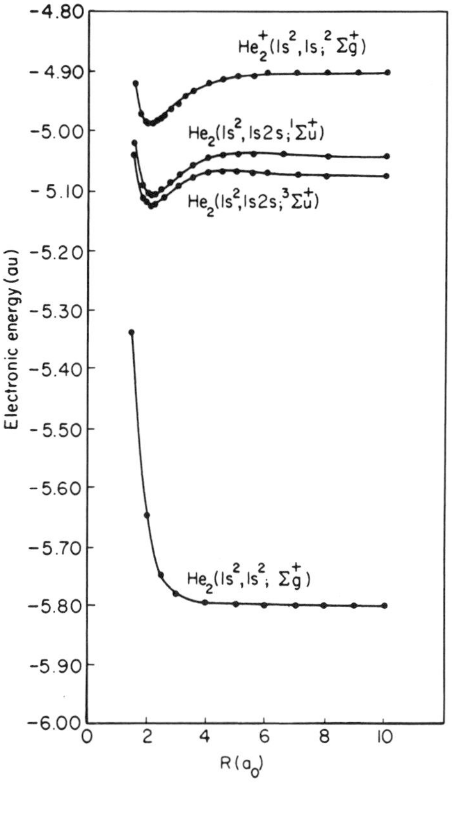

Fig. 7. Illustration of the energy spacing of the potential curves of the helium molecule and molecule ion.

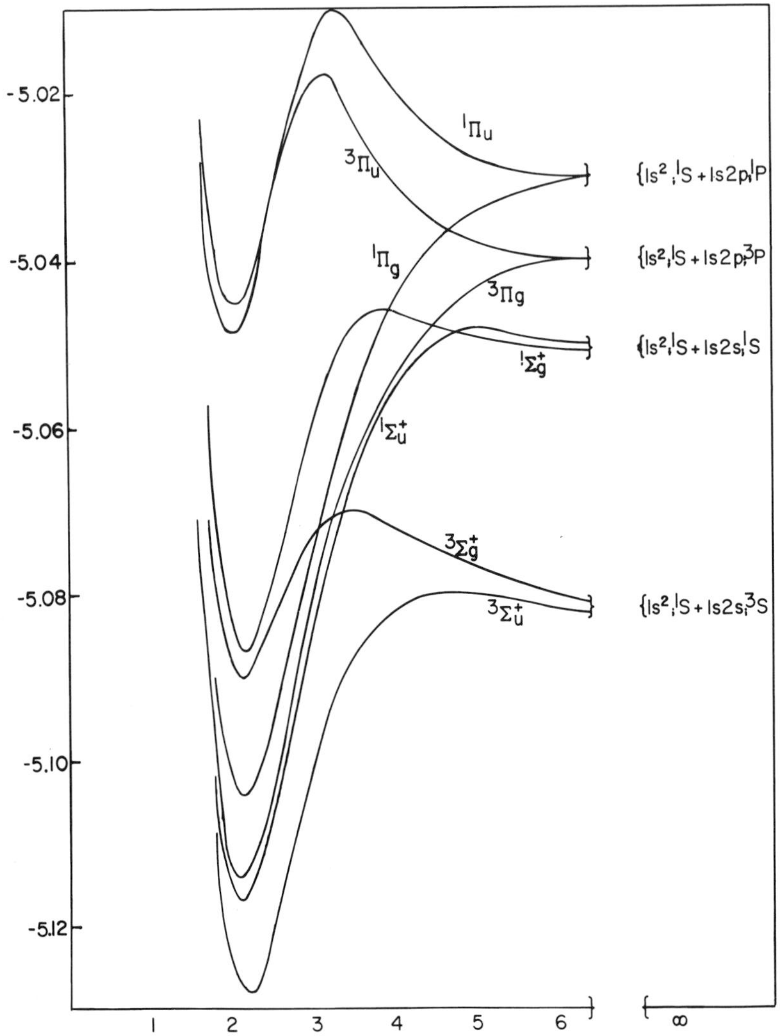

Fig. 8. This figure shows the richness and density of structure in the potential curves of the excited states of the helium molecule.

helium molecule. Both Mulliken's semitheoretical analysis and Ginter's analysis of his spectra depended on the magnitude of $D_e[He_2^+ ({}^2\Sigma_u^+)]$. If $D_e(He_2^+)$ were as large as 3 eV then a consistent analysis of the spectra could be carried out without invoking the occurrence of maxima, whereas a D_e much less than 2 eV would have suggested maxima markedly inconsistent with the rest of the data. Thus the electronic structures arising from the interpretation of these spectra rested crucially on the bounds for D_e provided by the accurate calculations of RBM. Additional corroborations of the existence of these rather large, 0.2–0.5 eV maxima were provided by 1965 calculations of Browne.[59,67] The existence of these maxima play a significant role in the understanding of a number of experiments, for example, dissociative recombination, excitation transfer, and diffusion cross sections for He(1s2s ^3S). Ginter[68] has continued his investigations and steadily more refined calculations have been made on some of these states.[52,69,70] Perhaps the most interesting result arises from Ginter and Battino[22] comparing their experimental potential curve for the $f^3\pi_u$ state with the calculated curve of Gupta and Matsen[70] for this state (Fig. 9). On the overlapping range of internuclear separations, the ab initio curve differs by not more than 0.02 eV from the experimental curve. Thus, since the theoretical curve is known to go to correct end points, it is fair to suggest that in this instance accurate calculations on a four-electron system have given a potential curve that is accurate to $\sim$0.02 eV from $O < R < \infty$, a conclusion which would have been greeted with derision only a few years previous.

G. Identification of Molecular Ions in Helium Discharges

Several authors[71,72] observed that there are present in helium discharges three ions of mobilities 10, 16, and 20 cm²/V·sec. Madson, Oskam, and

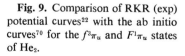

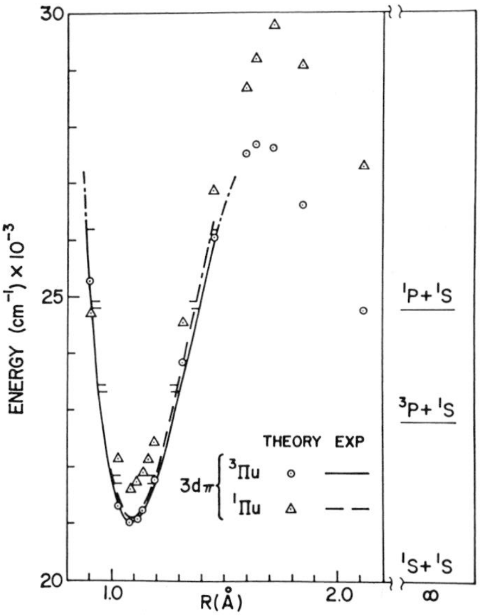

Fig. 9. Comparison of RKR (exp) potential curves[22] with the ab initio curves[70] for the $f^3\pi_u$ and $F^1\pi_u$ states of He_2.

Chanin[73] (1965) established that the ion of mobility 10 $cm^2/V \cdot sec$ was of mass 4, whereas the two ions of higher mobility were of mass 8. These authors confirm the ion of mass 8 and mobility 16 $cm^2/V \cdot sec$ as He_2^+ $(^2\Sigma_u^+)$ but could not assign the species of the other mass 8 ion in terms of known stable ions. Browne[74] (1966) showed with an accurate ab initio calculation that the metastable $^4\Sigma_u^+$ state of He_2^+ was stable with a binding energy of about 1 eV. Beaty, Browne, and Dalgarno[75] (1966), by combining experimental and theoretical constraints, established that the ion of mass 8 and mobility 20 $cm^2/V \cdot sec$ was the bound metastable $^4\Sigma_u^+$ molecular ion state calculated by Browne.

IV. SOME PROPERTIES OF LITHIUM HYDRIDE

A. The Dipole Moment of LiH

Lithium hydride is the simplest stable heteropolar molecule and, in consequence, is the simplest stable molecule that possesses a dipole moment. The dipole moments of most simple molecules were determined early in the century from the temperature coefficients of the electric polarization in the gaseous state. However, the dipole moment of lithium hydride could not be determined in this way since it is a solid at room temperature, which on heating yields mainly Li_2 and H_2 and small quantities of LiH.

In 1939 the first prediction of the dipole moment of lithium hydride was made by Pauling,[76] who predicted $\mu = 2.0\ D(Li^+H^-)$ on the basis of a semi-empirical theory. In 1957 Hurst, Miller, and Matsen[77] published the first ab initio prediction $\mu = 6.0\ D(Li^+H^-)$ employing a very modest basis set. In the next several years a number of calculations were carried out yielding results which ranged from $\mu = 5.6\ D$ to $6.1\ D(Li^+H^-)$. In 1960, three years after the first ab initio prediction, Wharton, Gold, and Klemperer[78] measured the dipole moment using the Stark effect on its microwave spectrum and obtained $\mu = 5.83$ $(v = 0)$. The dipole moment of lithium hydride was the first ab initio prediction of a property for a polyelectronic molecule and its success provided considerable encouragement to the *ab initioists*. Later and successively more accurate calculations by Kahalas and Nesbet,[79] Browne and Matsen,[80] Bender and Davidson,[81,82] and Brown and Shull[83] all give dipole moments in the range 5.8–5.9 D for LiH.

An excellent example of experimental accuracy forcing a reevaluation of a theoretical concept, in this case Brillioun's theorem, is found in the work of Bender and Davidson[84] on the dipole moment of LiH. Because of the Brillioun theorem it had been felt widely that SCF wave functions would yield accurate values of diagonal elements of one-electron operators since the SCF wave function does not contain first-order contributions from singly excited configurations. The case of the dipole moment of lithium hydride proves, however, to be an anomaly. The wave function of Cade and Huo,[85] which is a truly self-consistent wave function, yielded a dipole moment considerably above that of the experimental value (calculated $6.0D$; experimental $5.83D$). This anomaly was resolved by Bender and Davidson[82] who found that the contribution of singly excited configurations, due primarily to their strong interaction with the doubly excited configurations, were substantial contributors to the computed dipole moment in the case of LiH. This in effect says that second-order contributions to molecular properties determined by one-electron operators may, in fact, be significant. It is, therefore, not the case that one can always expect accurate values for one-electron properties from SCF wave functions. This is likely to be particularly true where the operator involved particularly weights a spatial region not closely similar to that region most heavily weighted by the energy operators.

B. The Quadrupole Moment of Li^7

In 1963 Kahalas and Nesbet[79] computed the electric field gradient at the lithium nucleus in LiH. They combined this with the quadrupole coupling constant determined by Wharton, Gold, and Klemperer[85] to obtain a value of $Q = 4.4 \times 10^{-26}$ cm^2 for the electric quadrupole moment of Li^7. In 1964 Browne and Matsen[80] recalculated the electric field gradient employing a very large basis set containing both atomic and elliptical orbitals and found complete agreement with the Kahalas and Nesbet result. The value of $Q = -4.4 \times 10^{-26}$ cm^2 has been referred to as the "experimental value."[86]

C. The Electronic Energy of LiH

The best (energy criterion) calculation by configuration interaction methods is by Bender and Davidson,[81] whose results lie 0.27 eV out of 216 V above the experimental value. Boys and Handy,[87] using their "trans correlated" wave function method (see Appendix A) have obtained an energy of 8.063 a.u. at $R = 3.04a_0$, in error by only 0.007 a.u.

D. The Ionization Potential of LiH

The ionization energy of LiH has not been measured. In 1959 Platas, Hurst, and Matsen[88] with a small basis set predicted that LiH^+ had zero binding energy. In 1961 Fraga and Ransil[89] predicted on the basis of an SCF–MO calculation that LiH^+ was bound and estimated ω_e (LiH^+, $^2\Sigma$)

to be 797 cm^{-1}. In 1963, Wilkinson[90] used a value of I.P.(LiH) of 6.5 $\pm$ 0.5 eV given by Price et al.[91] to obtain 1.3 $\pm$ 0.5 eV as an estimate of $D_e(LiH^+)$. In 1964 Browne[92] carried out a calculation with a large basis set and obtained 0.04 eV $\leq D_e(LiH^+) \leq$ 0.15 eV, from which one obtains 7.8 eV $\leq$ I.P.(LiH) $\leq$ 7.9 eV. This is the most accurate value of I.P. (LiH) to date. (Lin[93] has recently carried out a calculation which slightly improves the energies obtained by Browne.[92])

V. SUMMARY AND PROGNOSIS

In this article we have summarized some of the accomplishments of the *ab initioist* on three- and four-electron diatomic molecules. We have called attention to the close and highly profitable cooperation which has existed between the *ab initioist* and experimentalist functioning as co-investigators on many problems. The result of this collaboration has been the development of amazingly sophisticated and detailed knowledge of these simple systems and their relevant interactions.

Examples of calculations for large diatomic systems (four to twenty electrons) are becoming numerous. Schaefer and Harris[94] have carried out calculations on sixty-two states of the oxygen molecule. Although these calculations are not of high accuracy, they certainly reveal a great deal of structure which suggests a rich field for experiments in electron excitation, photodissociation, etc. Similar work has been done by Schaefer and O'Neil[95] on CO and by Michels[96] on N_2. Kouba and Ohrn[97] have carried out excellent calculations on C_2, and there are a number of other examples that could be quoted. As of this writing, the resolution in detail of the properties of the systems of more than five or six electrons are generally not in the same order of magnitude as that provided by experiment. It can reasonably be expected, however, that straightforward improvement in techniques and computing power will bring systems of this class into the range which can reasonably expect to obtain quite high resolution of detail by calculation. Polynuclear systems are inherently more difficult. There are fewer vanishing conditions on the integrals. Larger basis sets must be used to describe the more complex cusp and singularity structure of the wave function. Krauss[98] has given a good survey of recent calculations on the polynuclear systems. The key remark is that for four-electron systems the accuracy now being obtained for four-nuclei systems is roughly the same as that obtained ten years ago for the four-electron two-nuclei system. For detailed surveys of calculations see Krauss[99] (1967) or Allen[100] (1969).

Although the emphasis in this article has been on accurate ab initio computation, the less accurate (in energy) Hartree-Fock calculations have made significant contributions to our knowledge of electron densities, dipole moments, electric field gradients, and barrier heights. Highly sophisticated semiempirical procedures are also playing an important role in quantum chemistry. We believe that very active collaborations between experiment and theory at all levels represents the emerging patterns for the best and most productive research in chemistry in the coming decades.

Acknowledgment

The preparation of this manuscript and indeed much of the research herein described was supported by the Robert A. Welch Foundation of Houston, Texas.

APPENDIX A. NEW TECHNIQUES FOR COMPUTATION

There has been enormous activity in the field of quantum chemistry in the past ten years aimed at developing new methods of computation which will converge to answers of utilitian accuracy much more rapidly than the traditional techniques. The central concept of most of this work is that the so-called "correlation error" is essentially dominated by electron intrapair interactions. There are two historical threads leading into these new formalisms: the chemical bond pair concept and the formal perturbation theory developed for nuclear many-body problems. Sutcliffe and McWeeny,[101] Miller and Ruedenberg,[102] and Silver, Mehler, and Reudenberg[103] have reviewed the work originating in the chemical bond pair concept. There are at least three computational formalisms which have developed from the many-body perturbation theory. Sinanoglu[104] and his co-workers in a long series of papers have developed techniques for variationally computed pair functions. Nesbet[105] uses a systematic variational scheme for pairwise computation of correlation energies. Kelly[106] has developed the formal perturbation theory into a directly practical technique for atomic calculations. There is a very extensive literature on these subjects. The entire 1969 issue of *Advances in Chemical Phyiscs* is devoted to papers dealing with the correlation problem. Condon[107] and Nesbet[108] discuss the relations of the various theories. Browne[4] discusses the relation of these techniques to the more traditional methods of computation. Applications[109] of these techniques to calculations on atoms have been encouraging and have led to further developments[110] in the theories. Applications to molecules have been very sparse. Definitive bounds on the value of application of these techniques to molecules are not yet known. Bender and Davidson[111,112] have applied the Nesbet pair correlation theory to LiH and other first-row hydrides. Kelly[113] and Schulman and Kaufman[114] have applied Kelly's formalism to H_2 using a single-center wave function. Lee, Das, and Dutta[115] have applied Kelly's formalism to HF using a single-center wave function. In each case, the results were fairly good but not superior to calculations within traditional frameworks. However, the restriction to use of a single-center wave function makes comparison difficult. Two techniques which make computationally tractable the use of directly correlated wave functions for molecules with more than two electrons are currently in development and/or use. Conroy[116] uses the technique of minimizing the variance of the wave functions over a grid of points in configuration space

$$\text{Min}\,[U^2] = \left[\frac{\int (H\psi - E\psi)^2\, dx}{\int \psi^2\, dx}\right]$$

The integrations are done by purely numerical (random variate) methods. The trial wave function is chosen in a form which smoothes the electron-nuclear and electron–electron singularities in the Hamiltonian so that a smoothly distributed numerical integration grid of manageable size can be used. This method has been used to produce the most accurate potential surfaces yet available for H_3^+,[117] H_3,[118] and H_4.[119] It is not yet certain that application to systems of many electrons will be computationally attractive. Boys and Handy[120] have developed and implemented a means of calculation based on the method of moments whereby directly correlated wave functions can be obtained for polyelectron molecules.

The general form of the equations of the method of moments[121] can be written

$$\langle \psi_r | H_r - E | \sum c_i \Phi_i \rangle = 0, \; r = 1, \cdots, n$$

The set ψ_r is disjoint from the set Φ_i, and $\Phi = \sum c_i\Phi_i$ is an approximate solution of Schrodinger's equation. Boys and Handy choose

$$\psi_r = \frac{\partial(C^{-1}\Phi')}{\partial\beta_r} \quad \text{and} \quad \Phi_i = \frac{\partial(C\Phi')}{\partial\beta_r} \quad (\psi_1 = C^{-1}\Phi_1',\ \Phi_1 = C\Phi')$$

where C is a correlation function $C \equiv \prod_{i>j} f(r_{ij}, r_i, r_j)$. Φ' is an independent particle determinantal function and the β_r are the parameters upon which the C and Φ' depend. It is possible to choose C such that the most difficult integrals involve integrations of not more than nine dimensions and usually only six dimensions, instead of $3N$ where N is the number of electrons; the electron–electron singularities are removed from the integrals of the matrix elements. Boys and Handy evaluate the integrals of the matrix elements by direct numerical integration over six dimensions. Boys and Handy[87,122] have applied the formalism to the neon atom and to the equilibrium internuclear separation of the LiH molecule with encouraging results. Handy[123] has obtained excellent results for the Be atom. The primary drawbacks of this method are the absence of an upper bound relation for the energy eigenvalue and the necessity for dealing with unsymmetric matrix problems which sometimes have undesirable stability characteristics. It is clear that the method of moments offers the possibility of useful alternatives to the standard variational procedure. Note that the standard variational procedure is recovered if the sets ψ_r and Φ_i are identical.

References

1. W. J. Meath and J. O. Hirschfelder, *J. Chem. Phys.*, **44**, 3197 (1966).
2. H. A. Bethe and E. E. Salpeter, *Quantum Mechanics of One- and Two-Electron Systems*, Academic Press, New York, 1957.
3. R. T. Pack and J. O. Hirschfelder, *J. Chem. Phys.*, **49**, 4009 (1968).
4. J. C. Browne, *Advances in Atomic and Molecular Physics*, D. R. Bates and T. Estermann, Eds., Vol. 7, Academic Press, New York, 1971, p. 55.
5. G. H. Herzberg, *Spectra of Diatomic Molecules*, 2nd ed., Van Nostrand, Princeton, N. J., 1950.
6. F. T. Smith, *Phys. Rev.*, **179**, 111 (1969).
7. W. Lichten, *Phys. Rev.*, **131**, 229 (1963).
8. F. A. Matsen, *Advances in Quantum Chemistry*, P. O. Lowdin, Ed., Vol. 1, Academic Press, New York, 1964.
9. J. C. Slater, *Quantum Theory of Atomic Structure*, Vol. II, McGraw-Hill, New York, 1960, gives an excellent exposition of elementary SCF theory.
10. A. C. Wahl and G. Das, *Advances in Quantum Chemistry*, P. O. Lowdin, Ed., Vol. 5, Academic Press, New York, 1970, p. 261.
11. C. C. J. Roothaan, *Revs. Mod. Phys.*, **26**, 69 (1951).
12. G. G. Hall, *Proc. Roy. Soc. (London) Ser. A*, **205**, 541 (1951); R. K. Nesbet, *Proc. Roy. Soc. (London) Ser. A*, **230**, 312 (1955).
13. R. K. Nesbet, *Advances in Quantum Chemistry*, P. O. Lowdin, Ed., Vol. 3, Academic Press, New York, 1967, p. 1.
14. Ref. 10 gives a list of citations to this work and is a good source of references to papers on other MCSCF formalisms.
15. W. Kolos, *Int. J. Quantum Chem.*, **2**, 471 (1968).
16. J. O. Hirschfelder and W. Meath, "Intermolecular Forces," *Advances in Chemical Physics*, I. Prigogine and S. A. Rice, Eds., Vol. 12, Wiley-Interscience, New York, 1967.
17. J. C. Browne and F. A. Matsen, *J. Phys. Chem.*, **66**, 2332 (1962).
18. F. E. Harris and H. H. Michels, *Int. J. Quantum Chem.*, **35**, 461 (1970).
19. P. N. Reagan, J. C. Browne, and F. A. Matsen, *Phys. Rev.*, **136**, 304 (1963). This reference reviews the earlier calculated and experimental estimates of De($^2\Sigma_u^+$, He_2^+).
20. B. K. Gupta and F. A. Matsen, *J. Chem. Phys.*, **47**, 4860 (1967).
21. D. J. Klein, E. M. Greenawalt, and F. A. Matsen, *J. Chem. Phys.*, **47**, 4820 (1967).
22. M. L. Ginter and R. Battino, *J. Chem. Phys.*, **52**, 4469 (1960).
23. M. L. Ginter, and C. M. Brown, *J. Chem. Phys.*, **56**, 672 (1972).
24. K. M. Sando, *Mol. Phys.*, **21**, 439 (1971).
25. K. M. Sando and A. Dalgarno, *Mol. Phys.*, **20**, 103 (1970).
26. B. Liu, *Phys. Rev. Lett.*, **27**, 1251 (1971).
27. J. F. Franklin and F. A. Matsen, *J. Chem. Phys.*, **41**, 2948 (1964).

28. R. P. Marchi and F. T. Smith, *Phys. Rev. A*, **139**, 1025 (1965).
29. D. C. Lorents and W. Aberth, *Phys. Rev. A*, **139**, 1017 (1965).
30. P. E. Phillipson, *Phys. Rev.*, **125**, 1981 (1962).
31. D. C. Lorents, R. P. Marchi, and F. T. Smith, *Phys. Rev. Lett.*, **15**, 742 (1965).
32. J. C. Browne, *J. Chem. Phys.*, **45**, 2707 (1966). H. H. Michels made similar but more accurate calculations which reflected these same phenomena.
33. R. E. Olson and C. R. Mueller, *J. Chem. Phys.*, **46**, 3810 (1967).
34. M. Kennedy and F. J. Smith, *Mol. Phys.*, **16**, 131 (1969).
35. I. Amdur and A. L. Harkness, *J. Chem. Phys.*, **22**, 664 (1954). See also J. E. Jordan and S. O. Colgate, *J. Chem. Phys.*, **34**, 1525 (1961).
36. T. Wu, *J. Chem. Phys.*, **24**, 444 (1956).
37. W. R. Thorson, *J. Chem. Phys.*, **39**, 1431 (1963).
38. G. Matsumoto, C. F. Bender, and E. R. Davidson, *J. Chem. Phys.*, **46**, 402 (1967).
39. G. Barnett, *J. Phys.*, **45**, 137 (1967).
40. D. J. Klein, C. E. Rodriguez, J. C. Browne, and F. A. Matsen, *J. Chem. Phys.*, **47**, 4862 (1967).
41. J. E. Jordan and I. Amdur, *J. Chem. Phys.*, **46**, 165 (1967).
42. L. W. Burch and I. L. McGee, *J. Chem. Phys.*, **52**, 5884 (1970).
43. P. Bertoncini and A. C. Wahl, *Phys. Rev. Lett.*, **25**, 991 (1970).
44. H. F. Schaefer, III, D. R. McLaughlin, F. E. Harris, and B. J. Alder, *Phys. Rev. Lett.*, **25**, 988 (1970).
45. J. L. Nickerson, *Phys. Rev.*, **47**, 707 (1935).
46. R. A. Buckingham and A. Dalgarno, *Proc. Roy. Soc.* (*London*) *Ser. A*, **213**, 327 (1952).
47. A. V. Phelps and J. P. Molnar, *Phys. Rev.*, **89**, 1202 (1953).
48. A. V. Phelps, *Phys. Rev.*, **99**, 1307 (1955).
49. G. H. Brigman, S. J. Brient, and F. A. Matsen, *J. Chem. Phys.*, **34**, 589 (1961).
50. R. D. Poshusta and F. A. Matsen, *J. Chem. Phys.*, **132**, 307 (1963).
51. F. A. Matsen and D. R. Scott, *Quantum Theory of Atoms, Molecules, and the Solid State*, P. O. Lowdin, Ed., Academic Press, New York, 1966, pp. 133–155.
52. H. J. Kolker and H. H. Michels, *J. Chem. Phys.*, **50**, 1762 (1969).
53. F. D. Colegrove, L. D. Slearer, and G. K. Walters, *Phys. Rev.*, **135**, 353 (1964).
54. H. L. Richards and E. E. Muschlitz, *J. Chem. Phys.*, **41**, 559 (1964).
55. E. W. Rothe, R. H. Neynaber, and S. M. Trujillo, *J. Chem. Phys.*, **42**, 3310 (1965).
56. K. H. Ludlum, L. P. Larson, and J. M. Coffrey, **46**, 127 (1967).
57. W. A. Fitzsimmons, N. F. Lane, and G. K. Walters, *Phys. Rev.*, **174**, 193 (1968).
58. Y. Tanaka and K. Yoshino, *J. Chem. Phys.*, **39**, 3081 (1963).
59. J. C. Browne, *J. Chem. Phys.*, **42**, 2826 (1965).
60. D. C. Allison, J. C. Browne, and A. Dalgarno, *Proc. Phys. Soc.* (*London*), **89**, 41 (1966).
61. D. R. Scott, E. M. Greenawalt, J. C. Browne, and F. A. Matsen, *J. Chem. Phys.*, **44,** 2981 (1966).
62. Y. Tanaka and K. Yoshino, *J. Chem. Phys.*, **50**, 3087 (1969).
63. A. L. Smith and K. W. Chow, *J. Chem. Phys.*, **52**, 1010 (1970).
64. R. S. Mulliken, *Phys. Rev. A*, **136**, 962 (1964).
65. R. S. Mulliken, *J. Am. Chem. Soc.*, **86**, 3183 (1966); *ibid.*, **88**, 1849 (1967); *ibid.*, **91**, 4615 (1969).
66. M. L. Ginter, *J. Chem. Phys.*, **42**, 561 (1965).
67. J. C. Browne, *Phys. Rev. A*, **138**, 9 (1965).

68. See Ref. 22 for a summary and reference to the entire sequence of Ginter's papers on the spectra of He_2.
69. E. M. Greenawalt, Ph.D. Dissertation, University of Texas, 1967, unpublished.
70. B. K. Gupta and F. A. Matsen, *J. Chem. Phys.*, **50**, 3797 (1969).
71. M. A. Biondi and L. M. Chanin, *Phys. Rev.*, **94**, 911 (1954).
72. E. C. Beaty and P. L. Patterson, *Phys. Rev. A*, **137**, 346 (1965). These authors review the earlier measurements.
73. J. M. Madson, H. J. Oskam, and L. M. Chanin, *Phys. Rev. Lett.*, **15**, 1018 (1965).
74. J. C. Browne, *J. Chem. Phys.*, **45**, 2707 (1966).
75. E. C. Beaty, J. C. Browne, and A. Dalgarno, *Phys. Rev. Lett.*, **16**, 723 (1966).
76. L. Pauling, *The Nature of the Chemical Bond*, Cornell University Press, Ithaca, New York, 1940.
77. R. P. Hurst, J. Miller, and F. A. Matsen, *J. Chem. Phys.*, **26**, 1092 (1957).
78. L. L. Wharton, L. P. Gold, and W. A. Klemperer, *J. Chem. Phys.*, **33**, 1255 (1960).
79. S. L. Kahalas and R. K. Nesbet, *J. Chem. Phys.*, **39**, 529 (1963).
80. J. C. Browne and F. A. Matsen, *Phys. Rev. A*, **135**, 1227 (1964).
81. C. F. Bender and E. R. Davidson, *J. Phys. Chem.*, **70**, 2675 (1966).
82. C. F. Bender and E. R. Davidson, *J. Chem. Phys.*, **49**, 4222 (1968).
83. R. E. Brown and H. Shull, *Int. J. Quantum Chem.*, **2**, 663 (1968).
84. P. Cade and W. Huo, *J. Chem. Phys.*, **47**, 614 (1967).
85. L. Wharton, L. P. Gold, and W. A. Klemperer, *J. Chem. Phys.*, **37**, 2149 (1962).
86. R. D. Present, *Phys. Rev. B*, **139**, 300 (1965).
87. S. F. Boys, and N. C. Handy, *Proc. Roy. Soc. (London) Ser. A*, **311**, 309 (1969).
88. O. Platas, R. P. Hurst, and F. A. Matsen, *J. Chem. Phys.*, **31**, 501 (1959).
89. S. Fraga and B. J. Ransil, *J. Chem. Phys.*, **35**, 669 (1961).
90. P. G. Wilkinson, *Astrophys. J.*, **138**, 778 (1963).
91. W. C. Price, P. V. Harris, and J. R. Passmore, *J. Quant. Spectr. Radiant. Trans.*, **2**, 327 (1962); W. C. Price, T. R. Passmore, and D. M. Roessler, *Discussions Faraday Soc.*, **33**, 201 (1963).
92. J. C. Browne, *J. Chem. Phys.*, **41**, 3495 (1964).
93. C. S. Lin, *J. Chem. Phys.*, **50**, 2787 (1969).
94. H. F. Shaefer and F. E. Harris, *J. Chem. Phys.*, **48**, 4946 (1968).
95. S. V. O'Neil and H. F. Shaefer, *J. Chem. Phys.*, **53**, 3994 (1970).
96. H. H. Michels, *J. Chem. Phys.*, **53**, 841 (1970).
97. J. E. Kouba and Y. Ohrn, *J. Chem. Phys.*, **53**, 3923 (1970).
98. M. Krauss, *Ann. Rev. Phys. Chem.*, **21**, 39 (1970).
99. M. Krauss, "Compendium of *Ab Initio* Calculations of Molecular Energies and Properties," *Natl. Bur. Std. Tech. Note*, **438**, 1967.
100. L. C. Allen, *Ann. Rev. Phys. Chem.*, **20**, 315 (1969).
101. B. T. Sutcliffe and R. McWeeny, *Methods of Molecular Quantum Mechanics*, Academic Press, New York, 1969.
102. K. J. Miller and K. Ruedenberg, *J. Chem. Phys.*, **48**, 3414, 3444, 3450 (1968).
103. D. M. Silver, E. L. Mehler, and K. Ruedenberg, *J. Chem. Phys.*, **52**, 1174 (1970).
104. O. Sinanoglu, *Adv. Chem. Phys.*, **14**, 237 (1969). This is a survey paper with references to the earlier work in this series.
105. R. K. Nesbet, *Adv. Chem. Phys.*, **14**, 1 (1969). This is a survey paper with references to the earlier work in this series.
106. H. P. Kelly, *Adv. Chem. Phys.*, **14**, 129 (1969). This is a survey paper with references to the earlier work in this series.

107. E. U. Condon, *Rev. Mod. Phys.*, **40**, 872 (1968).
108. R. K. Nesbet, *Int. J. Quantum Chem.*, **45**, 117 (1971).
109. O. Sinanoglu and I. Oskutz, *Phys. Rev.*, **181**, 54 (1969); T. L. Barr and E. R. Davidson, *Phys. Rev. A*, **1**, 95 (1970). See also Refs. 104, 105, and 106.
110. H. F. Shaefer, III, and F. E. Harris, *Phys. Rev.*, **167**, 67 (1968).
111. C. F. Bender and E. R. Davidson, *J. Chem. Phys.*, **47**, 360 (1967).
112. C. F. Bender and E. R. Davidson, *Phys. Rev.*, **183**, 23 (1969).
113. H. P. Kelly, *Phys. Rev. A*, **2**, 1261 (1970).
114. J. D. Schulman and D. N. Kaufman, *J. Chem. Phys.*, **53**, 477 (1970).
115. T. Lee, N. C. Dutta, and T. P. Das, *Phys. Rev. Lett.*, **25**, 304 (1970).
116. H. Conroy, *J. Chem. Phys.*, **41**, 1327, 1331, 1336, 1341 (1964).
117. H. Conroy, *J. Chem. Phys.*, **51**, 3979 (1969).
118. H. Conroy and B. L. Bruner, *J. Chem. Phys.*, **47**, 921 (1967).
119. H. Conroy and G. Malli, *J. Chem. Phys.*, **50**, 5049 (1969).
120. S. F. Boys and N. C. Handy, *Proc. Roy. Soc. (London) Ser. A*, **309**, 209 (1969); *A*, **310**, 43 (1969).
121. N. C. Handy and S. J. Epstein, *J. Chem. Phys.*, **53**, 1392 (1970).
122. S. F. Boys and N. C. Handy, *Proc. Roy. Soc. (London) Ser. A*, **310**, 63 (1969).
123. N. C. Handy, *J. Chem. Phys.*, **51**, 3205 (1969).

PICOSECOND SPECTROSCOPY AND MOLECULAR RELAXATION

P. M. RENTZEPIS

Bell Telephone Laboratories, Incorporated, Murray Hill, New Jersey

CONTENTS

Abstract

Molecular relaxation, radiationless transitions, and most primary photochemical processes which have lifetimes of the order of 10^{-12} sec can be studied experimentally with picosecond pulses generated by mode-locked lasers. This paper reviews the application of picosecond pulses for the study of molecular relaxation. We describe several methods which we have utilized for the direct observation of radiative and radiationless transitions in liquids and gases. Simultaneous time-frequency resolved emission and absorption spectra methods have enabled us to measure emission and absorption processes with a time resolution limited by the time width of the picosecond pulse.

I. INTRODUCTION

The normal mode-laser with a relatively low power and long duration output was superseded by the *Q*-switched laser with peak power output of approximately $5 \times 10^{+8}$ W and duration limited to about 10^{-8} sec. The picosecond pulses are generated by the third "model" of lasers known as "mode-locked" lasers and are capable, at the present, of pulses with a time duration as short as $\sim 4 \times 10^{-13}$ sec and power of more than 10^{10} W. These ultrashort pulses have found extensive applications in optical nonlinear studies, plasma physics, optical radar and other physics and engineering disciplines, and also in molecular relaxation, which we have been studying in the past five years and which forms the basis of this chapter.

A. Mode-locking

The first mode-locking of a laser was achieved in 1963 with an acoustic–optic oscillator.[1] Thereafter solid-state lasers were mode-locked by means of a saturable absorber such as polymethene dyes,[2–6] cryptocyanene or other dyes which have a broad absorption band at the laser wavelength and a repopulation time shorter than the optical round-trip time of the laser cavity. Experimentally one can easily obtain a mode-locked laser pulse as shown in Fig. 1*a*. The laser rod provides the amplifying medium, and the mirrors form the Fabry-Perot cavity. Now upon insertion of a bleachable dye which has a much broader absorption band than the laser line, standing waves are set up with discrete optical frequencies defined by the Fabry-Perot modes

$$\nu_n = \frac{nc}{2L}$$

where L is the optical separation of the mirrors. For a typical laser where there are $\sim 10^6$ modes oscillating within the cavity, the adjacent mode separation $\Delta\nu$ is given by $\Delta\nu = \nu_n - \nu_{n-1} = C/2L$. In practice, however, the modes allowed have a spectral width $\Delta\nu$ which is a function of the mirrors and the effective aperture of the laser rod. The light beam passes through the bleachable dye and distorts the original pulse intensity distribution by increasing the peak intensities. In the cavity the beam passes back and forth many times, continuously sharpening the pulse and increasing the intensity, with the duration of the pulse theoretically limited by the spectral band width.

Computer calculations show that the pulse is sharpened after each successive pass with the maximum achieved after about 15 passes or so. Although the saturable absorber provides a method for sharpening the pulses and separates them by $C/2L$, no unified model provides sufficient information with respect to the formation of the initial sharp pulse, which will be amplified in the laser cavity faster than the broad envelope. Normally the modes of the laser oscillate in a random fashion without any set phase relationship. The locking of the phase is provided by the bleachable dye which modulates the frequencies $(\nu_0 + \Delta\nu)$, ν_0, and $(\nu_0 - \Delta\nu)$

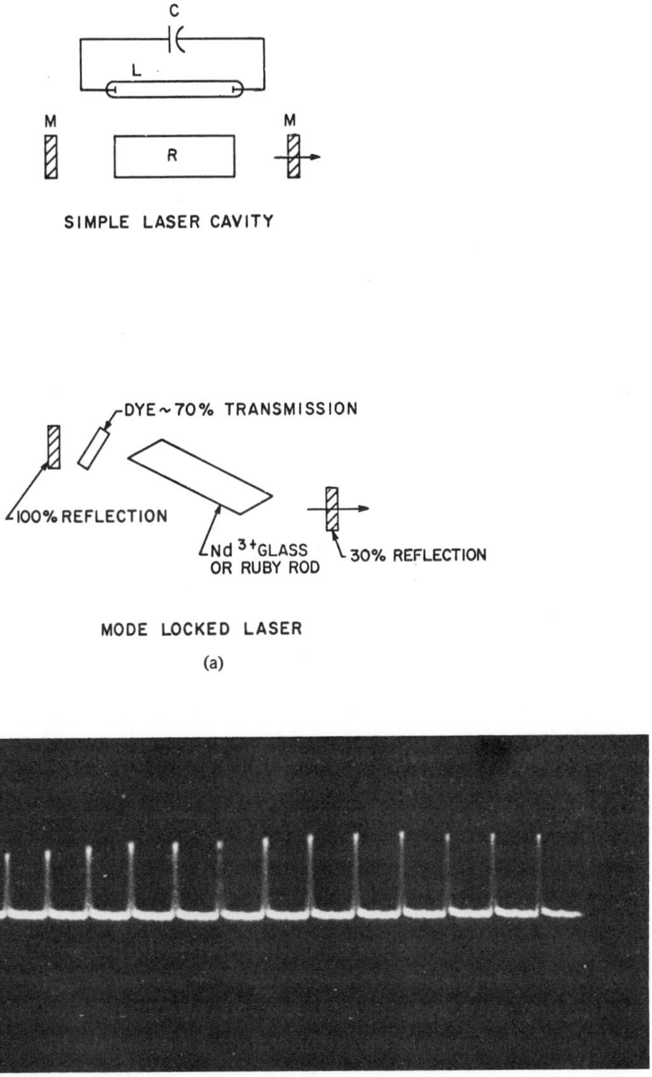

Fig. 1. (*a*) Schematic representation of a typical mode-locked Nd^{3+} glass laser. (*b*) Train of picosecond pulses detected by fast photodiode and displayed on 519 Tektronix oscilloscope.

(where $\Delta\nu = C/2L$) of the cavity interferometer and couples in new side bands to the first three. With each additional path new side bands are coupled until all the modes defined by the total band width are coupled. Experimentally mode-locking is achieved by ascertaining that all the optics (rod and Q-switch cell) are at the Brewster angle to eliminate back-reflections. The pulse intensity can vary from high intensities in short pulse trains to low intensities in long trains by adjusting the mirror reflectivity, length of the cavity, and concentration of the Q-switch dye.

Mode-locked laser experiments have used both Nd^{3+} glass and ruby, and YAG:Nd^{3+}, to generate pulses in the 10^{-12} sec range. The Nd^{3+} glass is most interesting because of the 100–200 cm^{-1} band width which provides the possibility of 10^{-13} sec pulses. In practice one usually generates pulses with duration of 2–7 psec utilizing Eastman-Kodak #4740 and #9860 dyes as the saturable absorber. In the case of the ruby, mode-locked pulses in the 40 psec range can be generated by using cryptocyanine (~25 psec), DDI (1,2-diethyl-2,2-dicarbocyanine iodide) (~8 psec), and dicyanine A (~50 psec).

A typical train of mode-locked picosecond pulses is shown in Fig. 1*b*. The pulses are separated by 8 nsec, which corresponds to a 1.2 m cavity. It is interesting to note that the pulse width is smaller than the resolution of either photodiodes or oscilloscopes of about 10^{-9} sec. One can obviously vary the separation of the pulses by changing the length of the cavity, and in the cases where small separation is desired one can utilize a front mirror with flat surfaces, thus creating pulse separations equivalent to twice the thickness of the mirror.

It seems appropriate to point out that gas lasers have also been mode-locked, as a matter of fact, even before other lasers, with acoustic–optic crystals; however, due to the small spectral widths the pulse duration is limited to $\sim 10^{-10}$ sec. In the case of argon, mode-locking has also been achieved and by frequency doubling we have obtained pulses at ~2550 Å as short as ~200 psec[8] with a separation rate of 1 pulse/3 nsec or longer. With the aid of fast electronics we are now able to monitor spectroscopic processes essentially continuously with a high signal to noise ratio. Lately also the CO_2 laset has been modified to generate picosecond pulses[9,10]; however, due to its low frequency its application is restricted to vibrational excitation of molecular systems.[11,77–79]

B. Single-Picosecond Pulse

It has been repeatedly shown that the pulses contained within a pulse train vary in both intensity and duration. It is very important therefore to utilize, when possible, a single pulse when relaxation rates are in the range of a few picoseconds. The first high-intensity single-pulse instrument was proposed by Vuylstoke.[12] He constructed a laser with mirrors having 100% reflectivity at low-incident light intensity, but near pulse intensity the front mirror switched rapidly to zero reflectivity, dumping the pulse within the optical cavity round-trip time and a single pulse of very high power.

An alternative method for extracting single pulses from a mode-locked train utilizes a spark gap charged to about 10 kV.[13] The gap is adjustable so that one can select the most intense pulse to initiate a discharge. A high-voltage pulse, generated by the discharge, is used to increase the voltage on a partially energized polarizing Pockels cell so that a 90° change of polarization is generated for about 5 nsec. A single pulse is thus transmitted wihch has a high intensity and short duration, ~2 psec. With the addition of a 25 cm long amplifier a single picosecond pulse with ~1 J of energy can be generated.[13]

II. MEASUREMENT OF PICOSECOND PULSES[7,14,15]

A. Two-Photon Method

The pulse duration of nanosecond pulses is usually measured directly by means of a photomultiplier–(or photodiode)oscilloscope combination. High-speed photodiodes have been found with a resolving power of ~0.3 nsec. However, present electronic techniques are not capable of measuring pulses in the 10^{-12} sec range; therefore nonlinear optical methods have been developed which are capable of displaying the pulse duration and shape directly. Another, as yet less well developed, possibility is offered by interferometric methods which determine the autocorrelation function of the pulse amplitude; this is in turn related to the Fourier transform of the power density spectrum, as measured by a spectrometer.

It is obvious, however, from the relationship $\Delta\omega\Delta\tau = 2\pi$, that the linear techniques provide only a lower limit for the pulse duration except in the case where the complete spectral width is contributing to the mode-locked pulse. In practice it is generally observed that when a high background level is observed in the oscilloscope trace, the modes are only partially locked and the spectrum of the pulse appears to be less wide, indicating that the inverse band width–time duration relationship is followed at least as a lower limit.

The method that has found most frequent use for the measurement of picosecond-duration pulses is the two-photon fluorescence method (TPF).[7,14,15] The simplicity, accuracy, and reliability of this method are the principal advantages over other methods, such as the second harmonic. The TPF measures the correlation of two pulses as illustrated in Fig. 2*a*.

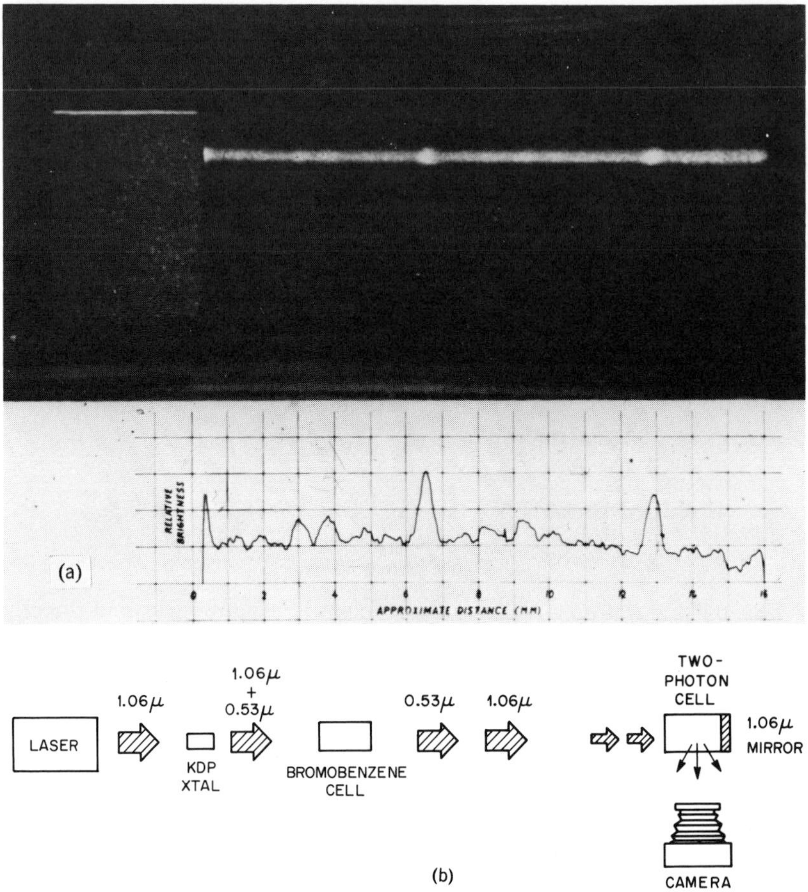

Fig. 2. Two-photon fluorescence technique (TPF). (*a*) The original apparatus for the direct measurement of picosecond pulses. (*b*) TPF technique using two picosecond pulses with different frequencies for improved contrast.

The first reported experiment on TPF made use of the second harmonic, 5300 Å of a train of mode-locked light pulses of a Nd^{3+} glass laser. After filtering the fundamental and pump light, the beam traverses a 20 cm cell containing a dibenzanthracene/benzene solution. Other fluorescent dyes have been used subsequently with this method and almost any dye is suitable if the solution does not absorb by one photon (it is not desirable, for example, to use methanol as the solvent with Nd^{3+} laser fundamental since it absorbs slightly at 1.06 μ), has large two-photon absorption cross section, and a high quantum yield for emission. The pulses of the train are

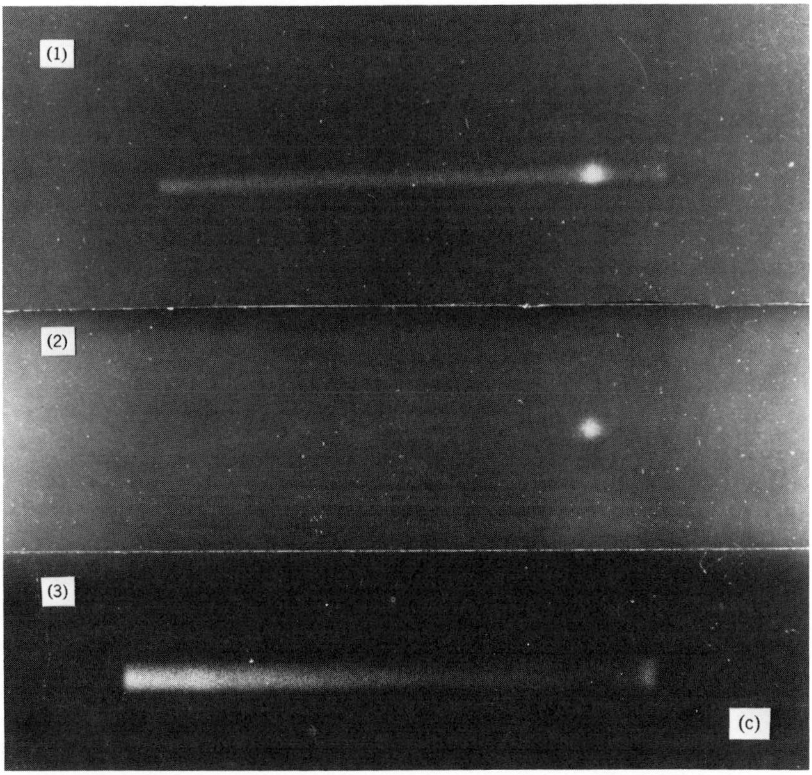

Fig. 2 (*continued*). (*c*) Photograph of picosecond pulses by (1) the original TPF technique, (2) TPF using two different frequencies, and (3) same as above but removing the bromobenzene cell and dissolving the TPF dye in a dispersing solvent.

reflected by a mirror situated at the end of the cell, then subsequently collide with the following members of the pulse train, inducing two-photon fluorescence at the point of overlap, Fig. 2*b*. The induced fluoresecence even from a single pulse is sufficiently intense to be directly displayed photographically (camera with $f = 2.8$, Polaroid film, 3000 speed).

The pulse duration, t, is simply calculated by $t = dn/\alpha c$ where d is the length of the two-photon bright spot at half-maximum intensity above background, α is a constant depending on the shape of the pulse, and n is the index of refraction of the solution. The separation of the pulses, t_2, displayed as the distance d_2 between bright spots is given by $ct_2/2n$. A typical display of picosecond pulses by the TPF technique is shown in Fig. 2*c*; in this case the pulses are 2–3 psec and separated by 50 psec.

Figure 2*c* clearly shows not only the fluorescence spot at the point where the pulses overlap but a continuous background induced by each pulse alone as it traverses the two-photon solution.

Subsequently it was shown[16–20] that the theoretical maximum contrast ratio of the TPF spot to the background for a two-photon virtual process is 3:1, whereas this ratio for the free-running (nonmode-locked) case is 1.5:1. The contrast ratios can be calculated assuming plane-polarized waves of frequency w reflected by a mirror of reflectivity R (100% in this case). The standing wave field strength in front of the mirror is given by

$$E_{0m} = E_0 \left[1 + R + 2R^{1/2} \cos\left(\frac{4\pi n d}{a - \delta}\right)\right]^{1/2}$$

where n is the refractive index, a is the wavelength, δ is the phase shift due to imperfect reflection, and d is the distance from the mirror. In the case of two-photon induced fluorescence the intensity, I_n, is proportional to $(E_{0m})^4$. Taking a running average over a wavelength,

$$I_n \sim E_0{}^4(1 + 4R + R^2)$$

which is valid for any pulse length reflected onto itself by the mirror. The background fluorescence emitted by the individual pulses acting alone is $1 + R^2$.

From this it is obvious that the ratio of the fluorescence intensity of the point where the two-photon mode-locked pulse meet to the background is 3:1 and in the free-running or random case 1.5:1.

The TPF method has the advantages of simplicity and accuracy; however, the low contrast ratio and inherent incapability of displaying the pulse symmetry are sufficient disadvantages to warrant the use of the simple and accurate three-photon method (3PF),[21,22] which circumvents the first of these disadvantages in a practical way and the second in principle.

B. Three-Photon Method

The three-photon method has all the attractive features of the TPF techniques including the direct photographic recording of the pulse and the simplicity of operation. In addition, it has the advantages of (1) a contrast ratio of 10:1 as opposed to 3 for TPF, and (2) the potential for measurement of the phase-intensity relationship. The first makes it possible to observe experimentally processes which were previously masked by the intense background.

The experiment is represented schematically in Fig. 3*a*. The solution contains a dye which has its first allowed state at three times the frequency

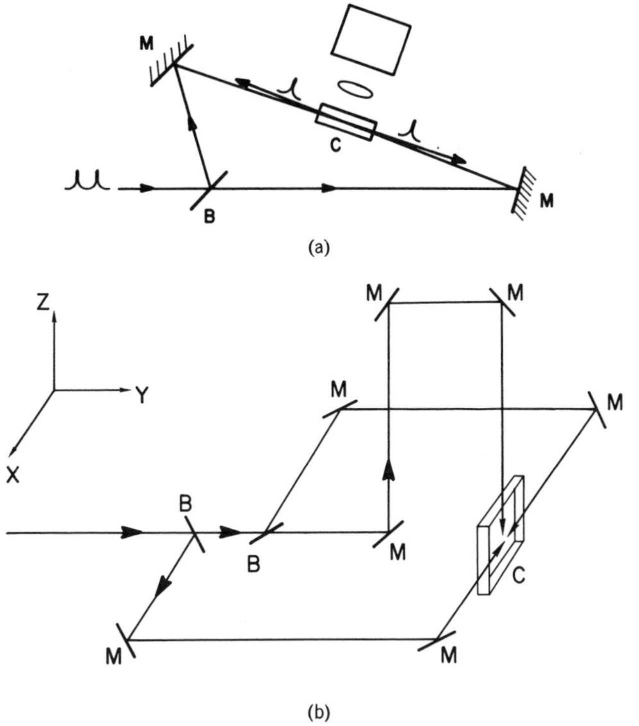

Fig. 3. Three-photon fluorescence method (3PF). (*a*) Schematic representation for the measurement of the duration of a single picosecond pulse. (*b*) Method for the simultaneous measurement of asymmetry and duration of a picosecond pulse.

of the laser and strongly fluoresces. In the first 3PF experiments we utilized 2,5-bis[5-terbutylbenzoxazolyl-(2)]thiophene (BBOT) in methylcycloxane and 1,4-bis[2-(4-methyl-5-phenyloxyazolyl)]benzene (dimethyl POPOP) in methylcyclohexane. The absorption cross section of this dye at the three-photon wavelength (3550 Å) of the Nd^{3+} glass laser is very high ($\epsilon_{3590\ \text{Å}} \approx$ 50,000) and the quantum yield for emission is approximately unity.

The Nd^{3+} glass laser was mode-locked and *Q*-switched in the normal manner by Eastman Kodak #9860 or #9740 bleachable dyes. The pulse traversed the optical paths shown in Fig. 3*a*. To eliminate reflection back into the laser cavity a set of $\lambda/4$ plates and polarizers were inserted in the path; one could achieve similar results by a slight misalignment of the back mirror, thus reflecting the beam away from the laser cavity. The data of Fig. 7 are typical of well mode-locked pulses with contrast ratio better than 9:1. The 3PF method, due to its large range of contrast ratios, has also the potential of depicting the degree of mode-locking much more clearly; in fact, studies show that as the degree of mode-locking decreases the pulses are broadened and the oscilloscope (519 Tektronix–photodiode combination) display has the appearance of noise between the pulse of the train and the 3PF intensity ratio decreases. We have observed the low-contrast ratio with either the TPF or 3PF methods is accompanied by a display of a "noisy" trace on the oscilloscope. Conversely, an oscillogram displaying a smooth signal without the background resulted in a relatively high ratio, that is, 9:1. We believe these observations are of practical importance because with them one need not process the film and measure the contrast ratio by photodensitometer of each photograph to verify that the laser does in fact generate mode-locked picosecond pulses. As mentioned previously, one of the major advantages of the 3PF method is that the third-order correlation function relates the 3PF intensity to the pulse symmetry.

$$G^{(3)}_{(\tau_1\tau_2)} = \int I(t)I(t + \tau_1)I(t + \tau_2)\, dt$$

and

$$P^3(\tau) = \int [I^2(t)I(t + \tau_1) + I^2(t + \tau_2)]\, dt$$

The analysis of such a third-order correlation function may give the relative phases of the pulse components and their amplitude, thus unveiling the shape of the pulse.[23]

The experimental arrangement of Fig. 3*b* shows the method employed to achieve the variable pulse delay ($t_1t_2t_3$) demanded by the $G^3{}_{(\tau_1\tau_2)}$ function. The apparatus consists of beam splitters B which divide the original pulse into three equal parts. Each pulse is reflected by the mirrors M and traverses the optical path shown. The prism P is situated on a moving platform providing the variable delay necessitated by G^3. The pulses recombine inside the cell c and the induced three-photon fluorescence is photographed by the camera situated normal to the path of the pulses.

We have previously observed that the pulse width within a train of picosecond pulses varies from ~2 psec to about 8 psec with the shorter pulses near the beginning of the train. Obviously it is difficult to observe the shape of a short pulse when one averages over the whole train; since the pulses are not continuously reproducible a single-pulse method is much more reliable for both the pulse width and shape determination and also for experiments in picosecond spectroscopy and molecular relaxation. Recently we have devised a method[22] which enables one to measure the pulse width and asymmetry with a single pulse. This method is a variation of the 3PF technique diagrammed in Fig. 3*a*. Refer again to Fig. 3*b*; the pulse is split into three components of equal amplitude while maintaining the original polarization. These now enter the dye cell along the x, y, and z axes, overlap, and generate the fluorescence pattern shown in Fig. 4. If the pulse is very asymmetric the fluorescence pattern would be asymmetric. Appropriate computer simulation shows that in many cases a well mode-locked laser has essentially a Gaussian shape and pulse width of ~2 psec.

We also mention here that several methods have been devised which essentially achieve pulse compression approaching the limit $\Delta\tau\Delta \approx 1$, and are capable of generating pulses with duration of $\sim 4 \times 10^{-13}$ sec.[25,26]

III. DIRECT MEASUREMENT OF RADIATIONLESS TRANSITIONS

In a general sense radiationless relaxation plays an important role in most photochemical processes. There are several theoretical treatments of radiationless transitions, Jortner and Berry,[27] Bixon and Jortner,[28,29] Freed,[24] Jortner et al.[30,31] and Robinson and Frosh,[32] Lim and Bershon,[33] and several reviews dealing with nuclear scattering which have a strong relevance to molecular systems.[34] The interpretation of these relaxation processes is based on the assumption that the system is a compound state. The system can be thought of as being composed of a set of zero-order states which, upon interaction, create the exact eigenstates of the Hamil-

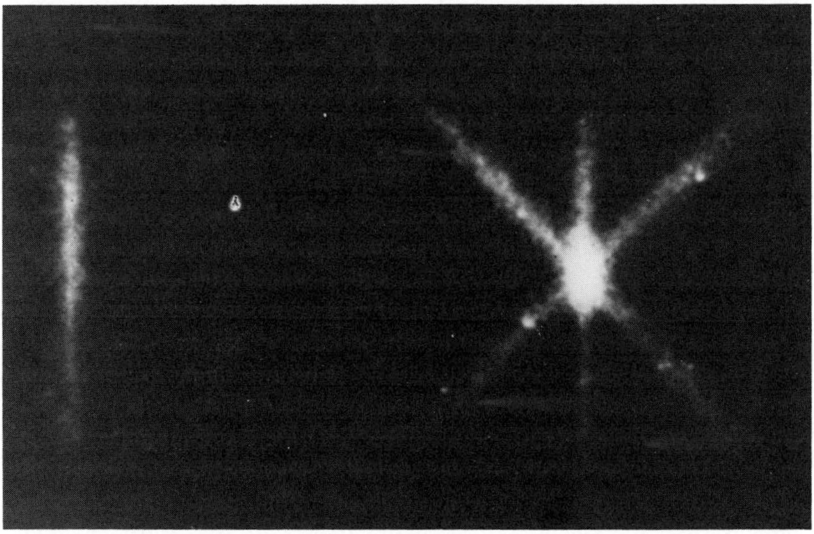

Fig. 4. Photographic display of picosecond duration and asymmetry of pulse(s) by 3PF.

tonian. As visualized by Jortner[35] the compound state can be subdivided into two zero-order states, a sparse and very dense system. The relaxation process then proceeds into the continuum when there is a superposition of the zero-order discrete and dense states. Since the zero-order states do not portray a physical real-life situation we can consider the relaxation of a state as a function of its total width, γ, which is composed of several components, such as nonradiative width Δ and radiative width Γ, or $\gamma = \Gamma + \Delta$. In this spirit the width of a state is zero until coupling with the radiative field of a state takes place, as in the case of coupling between stationary states, whereupon one observes a radiative transition with width Γ_i. Since nonradiative transitions are not allowed Δ is essentially zero and $\gamma = \Gamma = 1/\tau_R$, where τ_R is the radiative lifetime. Similarly, if the system is excited to a nonstationary state or, equivalently, the molecule is embedded in an inert matrix such as a nonreacting solvent, radiationless transitions are now allowed and the total width, γ, contains both components Γ and Δ. Δ can be constructed so as to include the collisional broadening and other deactivating processes, B,

$$\gamma = \Gamma_i + \Delta_i \qquad \text{and} \qquad \Delta_i = \delta + a + B_i$$

As an approximation, we will consider the excitation to a bound state which initially has zero width, and a width of certain magnitude appears as the decay channels develop.

In this discussion of the relaxation processes in large molecules we will consider three types of decay channels. The mode of decay when (*a*) the excited state is coupled to very large numbers ($\sim 10^8$) of levels resembling a quasicontinuum, (*b*) a sparse number of levels, and (*c*) the intermediate case where the energy gap between the excited state and a lower state is a few thousand wave numbers.

The theoretical treatment has been given by Bixon et al.,[28] who show that in the statistical limit the relaxation of the excited state molecule would exhibit similar characteristics in both the isolated molecule case and in solution. We present here experimental evidence which shows that a large molecule with a large energy gap relaxes nonradiatively from an excited electronic state at the same rate in both the "isolated molecule" case and when embedded in an inert matrix, that is, in solution.

The theoretical sparse level situation occurs in a real-life experiment when a zero-order state is strongly coupled to a small number of widely separated vibronic levels which belong to a lower electronic origin. Certain triatomic molecules[37] such as NO_2, SO_2, and CS_2 provide examples of a complex spectrum which may be due to the strong intramolecular interaction between the zero-order excited state and coarsely spaced levels.

The intermediate coupling provides a very interesting case which just lately has been discussed theoretically.[36,38] Some of the features predicted and partially exposed experimentally are (1) the nonexponential decay characteristics,[39,40] (2) lengthening of the lifetime of the excited state[41,42] (3) emission between highly excited states $S_2 \rightarrow S_1$,[43,44] and between S_2 and the ground manifold,[45] (4) deviations from the Stern-Volmer relationship,[46,47] and (5) quantum interference between levels in large molecules.[40–42]

A. Molecules Corresponding to the Statistical Limit—Azulene

The first experiments on radiationless relaxation[47,48] of large molecules with picosecond pulses were performed late in 1966; Jortner and his colleagues were concerned, independently, about the theoretical aspects of radiationless transitions of large molecules, focusing their attention on coupling with a dense manifold. As the first experiment ever performed on molecular relaxation with picosecond pulses,[48] we selected azulene because it exhibits a fast relaxation rate from the first excited singlet state ${}^1B_1$ by coupling either to the ground state or to a low-lying triplet.[49,50] The energy gap of 14400 cm^{-1}, ω_1, from the ground state to the "origin" of the first ${}^1B_1$ state could be covered conveniently by the fundamental of the ruby laser, and an upper vibronic level of the same excited state could also be populated by the second harmonic of a Nd^{3+} glass, at 18862 cm^{-1}, ω_2. The azulene was purified by sublimation and dissolved in spectroscopically pure methanol. The experimental arrangement is shown in Fig. 5. The fundamental 9431 cm^{-1} beam of a mode-locked Nd^{3+} glass laser generated

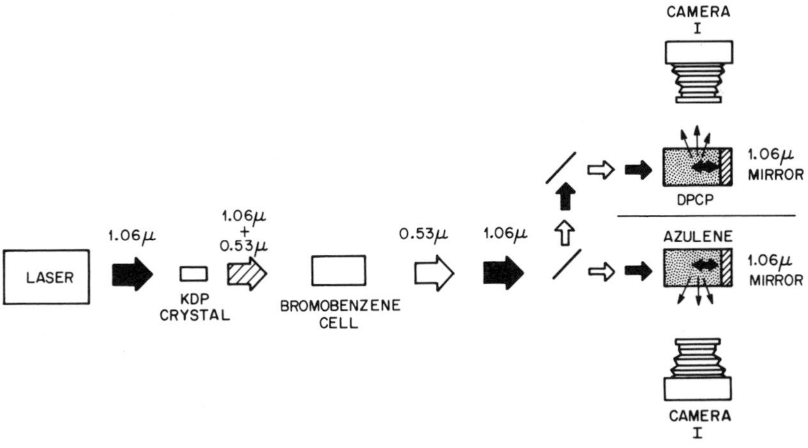

Fig. 5. Schematic version of the method for the direct observation of radiationless transition rate of azulene.

its second harmonic, 18862 cm^{-1} in a $1 \times 1 \times 3$ cm phase-matched KDP (potassium dihydrogen phosphate) crystal, with a conversion efficiency of ~12%. The two pulses propagated simultaneously and entered the dispersion cell containing bromobenzene. The fundamental pulse, ω_1, propagates faster than ω_2 and the separation of the ω_1 and ω_2 pulses can be calculated accurately by the group velocity relationship; the wave number of the two pulses and refractive index of bromobenzene at ω_1 and ω_2 are known. In order to measure simultaneously the pulse width and relaxation time of azulene in solution for every picosecond train, and thus avoid the fluctuation in pulse intensity and duration that occurs between successive laser shots, part of each beam was reflected into a two-photon cell in the manner shown in Fig. 6. The fluorescence induced in each cell was photographed by an *f*1.9 Polaroid camera. Specific care must be taken that the exposure of the film is within its linear range by using neutral density or other calibrated filters.

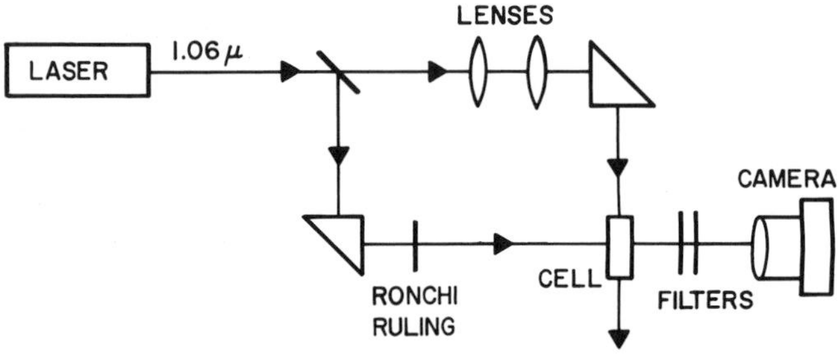

Fig. 6. The cross-beam method which enables one to measure in one shot essentially all the parameters of Fig. 9.

Pulse ω_1 enters the cell first. However, the photon energy is not sufficient to populate the first excited singlet state. Thus pulse ω_1 will not induce fluorescence from the emitting second singlet as it travels through the azulene solution. At the end of the cell the dielectric mirrors (Fig. 7) reflect ω_1 onto itself so it now travels in the opposite direction. Pulse ω_2 follows behind, at a predetermined distance which ensures that it will overlap with the returning ω_1 pulse inside the azulene cell. Pulse ω_2 has a frequency of 18862 cm^{-1} which is sufficient to populate an upper vibronic level of the azulene S_1. Azulene has been shown not to emit from the first singlet,[55,56] ($Q - 10^{-6}$);[57] thus no emission is recorded. To avoid biphotonic excitation by ω_2 to the second excited singlet state of azulene, which does emit, the intensity of ω_2 is reduced by detuning the KDP crystal, or by

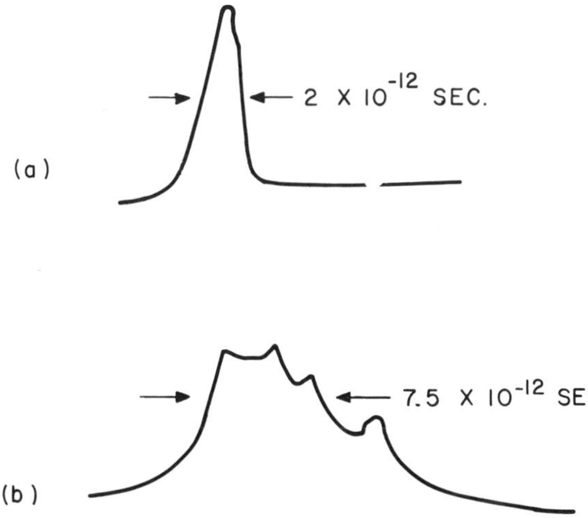

Fig. 7. Photodensitometer trace of (*a*) picosecond pulse and (*b*) lifetime of 18862 cm^{-1} excited level of azulene.

appropriate filters. Following the course of events we see that w_2 (Fig. 8) populates a vibronic level of S_1 situated at 18862 cm^{-1} above the ground state. At the time and space within the cell where the two pulses w_1 and w_2 overlap, the combined energy of the pulses is the equivalent of 28293 cm^{-1}, which is near the origin of the S_2 state of azulene. At the position of overlap the frequency and photon density are sufficient to gnerate observable fluorescence from S_2 to S_0 during the time of the pulse overlap. When the pulses separate ,the fluorescence should also disappear, unless the level which was populated by w_2 has a finite lifetime which is longer than the width of the pulses, $\sim$2 psec in this case. The w_1 pulse performs this "population interrogating" function as it travels against the direction of ω_2. It is evident that the distance which ω_1 traverses from the point of intersecting ω_2 and still encounters excited azulene corresponds to the direct measure of the relaxation time. The length of the fluorescence spot is simply transposed into time and the relaxation from the 18862 cm^{-1} vibronic level is measured, after the appropriate deconvolution, as $\sim$7 psec. This simple and accurate method made possible the first measurement of nonradiative molecular relaxation. This value of $\sim$7 psec is in very good agreement with the independent theoretical studies, based on the statistical limit model, of Jortner and Berry.[27] In a following section we shall discuss the relation of these experimental results with the theoretical predictions for the isolated molecule case.

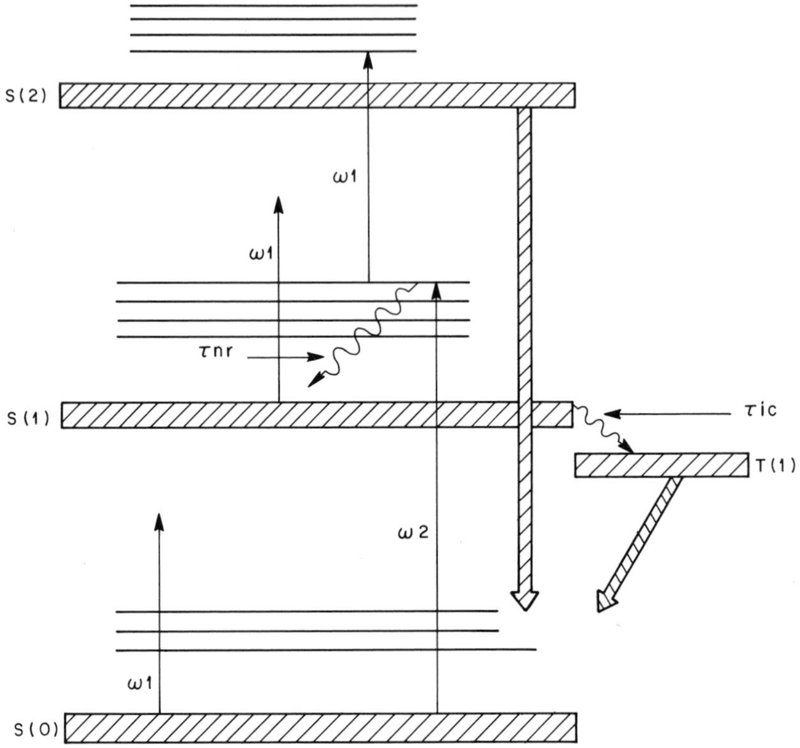

Fig. 8. Energy level diagram of azulene.

1. Repopulation of the Azulene Ground State

The previous experiment revealed that relaxation from the 18,862 cm^{-1} vibronic level of S_1 occurs with a rate of 1.3×10^{11} sec^{-1}; however, the experiment does not show a priori what is the mechanism of the relaxation. One cannot predict from this result whether the coupling takes place with the ground state or with the first excited triplet. To elucidate the mechanism we measured directly the relaxation rate to the ground state of azulene in solution[48] by exciting with 14400 cm^{-1} picosecond pulses from a mode-locked ruby laser. In the apparatus which we originally designed, shown in Fig. 9,[48,49] the picosecond pulses are partly reflected on photodiodes p_1 and p_3. The ratio of the intensities p_1:p_3 gives directly the number of molecules excited. The pulse traverses the azulene cell, then enters into the "two-photon liquid" located at the end of the apparatus. Here the length of the pulse is measured by the two-photon method by the fluores-

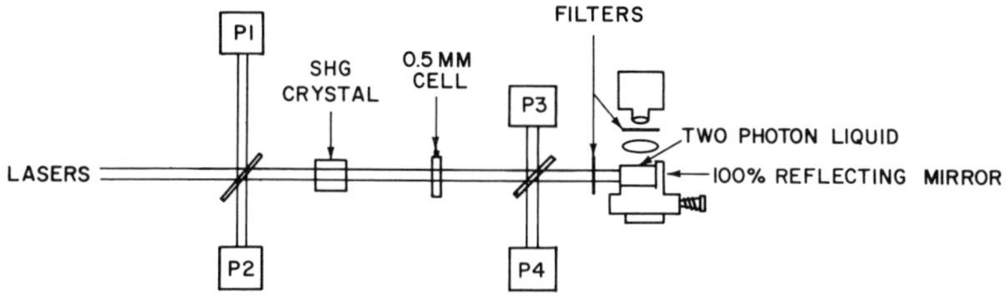

Fig. 9. The apparatus used for the rate of repopulation of the ground state of azulene.

cence spot recorded on the film in the above-described manner. The reflected pulse is attenuated by the filters shown in the figure and its intensity is measured again, before and after passage through the azulene cell, by p_4 and p_2. The returning pulse is attenuated to the extent that it is 95% absorbed when azulene is in the ground state. The rate of repopulation is directly measured by this apparatus in the following manner. The first pulse passes through the azulene and, upon reflection by the mirror in the "two-photon cell," returns into the azulene solution where it meets with the second pulse as it travels in the opposite direction. The ratio of the intensities at p_4:p_2 measures the concentration in the ground state. If the two pulses are synchronized and interesect each other within the 0.5 mm azulene cell the interrogating pulse measures the population of the ground state at the time of excitation, $t = 0$. To measure directly the repopulation after a few picoseconds, the mirror situated on the sliding micrometer platform at the right-hand end of the apparatus was translated so as to change the time elapsed between the excitation pulse and the arrival of the interrogating pulse. By this continuous translation one obtains a histogram of the repopulation to the ground state from S_1. This repopulation time is found to be $\sim 8 \times 10^{-12}$ sec. This result strongly supports the view that the $^1A_1$ state preferentially couples with the ground rather than a low-lying triplet state.

The method just described, which we originated for the study of azulene and polymethane dyes, has been utilized subsequently by several research workers in the studies of Q-switching dyes,[50] rotational relaxation,[51,52] and energy transfer in Rhodamine 6G.[53]

To confirm that the dissipation of the energy from the first excited singlet state manifold occurs via direct coupling with the dense manifold of the ground state ($\sim 10^8$ levels/cm^{-1} at ~ 14000 cm^{-1}), rather than through the low-lying triplet, we performed similar absorption experiments in the triplet state which indicate that the electronic transitions (intersystem crossing) of azulene, $S_1 \rightarrow T_1$, is approximately 10 times slower than the $S_1 \rightarrow S_0$ rate. Drent et al.[54] have reported the intersystem crossing time as 6 psec. Although their data are in excellent agreement with ours, they analyzed their data by considering the exciting pulse as having a delta width compared to the fluorescence spot of azulene. If the analysis of the data is performed by a deconvolution of the pulse, one obtains a 60×10^{-12} sec lifetime, as compared with 6×10^{-12} sec as in Ref. 54. If the analysis is performed in the same manner on both sets of data the results are in very good agreement. Further experiments in solution and gas phase azulene have supported direct coupling to the ground state and the statistical coupling model for the azulene molecule excited in the S_1 manifold.[44,45]

2. *Emission Between Excited States*[55]

Fluorescence between excited states of large molecules had not been observed previously, although these radiative decay processes are frequent phenomena in diatomic molecues. The difficulty in observing emission from higher excited states of large molecules rests mainly in the very efficient electronic relaxation. The lifetime of an excited state is a function of its radiative $\tau_r(j \rightarrow i)$ and nonradiative $\tau_{nr}(j)$ widths,

$$[\tau(j)]^{-1} = \left[\sum_{i<j} (J \rightarrow i)\right]^{-1} + [\tau_{nr}(j)]^{-1}$$

in most cases the nonradiative decay $\tau_{nr}(J)$ of an upper electronic state, that is, the second excited singlet ($j = 2$), is fast, $\tau_{nr}(2) \lll \tau_2(2 \rightarrow 1)$. The quantum yield $Y_{(2\rightarrow1)} = \tau_{nr}(^2t_{r(2\rightarrow1)}) \lll 1$, and is therefore very difficult to observe experimentally.

The anomalous nature of azulene[55–57] renders this molecule suitable for the experimental observation of emission between the first two excited singlet states due to the long lifetime of the $^1A_1$ excited state and the low symmetry of the molecule. The azulene solution was excited by the second harmonic of a mode-locked ruby laser and the emission observed with an S-1 type photomultiplier. The appropriate filters, beam splitting, and polarizing components were inserted for the elimination of pump light from the detecting instruments. The characteristics of the emission are as follows:

1. A strong band at ~7600 Å and a weaker one at ~7400 Å.[58]
2. Deuterated azulene enhances the quantum yield by ~20% compared to normal azulene. This observation is consistent with the deuterium effect

observed for the $S_2 \longrightarrow S_0$ fluorescence and contrary to the effect observed for the emission between the first singlet and the ground state.[59]

3. The quantum yield was found to be $>10^{-6}$.

4. Very recently,[60,45] azulene in the gas phase has been excited by a Q-switched ruby laser pulses (20 nsec, 100 MW peak power). The characteristics of the $S_2 \longrightarrow S_1$ emission from gaseous azulene coincide with the solution experiments except that the quantum yield for $S_2 \longrightarrow S_1$ is larger by $\sim$100.

This discrepancy can be due to either energy deactivation provided by the solvent or, since all other parameters are in good agreement, that is, spectra lifetimes, and isotope effect, the possibility of stimulated emission $S_2 \longrightarrow S_0$ induced by the high-power picosecond pulses. In the latter case the apparent decrease in $S_2 \longrightarrow S_1$ emission will result in the low quantum yield. This possibility is currently under study.[61]

The observation of the emission from the first singlet state to the ground state has been observed in solution and, recently, in the gas phase, by methods similar to those described for observation of the $S_2 \longrightarrow S_1$ transition. The gas phase and solution data are in good agreement and further validate the argument that azulene relaxes predominantly by direct coupling with upper vibronic levels of the ground state. It should be noticed also that the theoretical predictions of Jortner, Berry, and Freed, with regard to the relaxation of a molecule belonging to the statistical model, are confirmed by the experimental observation that azulene in both solution and gas phase relaxes with the same rate, and exhibits to a great extent the predicted spectroscopic properties of a statistical case. The data for the azulene molecule in solution are displayed in Fig. 10.

B. Small Electronic Energy Gap

We consider now the energy decay characteristics of large molecules and the effects of intramolecular nonadiabatic coupling between nearly degenerate zero-order vibronic levels of different electronic states. Strong nonadiabatic coupling between two electronic levels of large molecules is expected to occur when the two levels are in close proximity,[36] and in real-life situations this phenomenon will be met in an isolated large molecule which exhibits a small electronic energy gap (i.e., $\sim$3000 cm^{-1} between two states such as the first and second excited singlet states of 3,4-benzpyrene). In particular, we will consider the experimental consequences of the strong coupling between a vibronic level, φ_s of S_2 and a sparse manifold of the vibronic levels φ_l of S_1.

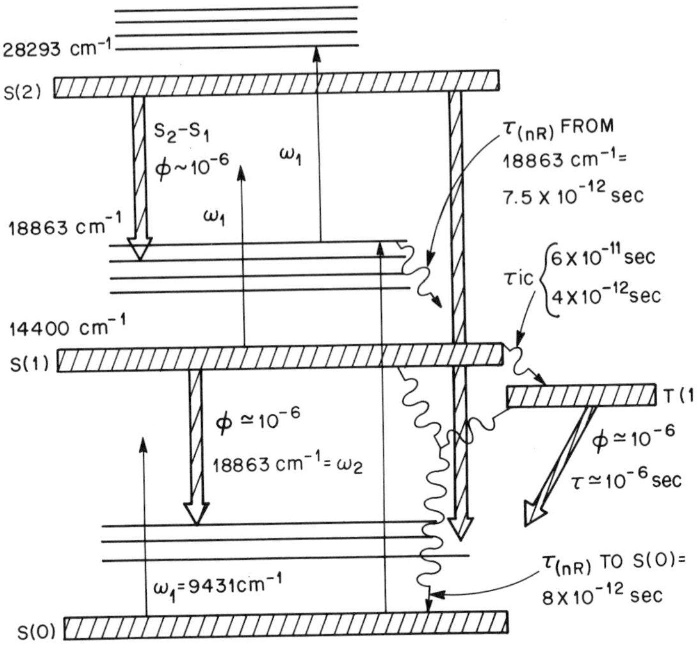

Fig. 10. Compilation of data for azulene by picosecond spectroscopy.

The intramolecular decay which occurs in the statistical case[27–30] is obviously not expected to be exhibited here but we would rather expect to see the small molecule characteristics portrayed by the following:

1. The absence of nonradiative electronic relaxation $\varphi_s' \sim\sim [\varphi_l]$.
2. Two distinct spectra regions in emission, one of which corresponds to the $\varphi_s \rightarrow \varphi_0{}^{0w}$ transition with the possible observation of resonance fluorescence from the S_2 state, and a second region which originates from the transition $\{\varphi_l\} \rightarrow \{\varphi_0{}^{lw}\}$.
3. The quantum yields, y_s, y_l, in the l and s states will be related to the pure radiative widths Γ_s and Γ_l by the ratio $R = (y_s/y_s)\Gamma_s / n\langle\Gamma_l\rangle$.
4. The experimental radiative decay in the s region will be lengthened compared to the calculated lifetime from oscillator strength measurements.
5. Under certain conditions molecular interference could be observed.

The two large molecules exhibiting these small molecule characteristics which we studied experimentally are 3,4-benzpyrene and naphthalene.

1. 3,4-Benzpyrene[42]

Pure 3,4-benzpyrene was heated to 150–230°C in a quartz cell. The excitation was provided by the second harmonic 28800 cm^{-1} generated, in a phase-matched KDP crystal, by mode-locked ruby laser (14400 cm^{-1}). The output consisted of a train of picosecond pulses with an average time width (FWHM) of 8 psec measured by 3PF and a separation of 7 nsec. Although picosecond lasers generate high-power pulses which provide an excellent means for a possible coherent excitation, the measured decay lifetime of 3,4-benzpyrene is longer than the separation of the pulses, thus inhibiting accurate measurements of the decay characteristics, including the possibility of molecular beats. The use of a *Q*-switch laser was also not convenient for the beat measurements and for the case of naphthalene the 20 nsec (FWHM) of the *Q*-switched pulse was also too long for comfort. To eliminate these disadvantages and retain the advantages of picosecond excitation, a new method was devised which proved to be both convenient and accurate.[41]

The laser beam was focused and collimated by a pair of 7 cm lenses. As the power of the picosecond pulses in the train increased an air spark was generated at the focal point of the first lens. The generated plasma absorbed the subsequent part of the laser train thus resulting in a sharp cut off within the pulse train as shown in Fig. 11. The fluorescence of 3,4-benzpyrene excited by this sharp cutoff light source was focused into the slit of a spectrometer, and detected by a 56 TVP photomultiplier.

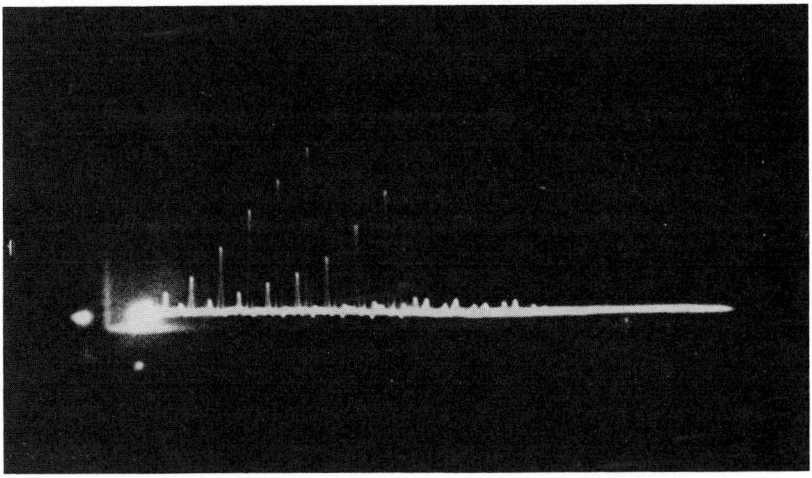

Fig. 11. Train of second harmonic ruby laser mode-locked pulses.

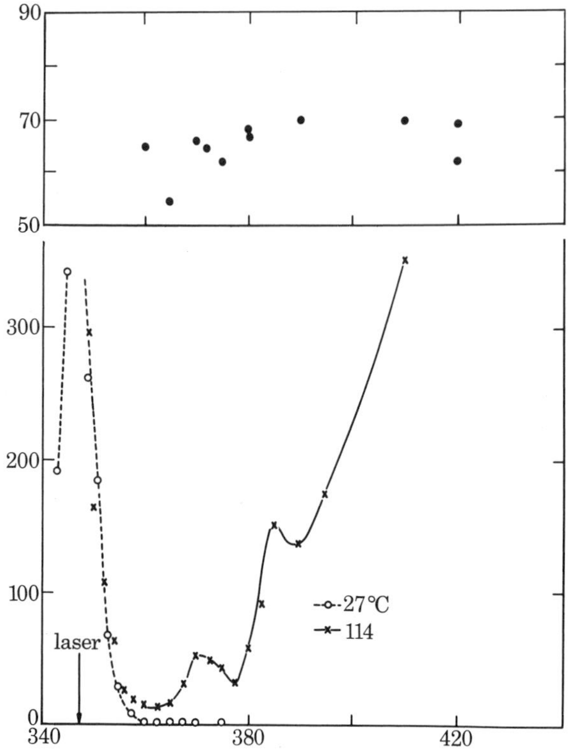

Fig. 12. Energy and time-resolved decay spectrum of 3,4-benzpyrene in the low-pressure gas phase excited by the second harmonic of the ruby laser. The room temperature (27°C) curve provides a measurement of the scattered light.

The predominant features are shown in the emission spectrum of Fig. 12 and the energy decay displayed in Fig. 13*a*. The decay spectrum clearly shows the $\varphi_s \rightarrow \varphi_0{}^{0w}$ transition with bands at 3700 and 3850 Å and the broad band with its origin at ~3900 Å which is assigned to the *l* region. This spectrum is in close agreement with that observed by Hoytink et al.[77] The time-resolved decay reveal the following. (1) The lifetime in the *s* region is larger by a factor of ~5 than the expected oscillator strength calculation for the $S_0 \rightarrow S_2$ transition. (2) The ratio of the emission for the *s* and *l* regions is ~0.1, whereas the ratio of the radiative widths $\Gamma_s/\langle\Gamma_l\rangle = 10$. Thus it is easily seen that the number of strongly coupled levels is ~100, which corresponds to the strong intramolecular coupling of the small molecule case.

The deviation from a straight line in the log *I* vs. t plot (Fig. 13*b*) in the oscillatory manner shown in very similar to that observed in atomic inter-

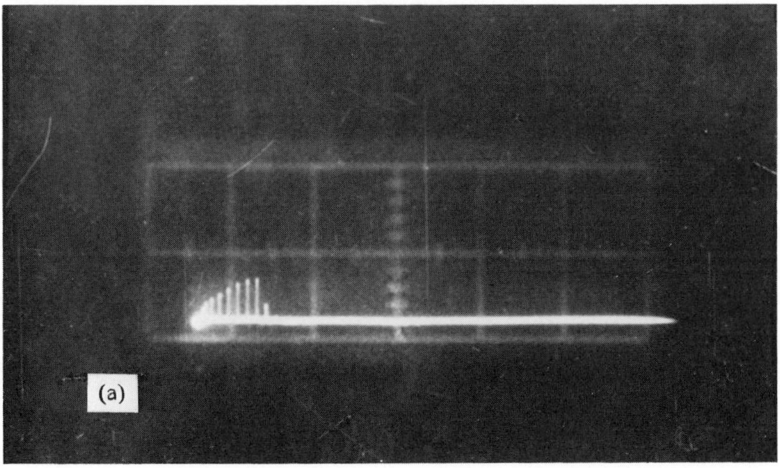

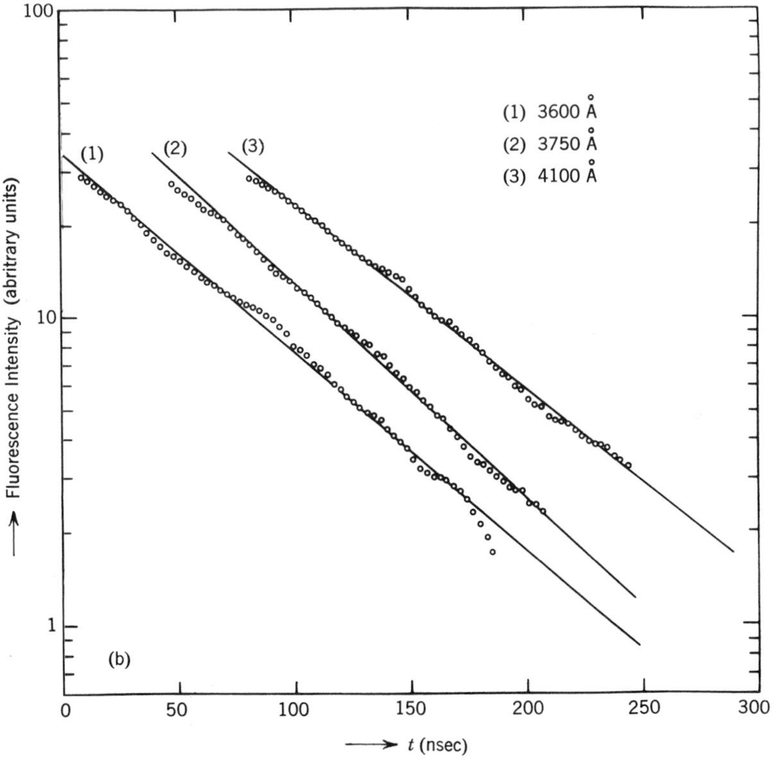

Fig. 13. (*a*) Typical energy resolved decay curves of 3,4-benzpyrene. Time scale 50 nsec/m, laser pulse. (*b*) Unimolecular energy resolved decay of 3,4-benzpyrene.

ference and possibly provides the first indication of molecular interference in large molecules.

2. *Naphthalene*, NO_2, SO_2

The second excited state of naphthalene provides an additional case for intermediate level spacing in a large molecule with an electronic energy gap of 3700 cm^{-1} between the $S_2(^1B_{2u})$ state with its origin at 32020 cm^{-1} and the S_2 $(^1B_{1u})$ at 35910 cm^{-1} and small molecule characteristics similar to the ones described for 3,4-benzpyrene. Although by conventional excitation sources previous investigators have not been able to observe resonance fluorescence from the S_2 state, by excitation with high-power mode-locked laser we have obtained the emission spectrum shown in Fig. 14.

The excitation was provided by the fourth harmonic triatomic molecules 37736 cm^{-1} of a mode-locked Nd^{3+} glass laser. The advantages over the conventional sources are the high power at a very short time and the

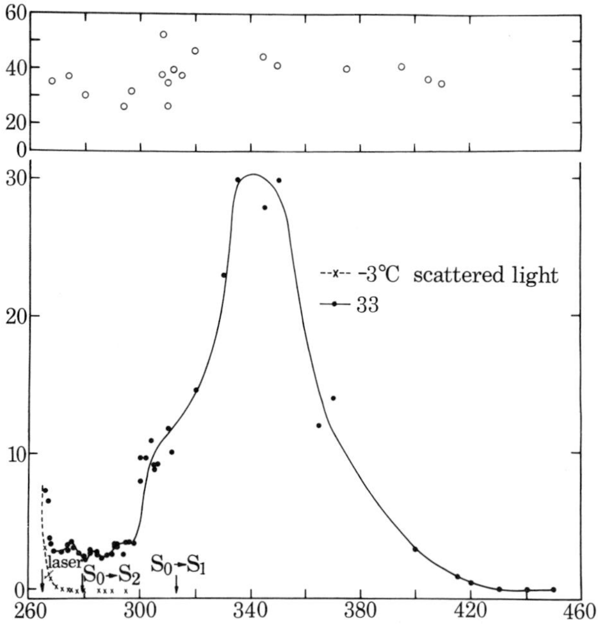

Fig. 14. Energy and time-resolved decay spectrum of naphthalene in the low-pressure gas phase excited by the fourth harmonic of Nd^{3+} laser. The data at $-3°C$ provide a measurement of the scattered light. The electronic origins of the two lowest excited singlet states[8] are indicated by arrows.

minimal scatter due to the collimation and monochromaticity of the laser pulse. The methods employed were similar to those described previously for the 3,4-benzpyrene experiments. The pulses were ~2–8 psec in duration and spaced apart by 6 nsec. The second and fourth harmonics were generated by phase-matched KDP and temperature-matched ADP crystals, respectively.

The naphthalene was heated to 33°C in a quartz cell (vapor pressure 0.2 torr) and excited by the 37736 cm^{-1} radiation of the fourth harmonic. The naphthalene emission was focused on a spectrometer slit, detected by an Amperex TUVP photomultiplier, and displayed on the oscilloscope. The energy and time-resolved spectra in Fig. 14 for the naphthalene molecule at low pressure exhibit a new emission between 2750 and 3000 Å (the time between collisions is longer than the decay lifetime). Three weak emission bands with ~1200 cm^{-1} spacing are shown at 2750 Å, 2840 Å, and 2920 Å. This fluorescence region is assigned to the *s* region whereas the broad band above 3000 Å is probably due to the l manifold. As in the case of 3,4-benzpyrene the $S_2 \rightarrow S_0$ decay is anomalously long with a lifetime an order of magnitude longer than the one calculated by oscillator strength data.

NO_2 and SO_2 are being studied with mode-locked short (~200 psec) laser pulses from an argon laser.[68] Nonexponential decays and anomalous emission have been observed by us in both these molecules, which seem to behave in a manner similar to 3,4-benzpyrene and naphthalene.

IV. TIME- AND FREQUENCY-RESOLVED SPECTRA

The methods by which we were first able to measure picosecond relaxation rates by absorption techniques, especially by the mirror translation method, have the following disadvantages:

1. The intensity of the pulses within a train varies.
2. The widths of the pulses in a single train vary widely.
3. The method is time consuming, requiring at least a laser shot per experimental point.
4. The pulse width and intensity vary widely from shot to shot.

In addition, wavelength resolution is totally absent. In the past few years we have utilized completely different experimental methods for picosecond spectroscopy and relaxation measurements which permit simultaneous time and frequency resolution with a single picosecond pulse.[62]

The time-frequency resolving method has been utilized for both absorption and emission measurements of relaxation processes.[62–64] This method,[62] which we believe to be the first simultaneous time-frequency resolving apparatus, is shown in Fig. 15. In this particular case a picosecond light pulse from a 1.06 μ (9431 cm^{-1}) Nd^{3+} glass laser generates the second harmonic for uses as interrogating pulse, although stimulated Raman or laser emission from a dye has been used for this purpose. The fundamental (9431 cm^{-1}) frequency pulse traverses the path shown in Fig. 15 and enters the optical shutter cell.[64] The shutter is a 0.1 $\times$ 20 mm CS_2 cell positioned between two polarizers crossed at 45° to the field of the 1.06 μ pulse. As this picosecond pulse propagates along the 20 mm length of the CS_2 cell it induces a birefringence in the CS_2, similar to that induced by the high electrical pulses of Kerr cell shutters. In the case of the picosecond pulse, the shutter transmits light through different points along the 20 mm cell of the shutter in step with the propagation of the opening pulse. The reorientation times for CS_2 is about 1.8 psec. Thus the time resolution at each point within the CS_2 is restricted, in this "Kerr liquid," to 1.8 psec or by the length of the pulse, which usually has a longer duration than 1.8 psec. Therefore, a spot of the shutter is "open," that is, transmits light through the cross polarizers, only during the orientation time and is then closed. Hence the light recorded by the spectrometer is dependent upon the time and point at which it arrives in the CS_2 cell. The second harmonic pulse is reflected by the mirror shown in Fig. 16 and induces the excitation, in this case, of Rhodamine 6G. The emitted radiation passes through the optical shutter during the time and at the point where the shutter is open for 2 psec. The path of the fundamental pulse along the length of the CS_2 cell provides the time-resolving mechanism. The CS_2 cell is imaged along the length of the slit of the spectrometer. The excitation pulse, corresponding to t = 0 with regard to fluorescence from the molecule under investi-

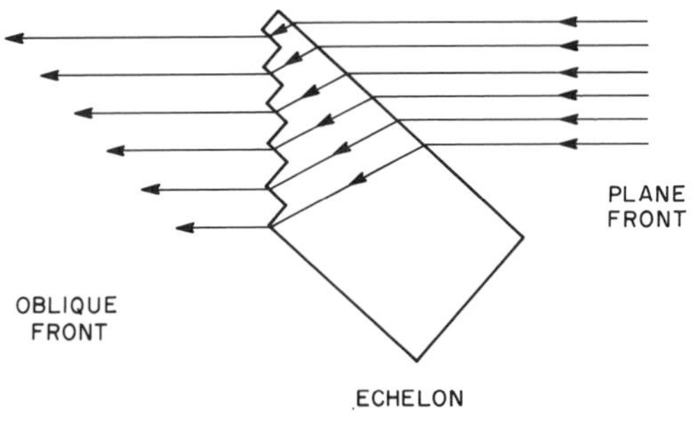

Fig. 15. The echelon: a single pulse passing through or reflected by the echelon is segmented into several pulses having any digitized separation equal intensity and the duration of the original pulse.

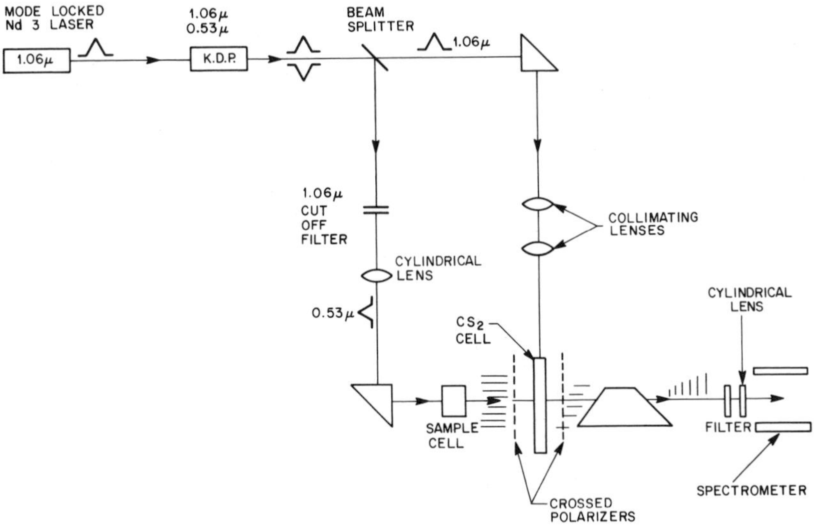

Fig. 16. Method for time-frequency resolving picosecond spectra.

gation, intersects the fundamental at the upper part of the CS_2 cell. Subsequent emission is transmitted through the shutter at various lower points of the CS_2 cell depending upon the time at which emission originates.

The light transmitted through the shutter is now time resolved, with time increasing from top to bottom in the CS_2 cell. Since the shutter cell is imaged along the length of the slit of the spectrometer with $t = 0$ on the top of the slit. The resulting spectrum now has two coordinates; one is the normal frequency display (ν), and the other is time in picoseconds (t).[63] This apparatus was first used for studies of vibrational relaxation in Rhodamine 6G.

Rhodamine 6G was excited by a 0.53 μ picosecond pulse, resulting in stimulated emission from the Rhodamine solution. The exciting 0.53 μ pulse was synchronized so that it arrives simultaneously with the shutter opening pulse and is transmitted at the upper point of the CS_2 cell corresponding to $t = 0$. The time-frequency plate (Fig. 17) shows the 0.53 μ pulse at $t = 0$. Later, a very small amount of spontaneous emission occurs from the upper vibrational levels of the first singlet[65] and, after a period of ~6 psec, the onset of the stimulated emission is observed with a duration of ~10 psec. The spectral resolution achieved by this method is strictly spectrometer limited. The Rhodamine 6G quantum yield for emission is almost unity, and since the stimulated emission occurs by a four-

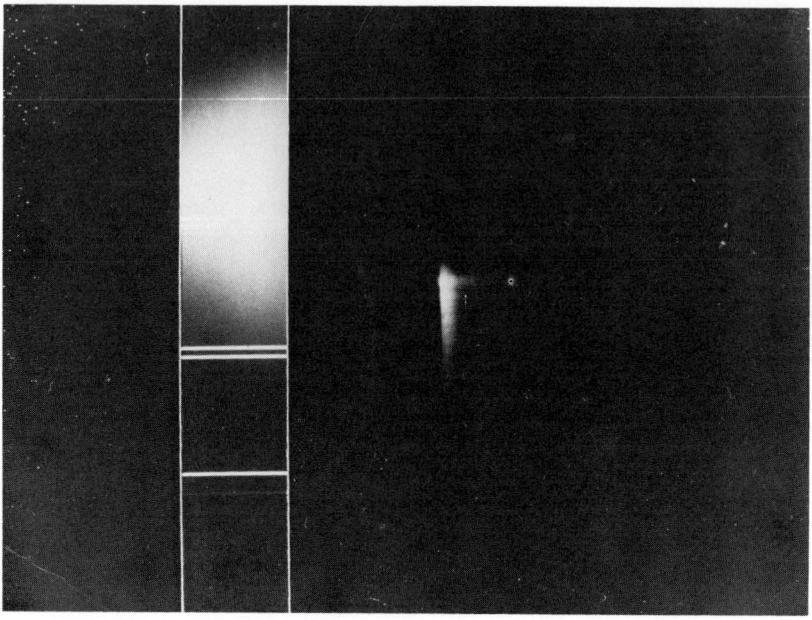

Fig. 17. Simultaneous time-frequency resolved spectrum of Rhodamine 6G.

level mechanism, the 6 psec time for the onset of stimulated emission represents the rate of vibrational relaxation within the electronically excited state.

The observed relaxation is due to either (1) intramolecular energy distribution by anharmonic coupling of the optically active modes with vibrational states resulting in the redistribution of the excess energy to intramolecular vibrational modes, or (2) dissipation of the energy by coupling to lattice modes. The vibrational relaxation is related to the width of the absorption bands by the Fourier transform and can be related to the solvent–solute "supermolecule" interaction consisting of the solute and the close-by solvent molecules.

By the same method the stimulated Raman emission[63] from the solvent was measured to be delayed by less than 3 psec, in agreement with theoretical postulations that molecular vibrations are only transiently excited by picosecond pulses. A further step in the development of high-resolution time-frequency spectroscopy is the utilization of an echelon,[65–67] which in practice generates picosecond pulses of identical duration and intensity, yet with selected constant separation (Fig. 15). This simple yet very reliable method is shown in Fig. 18. A train of pulses traverses the pockels cell

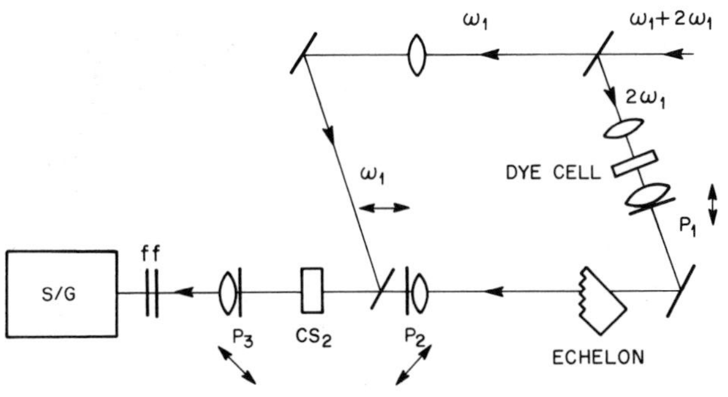

Fig. 18. Schematic of the echelon apparatus; this type with obvious variations has been used for both emission and absorption picosecond spectroscopy.

where a single pulse is extracted. This single pulse passes through the "stepped delay-echelon," which consists of a quartz block cut into several equal steps, or a stack of plates. As the pulse travels through, or is reflected by the echelon, it experiences variable delays corresponding to the length of that particular part of the echelon. The relative delay is simply a function of the refractive index, n, of the echelon medium, the thickness of the step d, and the angle of incident Θ.

$$\iota = \frac{d}{c}\left[n^2 - \sin^2 \Theta^{1/2} - \cos \Theta\right]$$

After the echelon, the output now consists of a train of pulses of equal intensity, identical duration, and separated selectively by one or more picoseconds. This set of pulses has been utilized as either the absorption or interrogation pulse placed before the sample cell, or as the time-resolving element in emission by locating the echelon after the sample cell. Two typical experimental arrangements are shown in Figs. 18 and 19; these have been applied in the study of relaxation processes. One application, for example, was to measure accurately the reorientation mechanism of molecules such as CS_2, which control the time resolution of the experiment, if the pulse duration is ultrashort.

To measure directly the reorientation time of CS_2, we placed the CS_2 cell between crossed polarizers as in a normal "shutter" and utilized the apparatus shown in Fig. 18. A single pulse is extracted from the train of picosecond pulses, and this is split into two parts, one with 90% of the intensity and the other with 10%. The intense pulse acts as the shutter opening pulse, whereas the weaker one passes through the echelon. The echelon output, which now consists of a set of picosecond pulses, enters

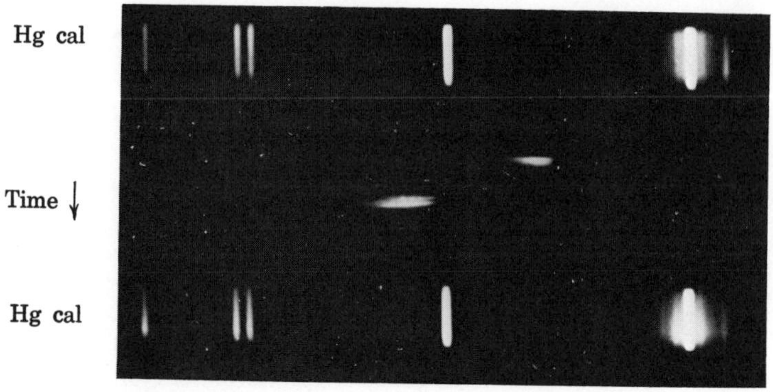

Fig. 19. The echelon method in emission. Stimulated emission from Rhodamine 6G in ethanol. Vertical echelon segments ≈ 3 psec per segment. Center broad emission at $\sim$545 nm. Structured segment at $\sim$7 psec and $\sim$558 nm.

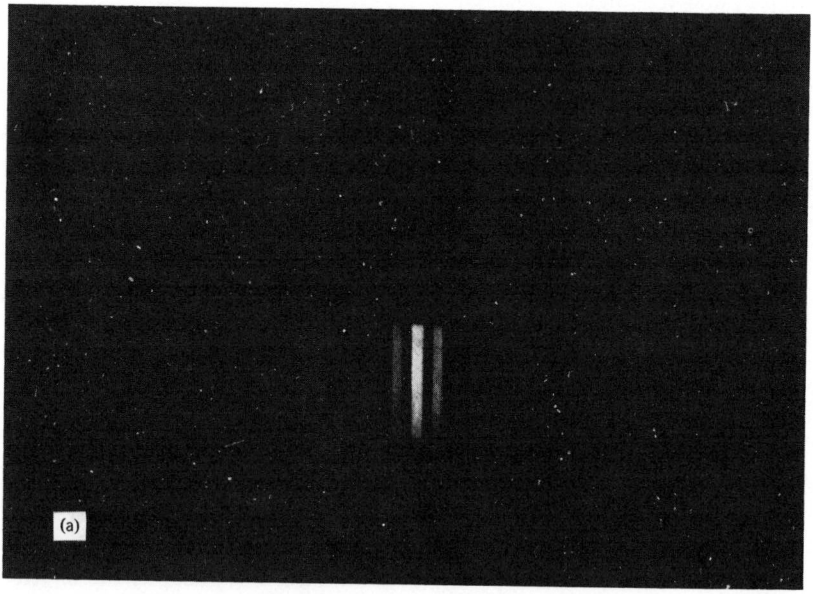

Fig. 20. (*a*) Photograph of the reorientational relaxation of CS_2 by the echelon method used as in absorption.

the CS_2 cell collinearly and simultaneously with the intense pulse. The transmitted segments of the echelon are then displayed on a photographic plate. Since each segment represents 1.2 psec, the number and intensity of the transmitted segments give directly the reorientation time of CS_2. Obviously, by minor variation of the apparatus, one can easily obtain the reorientation relaxation rate of other molecules in the excited and ground electronic states. The resulting absorption of the echelon-generated pulses is shown in Figs. 20*a* and *b* from which we measured the total opening

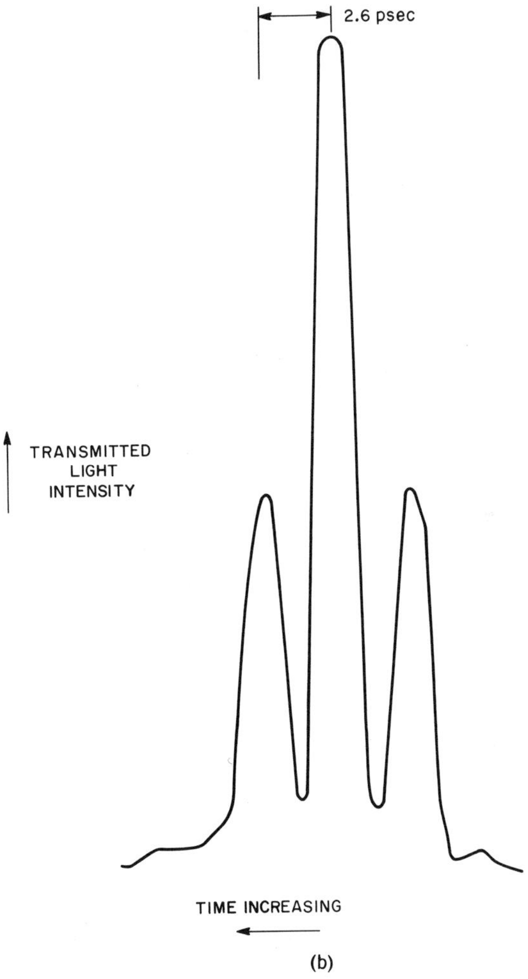

Fig. 20 (*continued*). (*b*) Photodensitometer trace of the echelon segments.

time of the shutter as 3.6 psec. This time includes both the CS_2 reorientation time and pulse width. Furthermore, we can not calculate the reorientation time of CS_2 as ~2 psec, which is in good agreement with light scattering data. It should be noted that although in CS_2 the rotational relaxation is fast and probably a dominant factor, at lower temperatures and in other liquids and solids, we have shown by similar direct methods[66] that the electronic polarizability and "rocking" mechanisms play the dominant roles. We have repeated by this method our 1968 experiment on the rotational relaxation of biphenylene in solution, and the relaxation of several organic molecules by absorption and emission, with greater reliability, accuracy, and ease than the original pulse train type of experiment.

Lately, by self-focusing a picosecond pulse a continuum is generated in several solutions which have several thousand wavenumbers bandwidth and duration on the order of or less than the pumping picosecond pulse. Utilizing this continuum in conjunction with the echelon we can study absorption and emission processes from ~1 psec to ~0.5 nsec with a single laser shot.

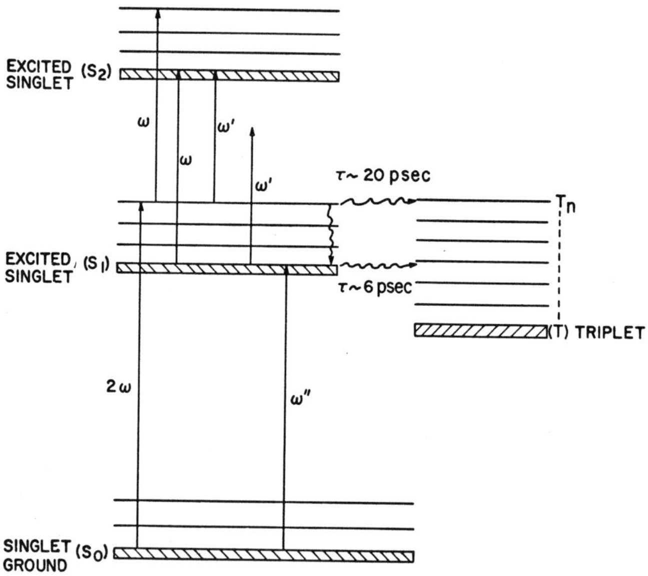

ENERGY LEVEL DIAGRAM FOR BENZOPHENONE

(a)

Fig. 21. (*a*) Electronic relaxation rates of benzophenone in solution from two vibronic levels of S_1.

V. BENZOPHENONE

The energy decay characteristics of benzophenone in solution and in gas phase provide an example of the utilization of the methods described for studies of radiative and nonradiative processes. The intersystem crossing rate of benzophenone in solution was studied at two photoselectively excited vibronic levels of the first excited singlet state. The effective energy gap between the lowest excited singlet and triplet states is $\sim$2000 cm^{-1}. The energy gap of benzophenone would cause it to be classified in the "small molecule case," and one would expect that electronic relaxation will not occur in the isolated molecule since the sparse triplet manifold cannot act as quasicontinuum. In solution, however, benzophenone exhibits irreversible electronic relaxation since the solvent provides a vibrational broadening mechanism, and the large-spin orbit coupling results in strong interstate singlet–triplet coupling which in solution is evident by the lack of fluorescence.

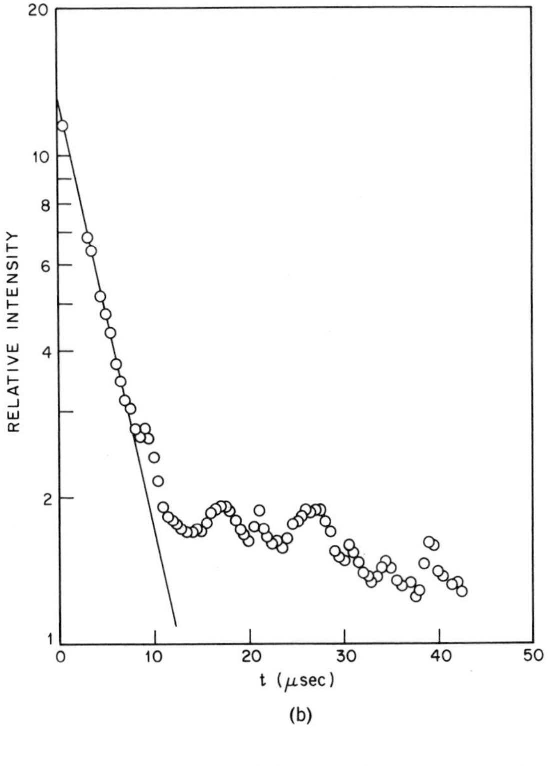

Fig. 21 (*continued*). (*b*) Nonexponential decay of benzophenone in the gas phase at 60°C and 4600 Å. The oscillator behavior is indicative of molecular interference.

Benzophenone in solution was excited at two vibronic levels of the first excited singlet: (1) near the electronic origin of the first excited singlet state 26,800 cm^{-1}, with a ($\Delta\nu = 200$ cm^{-1}) stimulated Raman line of the second harmonic of ruby 28,800 cm^{-1}; and (2) at a higher vibronic level located at 28,800 cm^{-1}. The energy decay from each of these levels was measured by the micrometer method in 1968 and more recently using the time-frequency resolving apparatus. The absorption (interrogating) beams (see Fig. 21*a*) were of the appropriate wavelength to correspond exactly to the energy gap between the level excited by the laser and the second excited singlet.[68]

The experimental data reveal that the upper vibronic level excited by the second harmonic 28,800 cm^{-1} has an electronic relaxation lifetime of $\sim$20 psec, whereas the lower vibronic level ot 26,800 relaxes with a rate approximately three times faster, or 6 psec. The high intersystem crossing rate is expected since benzophenone is characterized by the very high phosphorescence yield. However, the slower rate from an upper vibronic level compared to the lower level (at the origin) is in contrast to the normal behavior of large molecules corresponding to the "statistical case." Recent calculations[69] show that for small energy gap the major channel for the decay of the upper electronic level is electronic relaxation rather than the vibrational relaxation within the singlet electronic manifold. These experimental observations indicate that benzophenone in the "isolated molecule" case will exhibit the decay characteristics of a small molecule, and the fluorescence lifetime measured is longer than that calculated from the oscillator strength. 3,4-Benzpyrene[41] and naphthalene[42] excited near the electronic origin of the second singlet state exhibit this "small molecule" behavior, and their emission lifetime is larger by a factor of three or more than the lifetimes obtained from oscillator strength calculating.

Experiments performed[67,68,70] in gaseous benzophenone at ~ 1 μ pressure not only validate these expectations, but in addition, we observed nonexponential decay and, possibly, quantum interference. The nonexponential behavior can be summarized as being composed of two predominant decay rates. One is the result of strong coupling and has a lifetime of $\sim 10 \times 10^{-6}$ sec, in contrast to the 1×10^{-6} sec lifetime expected from the oscillator strength data. The tenfold lengthening of the benzophenone lifetime is due to the dilution of the oscillator strength excited singlet level by an average of ten strongly coupled levels of the triplet manifold. The second decay, with a lifetime of $\sim 200 \times 10^{-6}$ sec, is due to the weakly coupled levels. The characteristic decay mechanisms of benzophenone in the gas phase should be a general phenomenon for large molecules corresponding to the intermediate density level case, and experimental data should provide further insight into the coupling mechanism of the various levels and enhance our understanding of molecular relaxation processes.

Several other chemical processes have been studied by us which are not described in this chapter. They include (1) the energy transfer mechanism within single long-chain molecules, such as polymers[71] and biological molecules;[72,73] (2) the relaxation mechanism of dyes such as Rhodamine 6G, BBOT, and polymethanes;[74] (3) cis–trans isomerization[51] of molecules such as Rhodapsin[75]; and (4) dynamic reorientation of large molecules.[63] Other current studies include the cage effect[76] and several other fast processes which we believe will provide information concerning the mechanism and characteristics of ultrafast molecular relaxation via picosecond spectroscopy.

Acknowledgments

I am thankful for the benefits derived from the collaboration and discussions with Dr. D. C. Douglass. Also, the competent technical assistance of Mr. R. P. Jones is appreciated.

References

1. L. E. Hargrove, R. L. Fork, and M. A. Pollack, *Appl. Phys. Lett.*, **5**, 4 (1964).
2. H. W. Mocker and R. J. Collins, *Appl. Phys. Lett.*, **9**, 270 (1965).
3. D. A. Stetser and A. J. DeMaria, *Appl. Phys. Lett.*, **9**, 118 (1966).
4. N. Bloembergen, *Comm. Solid State Phys.*, **1**, 37 (1968).
5. J. A. Amstrong, *Appl. Phys. Lett.*, **10**, 16 (1967).
6. P. M. Rentzepis, *Chem. Phys. Lett.*, **2**, 117 (1968).
7. J. A. Giordmaine, P. M. Rentzepis, S. L. Shapiro, and K. W. Wecht, *Appl. Phys. Lett.*, **11**, 216 (1967).
8. G. E. Busch and P. M. Rentzepis, "Picosecond Molecular Relaxation with a Continuous Mode-locked Laser," *Chem. Phys. Lett.*, to be published.
9. C. K. N. Patel, *Appl. Phys. Lett.*, **18**, 25 (1971).
10. O. R. Wood, R. L. Abrams, and T. J. Bridges, *Appl. Phys. Lett.*, **17**, 376 (1970).
11. See for example, E. D. Shaw and C. K. N. Patel, *Appl. Phys. Lett.*, **18**, 215, 274 (1971).
12. A. A. Vuylstoke, *J. Appl. Phys.*, **34**, 1615 (1963).
13. A. J. DeMaria, *Proc. IEEE*, **57**, 1 (1969).
14. P. M. Rentzepis and M. A. Duguay, *Appl. Phys. Lett.*, **11**, 218 (1967).
15. H. P. Weber, *Phys. Lett.*, **27**, 321 (1968).
16. J. R. Klauder, M. A. Duguay, J. A. Giordmaine, and S. L. Shapiro, *Appl. Phys. Lett.*, **13**, 174 (1968).
17. R. J. Harrach, *Phys. Lett.*, **28**, 393 (1968).
18. H. P. Weber and R. Dandliker, *IEEE J. Quantum Elec.*, **4**, 1009 (1968).
19. K. H. Drexhage, *Appl. Phys. Lett.*, **14**, 318 (1969).
20. T. I. Kuznetosova, *Soviet Phys. JETP*, **28**, 1303 (1969).
21. P. M. Rentzepis, C. J. Mitschele, and A. C. Saxman, *Appl. Phys. Lett.*, **17**, 122 (1970).

22. P. M. Rentzepis and D. C. Douglass, "Measurement of Pulse Shape by the Three Photon Fluorescence Method," to be published.
23. E. I. Blount and J. R. Klauder, *J. Appl. Phys.*, **40**, 2874 (1966).
24. E. B. Treacy, *Appl. Phys. Lett.*, **17**, 14 (1970).
25. S. L. Shapiro and M. A. Duguay, *Phys. Lett. A*, **28**, 698 (1969).
26. D. J. Bradley, C. H. C. New, and S. J. Laughey, *Phys. Lett. A*, **30**, 78 (1969).
27. J. Jortner and R. S. Berry, *J. Chem. Phys.*, **48**, 2757 (1968).
28. M. Bixon and J. Jortner, *J. Chem. Phys.*, **48**, 715 (1968).
29. M. Bixon and J. Jortner, *J. Chem. Phys.*, **50**, 4061 (1969).
30. K. Freed and J. Jortner, *J. Chem. Phys.*, **50**, 2916 (1969).
31. A. Nitzan, J. Jortner, and P. M. Rentzepis, "Internal Conversion of Large Molecules," *Mol. Phys.*, **22,** 585 (1971).
32. G. W. Robinson and R. P. Frosh, *J. Chem. Phys.*, **38**, 1187 (1964).
33. S. H. Lin and R. Bersohn, *J. Chem. Phys.*, **48**, 2732 (1968).
34. B. Block and H. Feschbach, *Ann. Phys.* (*N.Y.*), **23**, 47 (1963).
35. J. Jortner, *J. Chim. Phys. Paris*, **9**, 1970 (1969).
36. A. Nitzan, J. Jortner, and P. M. Rentzepis, "Intermediate Level Structure in Highly Electronic States of Large Molecules," *Proc. Roy. Soc.* (*London*), in press.
37. A. E. Douglas, *J. Chem. Phys.*, **45**, 1007 (1966).
38. A. Nitzan and J. Jortner, to be published.
39. G. E. Busch, P. M. Rentzepis, and J. Jortner, *J. Chem. Phys.*, **56,** 361 (1972).
40. G. E. Busch, P. M. Rentzepis, and J. Jortner, *Chem. Phys. Lett.*, **11,** 437 (1971).
41. P. G. Wannier, P. M. Rentzepis, and J. Jortner, *Chem. Phys. Lett.*, **10**, 102 (1971).
42. P. G. Wannier, P. M. Rentzepis, and J. Jortner, *Chem. Phys. Lett.*, **10**, 193 (1971).
43. P. M. Rentzepis, J. Jortner, and R. P. Jones, *Chem. Phys. Lett.*, **4**, 599 (1970)
44. D. Huppert, J. Jortner, and P. M. Rentzepis, *J. Chem. Phys.*, **56,** 4826 (1972).
45. D. Huppert, J. Jortner, and P. M. Rentzepis, *Chem. Phys. Lett.*, **13**, 225 (1972).
46. A. Nitzan, J. Jortner, E. Drent, and J. Kommandeur, *Chem. Phys. Lett*, in press.
47. E. Drent, Ph.D. Thesis, University of Groningen, The Netherlands, 1971.
48. P. M. Rentzepis, *Chem. Phys. Lett.*, **2**, 117 (1968).
49. P. M. Rentzepis, *Chem. Phys. Lett.*,**3**, 717 (1969).
50. R. I. Scarlet, J. F. Figueria, and H. Mahr, *Appl. Phys. Lett.*, **13**, 71 (1968).
51. P. M. Rentzepis, *Photochem. Photobiol.*, **8**, 579 (1968).
52. K. B. Eisenthal and K. H. Drexhage, *J. Chem. Phys.*, **51**, 5720 (1969).
53. D. Rehm and K. B. Eisenthal, *Chem. Phys. Lett.*, **9**, 387 (1971).
54. E. Drent, G. M. Van Der Deije, and P. J. Zandstra, *Chem. Phys. Lett.*, **2**, 526 (1968).
55. M. Beer and H. C. Longuet-Higgins, *J. Chem. Phys.*, **23**, 1390 (1955).
56. G. Viswanath and M. Kasha, *J. Chem. Phys.*, **24**, 574 (1955).
57. P. M. Rentzepis, *Science*, **169**, 239 (1970).
58. P. M. Rentzepis and C. J. Mitschele, *Anal. Chem.*, **42**, 20 (1970).
59. M. M. Malley and P. M. Rentzepis, *J. Lumin.*, **1,2** 448 (1970).
60. D. Huppert, Ph.D. Thesis, University of Tel-Aviv, Israel, 1972.
61. G. Busch, J. Jortner, and P. M. Rentzepis, unpublished data.
62. P. M. Rentzepis, M. R. Topp, R. P. Jones, and J. Jortner, *Phys. Rev. Lett.*, **25**, 1742 (1970).
63. P. M. Rentzepis and M. R. Topp, *Ann. N.Y. Acad. Sci.*, **33**, 284 (1971).
64. M. R. Topp and P. M. Rentzepis, *Chem. Phys. Lett.*, **4**, 1 (1971).
65. M. R. Topp, P. M. Rentzepis, and R. P. Jones, *J. Appl. Phys.*, **42**, 3415 (1971).
66. M. R. Topp and P. M. Rentzepis, *J. Chem. Phys.*, in press (1972).
67. P. M. Rentzepis and G. E. Busch, *Mol. Photochem.*, in press (1972).

68. G. E. Busch and P. M. Rentzepis, to be published.
69. A. Nitzan, J. Jortner, and P. M. Rentzepis, *Chem. Phys. Lett.*, **8**, 445 (1971).
70. P. M. Rentzepis, *Photochem. Photobiol.*, to be published (1972).
71. J. Jortner and P. M. Rentzepis, to be published.
72. P. M. Rentzepis, ref. 57.
73. M. M. Malley and P. M. Rentzepis, *Chem. Phys. Lett.*, **3**, 534 (1969); **7**, 57 (1970).
74. M. R. Topp and P. M. Rentzepis, *Phys. Rev. A*, **3**, 538 (1970).
75. In collaboration with Drs. A. A. Lamola and G. E. Busch.
76. In collaboration with Prof. M. Szwarc, *Proc. Natl. Ac. Sci.*, Oct. (1972).
77. P. A. Geldof, R. P. H. Rettschnick, and G. J. Hoytink, *Chem. Phys. Lett.*, **4**, 59 (1969).

SOME MODERN ASPECTS OF EXCITON THEORY*

MICHAEL R. PHILPOTT†

Institute of Theoretical Science and Department of Chemistry, University of Oregon, Eugene, Oregon

CONTENTS

* Supported in part by the National Science Foundation.
† Alfred P. Sloan Fellow.

I. INTRODUCTION

Experimental and theoretical studies of excitons in polymers and molecular crystals made striking advances in the last decade. In the fifties most of the experimental work reported consisted of the measurement of polarized absorption and fluorescence spectra. In the sixties these techniques were refined and extended to include phosphorescence, isotopic impurity, and hot band spectroscopy with great emphasis placed on obtaining high-resolution spectra of crystals of carefully controlled composition at very low temperatures. Nor has the pace slowed during the last two years, for already resonance pair and reflection spectroscopy are supplying new information about excitons in molecular crystals. On the theoretical side new mathematical tools such as second quantized and Greens operator techniques are supplanting the use of perturbation and finite basis theories.

Reviews and progress reports on theoretical and experimental aspects of exciton phenomena are fairly numerous.[1–29] However, apart from Davydov's latest book,[28] there has been no major review of theory since 1966.[17] It is not possible in one article to do justice to all new developments and so the focus has been narrowed to the theory and experiment of single-exciton states with zero spin (singlets). Omitted from the discussion are triplet excitons and multiparticle processes, for example, interaction of excitons with lattice and molecular vibrations, charge transfer states, and conduction bands. Some of these areas are presently undergoing rapid evolution and deserve a review at a later date when some of the ferment has subsided.

The term phonon is used exclusively to describe the vibrational modes of the lattice in which the molecules move as rigid bodies. These modes consist of acoustical, optical, and torsional oscillations. The term vibrational exciton is reserved for states arising from the intramolecular vibrational motions of the molecules.

In preparing this article one objective has been to organize a coherent account of exciton theory that is applicable to crystal and polymer transitions of low, medium, and high intensity. None of the calculations performed so far for a transition of medium intensity, like that at 4000 Å in crystalline anthracene, are satisfactory, for reasons explained in Section IV.

A complete description of strong transitions requires a multiparticle theory; however, some major features of the spectra can be understood using the single-particle (single-exciton) theory, as will be shown. A theory of strong transitions has many applications to, for example, the higher singlet transitions of aromatic hydrocarbon crystals, biopolymers, synthetic monomolecular layers, and the chromophores in the lamellae of chloroplasts.[30] At a rather speculative level it is of interest to note that in Little's model[31,32] of a superconducting molecule the exciton states of the cyanine dye side groups play a role similar to phonons in the conventional BCS theory.[33]

Nearly all the advances made in the last decade have been for the very weak crystal transitions for which dipole interactions are negligible since the molecular transition dipole moments are less than 0.1 Å. For strong transitions, however, the dipole interactions provide the major component of the exciton interaction and their long range renders useless the restricted Frenkel approximation[34] which now forms part of the foundation of weak transition theory. Three-dimensional lattice sums of dipole–dipole interactions are conditionally convergent and must be performed in a manner consistent with any boundary conditions imposed by the radiation field used to excite the transition. It turns out that the dipole sums required in exciton calculations depend on the direction of the exciting radiation and are discontinuous at the center of the Brillouin zone ($\mathbf{k} = 0$ region). Around this point ($\mathbf{k} = 0$) exciton energies depend on the direction, $\hat{\mathbf{k}}$, of the wave vector, and so in crystals with several molecules per unit cell the factor group splitting depends on the crystal face used in the measurement.

The theory of single-exciton states in the dipole approximation is most easily developed from the equations of classical electrodynamics. The basic equations determining exciton frequencies in the classical dipole theory have exactly the same form as those obtained from the more general, but considerably more complicated, quantized theory when the latter is restricted to dipole interactions. Since it has been the author's experience that dipole theories can serve as a useful guide in formulating more general theories based on second quantized techniques, a complete section has been devoted to classical dipole theory.

Molecular transitions will be called strong, medium, or weak according to the following classification of oscillator strength[35]:

$$f \gtrsim 3.0 \quad \text{Strong}$$
$$f \simeq 0.3 \quad \text{Medium}$$
$$f \lesssim 0.03 \quad \text{Weak}$$

Transitions with oscillator strengths $f \simeq 0.01$ are the ones referred to as very weak. Note that these are not solution f values.[35]

A. Excitons in Molecular Crystals

The concept of the exciton was first introduced by J. Frenkel in 1931,[36–38] in papers aimed at explaining how, in atomic crystals, electronic excitation energy resulting from photon absorption is degraded into heat. Frenkel realized that electronic excitation initially localized on a particular atom would, under the influence of the interatomic interactions, migrate to neighboring atoms. This migrating packet of electronic excitation energy he termed an exciton.

In crystals of aromatic hydrocarbons the lowest excited electronic states appear to be tightly bound Frenkel excitons. This is supported by calculations that have revealed that there is poor overlap between excited state orbitals on different molecules.[39,40] The reverse is true in rare gas crystals where overlap effects are very important.

If the orbital of the excited electron encompasses several molecules then the electron and its hole (vacant ground-state orbital) effectively occur on different sites. This type of exciton is qualitatively different from the Frenkel exciton where electron and hole occur on the same molecule. It is called a Wannier exciton[41] and has been detected spectroscopically in many inorganic insulators by its hydrogenic spectrum. In these solids the electron can orbit its hole at greater and greater distances, just as in an electronically excited hydrogen atom.

At present there is no reliable spectroscopic evidence for the existence of Wannier excitons in crystals of aromatic hydrocarbons.[42] The presence of Wannier excitons (also called charge-transfer and ion-pair states) and conduction states has been inferred from less direct experiments.[43] It has been suggested that the factor group (Davydov) splittings of the first singlet transitions of crystalline benzene and naphthalene arise because of a coupling between the neutral Frenkel exciton and close-lying charge-transfer states.[17,44,45,47,48] All the accumulated experimental evidence is against this interpretation, however.[46,49,50]

Nowadays it is customary to use the term exciton for almost any migrating packet of molecular excitation energy. However, we here restrict usage to purely electronic, vibronic, and vibrational excitations. An electronic exciton corresponds to the transfer of purely electronic excitation; there is no change in the vibrational state of the molecule. The concept of an electronic exciton is important for the understanding of intense transitions in molecular aggregates. For strong transitions the purely electronic excitation moves too rapidly between sites for the molecules to adjust their vibrational motions. Vibronic excitons are associated with weak- and medium-intensity transitions; they correspond to the simultaneous migration of electronic and vibrational energy. Finally, vibrational excitons are migrating intramolecular vibrational excitations of the ground electronic state of the molecules.

For transitions to the single-exciton states from the crystal ground state the selection rule[51] on the exciton wave vector is

$$\mathbf{k} = \mathbf{q} \tag{1}$$

where $\mathbf{q}$ is the wave vector of the incident photon. For transitions between single-exciton states the rule is

$$\Delta\mathbf{k} = \mathbf{k}_f - \mathbf{k}_i = \mathbf{q} \tag{2}$$

where $\mathbf{k}_f$ and $\mathbf{k}_i$ refer to the final and initial crystal states, respectively. Since $|\mathbf{q}|$ is extremely small for transitions in the optical and near-ultraviolet it is customary to write these selection rules in the form

$$\mathbf{k} = 0\,, \qquad \Delta\mathbf{k} = 0 \tag{3}$$

For weak transitions, where dipole interactions are negligible, no difficulties result from this approximation since the exciton energies are essentially analytic functions of wave vector. However, for crystal transitions of medium and high intensity, where the exciton energies are nonanalytic functions of $\mathbf{k}$ at the center of the Brillouin zone, the selection rules should be written

$$|\mathbf{k}| = 0\,, \qquad \hat{\mathbf{k}} = \hat{\mathbf{q}} \tag{4}$$

for the first case, and

$$|\mathbf{k}_f| = |\mathbf{k}_i| \tag{5}$$

$$\hat{\mathbf{k}}_f - \hat{\mathbf{k}}_i = \frac{q}{k}\hat{\mathbf{q}} \tag{6}$$

for the second case. (In these last equations $\hat{\mathbf{k}}$ is the unit vector in the direction of $\mathbf{k}$). *These rules preserve the direction of the absorbed photon.*

B. Excitons in Polymers

The electronic properties of many polymers of biological importance have been interpreted using molecular exciton theory. This was a task first started by Moffit[52] who used exciton theory to develop a treatment of the optical rotation (OR) of helical polypeptides.[53–55] The theory has since been extended to treat the optical rotation and circular dichroism (CD) of many helical oligomers and polymers.[56–60] Since the monomers generally have little or no symmetry the spectra are congested (nearly all vibrations are totally symmetric) and consequently appear as broad envelopes. A little information about the magnitude of exciton interactions has been gleaned from the absorption and fluorescence spectra. Much more information has been obtained from the measurement of the optical rotatory dispersion (ORD) and CD spectra.

Many of the cyanine and related dyes aggregate in concentrated solution forming long threadlike polymers.[61,62] The absorption spectrum of aggregate formed by the cationic dye pseudoisocyanine displays a unique "*J*-band" about 1500 cm^{-1} to the red of the monomer spectrum.[63–65] This *J*-band is extremely narrow,[66] about 50 cm^{-1}, and appears to carry an appreciable fraction of the oscillator strength of the monomer transition. Other dye aggregates are known in which there is a single intense blue shifted transition.[67] These spectacular spectral changes accompanying aggregate formation point strongly at the presence of exciton bands in these systems. Since the isolated dye transitions are so intense the exciton bands in these systems are expected to be excellent examples of strong vibronic coupling.[68,69]

The exciton states of a cyclic helical polymer with $N(\rightarrow\infty)$ residues are conveniently classified by the N irreducible representations of the Abelian group of screw operations.[70] A one-dimensional wave vector k can be defined that points along the helix axis z. For singlet states time-reversal invariance ensures the degeneracy of levels with wave vectors $\pm k$. If q_z is the projection on the helix axis of the photon wave vector $\mathbf{q}$, then for a transition from the polymer's ground state, the selection rules are

$$k = q_z \tag{7}$$

$$k = q_z \pm S \tag{8}$$

where S is a wave vector characteristic of the angle of twist φ between adjacent monomers and the step-height t

$$S = \frac{\varphi}{t}$$

The first selection rule holds for helix transitions polarized along the z axis and the second for those polarized perpendicular to the axis.

For transitions in the optical and near-ultraviolet regions the photon wave vector $q_z \simeq 0$ so that the selection rules reduce to

$$k = 0\,, \qquad k = \pm S \tag{10}$$

This approximate selection rule must be applied with caution when calculating the contributions to ORD and CD spectra of electric dipole transitions.

C. Excitons in Molecules

To those accustomed to regarding excitons as transfer of excitation energy between molecules the idea of excitons in molecules may sound nonsensical. However Raymonda and Simpson[71] have analyzed a large number of alkane spectra in terms of CC σ-bond excitations, and more recently Partridge has extended their ideas to saturated polymers.[72,73] The spectra of some other large molecules have also been analyzed in terms of bond excitons.[74] Although there can be no doubting the regularities detected in the absorption spectra much remains to be considered, especially to what extent excitation of the CH bonds participate.[75,76]

The bond excitons of saturated molecules may be regarded as a type of Frenkel excitation. In unsaturated molecules containing extended π-electron systems like the dyes there is the possibility of having excitations of the Wannier type, and an attempt to analyze some dye spectra along these lines has been made by Tric and Parodi.[77]

Finally in passing we note that the spectra of weakly bound dimers,[78–80] trimers,[81] and chelated ions[82] are often analyzed satisfactorily using the exciton picture; however, these examples are more properly considered along with the polymers of the last section.

II. THEORIES OF SINGLE EXCITON STATES

Regardless of the intensity of the molecular transition an analysis of the crystal or polymer spectrum in terms of single-exciton states is generally a necessary preliminary. In the case of the weak 0–0 band of the first excited singlet in crystalline benzene and naphthalene, a complete quantitive analysis[26,49,83,84] has been made, implying the presence of practically isolated bands of single-exciton states. The higher vibronic transitions of the same electronic states display the effect of coupling to close-lying two-particle bands containing vibronic and vibrational excitons.[85,86] For very strong transitions the structure of the electronic exciton band can be calculated using single-particle theory. It is also of interest to determine the wave vectors of the electronic states occupying the unmasked region of single-particle excitations.[87]

In passing it should be mentioned that the phenomena of excimers and exiplexes[27] is outside the scope of single-exciton theory since MO calculations indicate that charge transfer interactions contribute greatly to the stabilization of these excitations.[88,89]

Early calculations of exciton energies and intensities used perturbation theory (up to second order). The next refinement consisted in actually diagonalizing the secular matrix instead of approximating it. This second method is here referred to as the finite basis approach. It has found wide use and seems perfectly adequate for calculations concerning weak and very weak transitions. For systems containing coupled transitions of intermediate and higher intensity the finite basis approach (which is nothing more than variation theory applied to a limited basis set or degenerate perturbation theory applied to nondegenerate states) can break down and what is required is a full many-body solution of the problem. This is obtained using the classical dipole theory (Section II.B) or better still the second quantized exciton theory (Section II.C).

A. Perturbation and Finite Basis Theories

All theories of the electronic spectra of molecular aggregates assume the availability of some free molecule data, usually the excitation energies and dipole moments relative to the principal axes. From these data and the crystal or polymer structure the properties of a useful reference system, the oriented gas model, can be calculated.

The Schrödinger equation for an isolated oriented molecule fixed on site s is written

$$H_s \varphi_s^u = \epsilon_s^u \varphi_s^u \tag{11}$$

where H_s is the molecular Hamiltonian, φ_s^u the uth excited state wave function, and ϵ_s^u the energy eigenvalue. The superscript u is omitted for the ground level ($u = 0$) except when needed for emphasis. For any aggregate of molecules (monomers or residues) on sites s the Hamiltonian is

$$\mathcal{H} = \sum_s H_s + \tfrac{1}{2} \sum_s \sum_{s'}{}' V_{ss'} \tag{12}$$

where $V_{ss'}$ is the potential energy of interaction between molecules on sites s and s'.

Before proceeding further it is worth emphasizing that the label u stands for any excited state including vibrational levels of the ground electronic state, as well as the vibronic sublevels of the excited electronic states. Rotational and librational motions are excluded since the molecules are assumed oriented at the lattice sites (rigid lattice model).

The eigenfunctions of the aggregate are linear combinations of the following one, two, . . ., localized exciton states:

$$\Psi_G = \prod_s \varphi_s^0 \tag{13}$$

$$\phi_s^u = \varphi_s^u \Psi_G (\varphi_s^0)^{-1} \tag{14}$$

$$\phi_{ss'}^{uu'} = \varphi_s^u \varphi_{s'}^{u'} \Psi_G (\varphi_s^0 \varphi_{s'}^0)^{-1} \tag{15}$$

$$\phi_{ss's''}^{uu'u''} = \varphi_s^u \varphi_{s'}^{u'} \varphi_{s''}^{u''} \Psi_G (\varphi_s^0 \varphi_{s'}^0 \varphi_{s''}^0)^{-1} \tag{16}$$

The unperturbed ground state function Ψ_G is sometimes referred to as the excitation vacuum state.

In practically all the aggregates of interest (mainly molecular crystals and biopolymers) the gap between the ground and excited electronic states is large ($\simeq$3 eV for singlets) and so *in perturbation and finite basis theories it is assumed that Ψ_G does not mix with any excited electronic state.* Thus Ψ_G becomes the ground-state eigenfunction with energy

$$E_G = \langle \Psi_G | \mathcal{H} | \Psi_G \rangle \tag{17}$$

If electron exchange between sites is ignored (tight binding approximation) E_G is simply a sum of monomer ground energies and two-body interactions

$$E_G = \sum_s \epsilon_s^{\,0} + \tfrac{1}{2} \sum_s \sum_{s'}{}' (\varphi_s\varphi_{s'} | V_{ss'} | \varphi_s\varphi_{s'}) \tag{18}$$

The second term represents the van der Waals binding energy due to the interaction of static dipole, quadrupole, octopole, etc., moments. Electron exchange provides the balancing repulsive interactions and may be included if Ψ_G is antisymmetrized and normalized by procedures like the one described by Löwdin.[90]

The next approximation consists of ignoring the states containing two or more excitons so that, with the previous exclusion of Ψ_G, the excited states are assumed to have the form

$$\Psi = \sum_s \sum_u a_{su} \phi_s^{\,u} \tag{19}$$

where the coefficients are found by solving the secular equations

$$\sum_{s'} \sum_{u'} [\langle \phi_s^{\,u} | \mathcal{H} | \phi_{s'}^{\,u'} \rangle - E\delta_{ss'}\delta_{uu'}] a_{s'u'} = 0 \tag{20}$$

Thus in finite basis theories the exciton energies are obtained by diagonalizing

$$\det | \langle \phi_s^{\,u} | \mathcal{H} | \phi_{s'}^{\,u'} \rangle - E\delta_{ss'}\delta_{uu'} | = 0 \tag{21}$$

The simplest optical properties of aggregates are related to the total dipole moment

$$\mathbf{D_q} = \sum_s \boldsymbol{\mu}_s \exp(-i\mathbf{q}\cdot\mathbf{r}_s) \tag{22}$$

where $\boldsymbol{\mu}_s$ is the electric dipole operator of molecule s. The modulation arises from a photon or some other external field, responsible for the transition $\Psi_G \rightarrow \Psi$, and the matrix element is

$$\mathbf{M_q} = \langle \Psi_G | \mathbf{D_q} | \Psi \rangle \tag{23}$$

In detail

$$\mathbf{M}_{\mathbf{q}} = \sum_{s} \sum_{u} \boldsymbol{\mu}_s{}^u a_{su} \exp(-i\mathbf{q}\cdot\mathbf{r}_s) \tag{24}$$

where

$$\boldsymbol{\mu}_s{}^u = (\varphi_s | \boldsymbol{\mu}_s | \varphi_s{}^u) \tag{25}$$

is the matrix element for the electric dipole transition within the isolated molecule.

The major approximations in setting up the secular equations are as follows:

1. Neglect of Ψ_G.
2. Neglect of multiple excitations.
3. Neglect of overlap (tight binding).

Of these three the last is often removed by antisymmetrizing the wave functions—an essential step in the calculation of triplet exciton interactions. The implications of the first two approximations seem not to have been fully explored. It would be interesting to examine these approximations from the viewpoint of the infinite Brillouin-Wigner perturbation series and the linked cluster expansion.[91]

1. Excitons in Pure Crystals

For crystals with more than one molecule per unit cell the site index $s = n\alpha$, where n labels the unit cell and α the sites within a cell. Translational symmetry dictates the form of the coefficients a_{su} in infinite periodic crystals containing N unit cells per periodic volume

$$a_{su} = N^{-1/2} \exp(i\mathbf{k}\cdot\mathbf{r}_{n\alpha}) a_{\alpha u}(\mathbf{k}) \tag{26}$$

The exciton wave vectors $\mathbf{k}$ lie in the first Brillouin zone. The secular equations are diagonal in $\mathbf{k}$ and given by

$$\sum_{\beta} \sum_{v} [\langle \Phi_\alpha{}^u(\mathbf{k}) | \mathcal{H} | \Phi_\beta{}^v(\mathbf{k}) \rangle - E\delta_{uv}\delta_{\alpha\beta}] a_{\beta v}(\mathbf{k}) = 0 \tag{27}$$

where

$$\Phi_\alpha{}^u(\mathbf{k}) = N^{-1/2} \sum_{n} \exp(i\mathbf{k}\cdot\mathbf{r}_{n\alpha}) | \phi_{n\alpha}{}^u \rangle \tag{28}$$

are Blochlike sums of single-site excitation functions for molecules with orientation α. For crystals with a center of symmetry

$$\langle \Phi_\alpha{}^u(\mathbf{k}) | \mathcal{H} | \Phi_\beta{}^v(\mathbf{k}) \rangle = [E_G + (\epsilon^u - \epsilon^0)]\delta_{uv}\delta_{\alpha\beta} + D_\alpha{}^{uv}\delta_{\alpha\beta} + I_{\alpha\beta}{}^{uv}(\mathbf{k}) \tag{29}$$

where

$$D_\alpha{}^{uv} = \sum_{m\beta}{}' [(\varphi_{n\alpha}{}^u\varphi_{m\beta}|V_{n\alpha,m\beta}|\varphi_{n\alpha}{}^v\varphi_{m\beta}) - \delta_{uv}(\varphi_{n\alpha}\varphi_{m\beta}|V_{n\alpha,m\beta}|\varphi_{n\alpha}\varphi_{m\beta})] \tag{30}$$

is the gas to crystal site shift matrix[92,93] and

$$I_{\alpha\beta}{}^{uv}(\mathbf{k}) = \tfrac{1}{2}\sum_{m}{}' \cos[\mathbf{k}\cdot(\mathbf{r}_{m\beta} - \mathbf{r}_{n\alpha})] [(\varphi_{n\alpha}{}^u\varphi_{m\beta}|V_{n\alpha,m\beta}|\varphi_{n\alpha}\varphi_{m\beta}{}^v) + (\varphi_{m\alpha}{}^u\varphi_{n\beta}|V_{m\alpha,n\beta}|\varphi_{m\alpha}\varphi_{n\beta}{}^v)] \tag{31}$$

is the matrix of exciton transfer interactions.[92] The primed sum means that the origin is omitted.

For general points **k** of the Brillouin zone the secular equations cannot be simplified any further. For certain points and directions of high symmetry, however, it is possible to reduce the order of the secular matrix. Several excellent accounts of the representations of space groups and applications to exciton theory are avilable. A particularly useful tool is the concept of interchange symmetry discussed lucidly in Kopelman's papers.[94] Maximum reduction occurs if the $\mathbf{k} = 0$ wave vector corresponds to a uniquely defined state as it does for weak and very weak transitions (recall that the exciton energies are analytic at $\mathbf{k} = 0$ in this case). The secular determinant for $\mathbf{k} = 0$ becomes

$$\det|\langle\Psi^u(\mathbf{k}\zeta)|\mathcal{H}|\Psi^v(\mathbf{k}\zeta)\rangle - E\delta_{vv'}| = 0 \tag{32}$$

where

$$\Psi^u(\mathbf{k}\zeta) = (h)^{-1/2}\sum_\alpha c_\alpha(\zeta)|\Phi_\alpha{}^u(\mathbf{k})\rangle \tag{33}$$

The wave vector **k** has been retained because these equations also hold (at least approximately) in the neighborhood of $\mathbf{k} = 0$. The coefficients $c_\alpha(\zeta)$ are found using the character table of the interchange group. For naphthalene there are two molecules per unit cell $h = 2$, so $\alpha = 1,2$ are the site labels. Let the $\alpha = 2$ molecules be generated by the screw operation; then the (proper) interchange group is C_2, the symmetry label is $\zeta = A,B$, and

$$\begin{aligned} c_\alpha(A) &= 1 \\ c_\alpha(B) &= (-1)^{1+\alpha} \end{aligned} \tag{34}$$

For transitions of intermediate and high intensity the $\mathbf{k} = 0$ wave functions described above cannot be used. However, nature is often kind, and experiments usually involve a direction of symmetry, in which case the group of the wave vector **k** is nontrivial and may be used to reduce the order of the secular matrix. For an interesting account of such a case (crystalline pyrene) see Ref. 23.

In the perturbation theory the interactions between exciton bands are presumed to be smaller than the energy separations. For definiteness the case of crystalline naphthalene will be considered. The exciton energies, relative to the ground energy E_G, and correct to the first order are ($\mathbf{k} = 0$ region, weak transitions)

$$E^u(\mathbf{k},\pm) = [(\epsilon^u - \epsilon^0) + D^u] + I_{11}{}^{uu}(\mathbf{k}) \pm I_{12}{}^{uu}(\mathbf{k}) \tag{35}$$

The factor group or Davydov splitting is $2I_{12}{}^{uu}(\mathbf{k})$, to first order.

The first singlet transition of naphthalene is weak but not so the higher singlets. Therefore the $\mathbf{k} = 0$ wave functions cannot be used for calculating the matrix element for an arbitrary direction in the Brillouin zone. However, for $\mathbf{k}$ close to the origin and lying in the *ac* crystal plane or pointing along the *b* crystal axis the states are given by $\mathbf{k} = 0$ formulas.[95] The second correction to the exciton energy is

$$\Delta E^u(\mathbf{k},\pm) = \sum_{v(\neq u)} \frac{[D^{uv} + I_{11}{}^{uv}(\mathbf{k}) \pm I_{12}{}^{uv}(\mathbf{k})]^2}{E^u(\mathbf{k},\pm) - E^v(\mathbf{k},\pm)} \tag{36}$$

and the state vector correct to the first order is

$$\Psi = \Psi^u(\mathbf{k},\pm) + \sum_{v(\neq u)} \frac{[D^{uv} + I_{11}{}^{uv}(\mathbf{k}) \pm I_{12}{}^{uv}(\mathbf{k})]}{E^u(\mathbf{k},\pm) - E^v(\mathbf{k},\pm)} \Psi^v(\mathbf{k},\pm) \tag{37}$$

Only a few calculations retain the nondiagonal site shift $(1 - \delta_{uv})D^{uv}$; for molecules with at least D_{2h} symmetry there are no static dipole moments in any state and in the dipole approximation the diagonal site shifts are zero. However, this is hardly sufficient reason to ignore them since gas to crystal shifts are around -500 cm^{-1} for the first singlets and higher still for the second singlet transitions of aromatic hydrocarbons.

The lowest transitions of aromatic hydrocarbon crystals are generally hyperchromic. This behavior contrasts sharply with the typical helical polynucleotide system where the transitions are hypochromic. To calculate the transition probability, which gives a measure of this effect, the crystal wave function must be substituted into (23).

Mention has already been made of the restricted Frenkel approximation introduced by Colson, Kopelman, and Robinson and subsequently put to great use in the analysis of the 0–0 exciton bands of benzene and naphthalene. The interactions for the $^1B_{2u}$ exciton band of benzene and naphthalene are of very short range, so short in fact that only nearest-neighbor interactions seem to be important in the sums of translationally equivalent interactions. Thus the sums $I_{\alpha\alpha}{}^{uu}(\mathbf{k})$ are effectively the same for all α, and if the exciton band is assumed to be isolated the wave functions are given by (33) for all **k**. A simple calculation reveals that the probability of an optical (**q** = 0) transition from any level of a ground vibrational exciton band is independent of **k**. Consequently hot transitions from narrow vibrational exciton bands have an intensity proportional to the density of states in the upper vibronic band. The density of states found in this way checks well against those obtained by less direct measurements such as variation of the energy denominator (mixed crystal absorption spectra) or from resonance pair spectra.[96–98]

2. *Dilute Mixed Crystals with Isotopic Guests*

In mixed crystals a proportion of the host (or solvent) molecules are substitutionally replaced by guest (solute) molecules. For chemically different molecules mixed crystals exist over only very limited concentration ranges; however, for guests that are simply isotopically labeled hosts the full range of mixed crystals can be prepared. Mixed crystals with chemical impurities were the first to be studied, originally to determine the symmetry of excited guest states without interference from the exciton resonance interactions and intensity transfers found in pure crystals. For weak transitions isotopic impurities have long since eclipsed chemical impurities in importance. One of the difficulties with chemical guests is the extent to which they disturb the host lattice; on geometric grounds alone severe distortion is expected around a guest like tetracene in a naphthalene host crystal. There are no such problems with isotopic and geometrically similar impurities; in fact, there is experimental evidence, from ESR measurements on the first triplet state, of complete alignment of guests with hosts.[100]

In the aromatic hydrocarbons replacement of C^{12} with C^{13} generally results in minute frequency shifts detectable in the 0–0 vibronic exciton bands of benzene.[99] Much greater frequency shifts are obtained by replacing hydrogen with deuterium. For the first singlet transition the maximum difference in guest and host excitation energies is around 200 cm^{-1} (benzene). Other aromatic hydrocarbons have frequency shifts that drop rapidly with increasing size of the molecule: naphthalene[101,102] (120 cm^{-1}), phenanthrene[103] (60 cm^{-1}), and anthracene[104,105] (60 cm^{-1}).

The theory of isotopic impurities in molecular crystals was first developed by Rashba[106–108] with subsequent contributions from many other workers who developed alternative viewpoints and performed practical calculations.[109–113] In all these works it is assumed that the guest–host interaction is identical with the host–host interaction and that the difference in guests and hosts is solely a difference in excitation energies. This approximation is equivalent to assuming that the Franck-Condon factors are the same for guest and host and is probably a very good one for the 0–0 transition.

The Hamiltonian of a mixed crystal, with guests on certain sites designated p, is

$$\mathcal{H} = \mathcal{H}_0 + \sum_p (H_p' - H_p) \tag{38}$$

where $\mathcal{H}_0$ is the pure (or neat) crystal Hamiltonian and H_p' the Hamiltonian of the guest on site p. The subtraction of the displaced host Hamiltonian is symbolic. In the localized one-exciton basis the representatives of $\mathcal{H}$ may be divided into a pure crystal part and a "perturbation" due to the guests. We may define the perturbation matrix by

$$\langle \phi_s{}^u | \sum_p (H_p' - H_p) | \phi_{s'}{}^v \rangle = \delta_{ss'}\delta_{uv}[(\epsilon_g{}^0 - \epsilon^0) \sum_p (1 - \delta_{sp}) + (\epsilon_g{}^u - \epsilon^u) \sum_p \delta_{sp}] \tag{39}$$

where $\epsilon_g{}^u$ are the energies of an isolated guest molecule. Actually this matrix contains a ground-state term that should be added to E_G, the ground energy of the pure crystal. The perturbation matrix is then redefined by

$$\langle \phi_s{}^u | \sum_p (H_p' - H_p) | \phi_{s'}{}^v \rangle = \delta_{ss'}\delta_{uv}[(\epsilon_g{}^u - \epsilon_g{}^0) - (\epsilon^u - \epsilon^0)] \sum_p \delta_{sp} \tag{40}$$

where

$$\delta = (\epsilon_g{}^u - \epsilon_g{}^0) - (\epsilon^u - \epsilon^0) \tag{41}$$

is called the trap-depth.

If it is assumed that the host exciton band of interest is an isolated one, that is, not coupled to any other, then the secular equations are readily solved for the case of noninteracting guests (infinite dilution). For trap-depths δ large enough to split a level away from the host band, the secular determinant can be reduced to the following equation

$$1 + [\delta/(hN)] \sum_{\zeta k} [E(\mathbf{k}\zeta) - E]^{-1} = 0 \tag{42}$$

The amplitude for dipole transitions from the ground state is

$$\mathbf{M} = \sum_\zeta \frac{\mathbf{M}(\zeta)}{E - E(\mathbf{q}\zeta)} \left\{ \sum_{k\zeta} [E(\mathbf{k}\zeta) - E]^{-2} \right\}^{-1} \tag{43}$$

where $\mathbf{M}(\zeta)$ is the amplitude for the optical transition with symmetry ζ in the pure crystal and $\mathbf{q}$ is the photon wave vector. From this last equation a formula describing the Rashba effect can be obtained. For light polarized parallel and perpendicular to the b crystal axis and propagating along the normal of the (001) face, the polarization ratio of the guest transition with energy E is

$$\mathcal{P}'(E) = \left(\frac{E(\mathbf{q},a) - E}{E(\mathbf{q},b) - E}\right)^2 \mathcal{P}\left(\frac{b}{a}\right) \tag{44}$$

where $\mathcal{P}(b/a)$ is the polarization ratio of the optical transitions in the pure crystal.

By decreasing the amount of isotopic labeling the trap-depth δ can be reduced and the energy of the guest level be made to approach the host band. The intensity of this level will then either "flare up" if a host optical level lies at the closest band edge, or fade away if the band edge is not an optically accessible state. In the 3200 Å naphthalene absorption band system the *ac*-polarized transition lies at the bottom and the *b*-polarized transition lies near the top of the 0–0 exciton band. Thus the polarization ratios for traps below the band fall to zero as the band edge is approached, whereas the ratios for traps above the band soar to "infinity." This phenomenon is called the Rashba effect.

In three-dimensional crystals there is a critical trap-depth that must be exceeded before a guest state can separate from the host band. For trap-depths less than critical the impurity may still localize (but not trap) a host exciton. The secular equations can be solved for this case too, in which energy E lies inside the host exciton band. There now occurs scattering of the excitons of the host band instead of the formation of a bound state. The theory of this process is outlined later in Section II.C where calculations based on three model density of states functions are described.

There is one simplification of the secular determinant of great practical and theoretical importance. Let $\rho(\epsilon)$ be the normalized density function of host exciton states (all branches); then

$$1 + \delta \int \frac{\rho(\epsilon)\, d\epsilon}{\epsilon - E} = 0 \tag{45}$$

where the integral runs over the extent of the host exciton band. This relation holds only for guest states with excitation energy E lying outside the exciton band. If we expand the denominator, (45) can be expressed in terms of the moments S_l of the exciton distribution

$$1 - \frac{\delta}{E} \sum_{l=0}^{\infty} \frac{S_l}{E^l} = 0 \tag{46}$$

where

$$S_l = \int d\epsilon\, \epsilon^l \rho(\epsilon) \tag{47}$$

Given a series of experimentally determined guest level energies E and the corresponding trap-depths δ, these last equations can be solved for the first few moments. These moments can be related to intermolecular interactions by expanding the denominators of (42) directly, without going through the density function.

3. Dilute Mixed Crystals with Chemical Guests

At one time mixed crystals containing chemical guests were a popular way to observe the absorption spectrum of the guest without too much interference from exciton resonance effects. However, since the discovery of the Shpolskii effect, aromatic hydrocarbons have been largely superseded by *n*-alkanes as hosts. For the larger aromatic hydrocarbons, like anthracene, the maximum trap-depth available from isotopic labeling is around 60 cm^{-1}. Most of the larger hydrocarbons have first singlet transitions in the intermediate-intensity range (e.g., anthracene $f = 0.3$, perylene $f = 1.1$) and the exciton bands are wider than those observed for the benzene and naphthalene first singlets. Therefore 60 cm^{-1} is probably too close to the critical value for isotopic guests to be useful probes of exciton band structure for the larger hydrocarbons. This leaves chemical guests as the only probes available.

A theory for the exciton states of a crystal containing chemical guests can be obtained for the dilute concentration limit provided certain interactions, called secondary traps, are neglected.[114–122] The role of secondary traps is to provide additional perturbations that may result in more discrete states being split from the exciton band of host crystal.[117] For simplicity it is assumed that the host crystals have one molecule per unit and that all molecules, guest and host, have only one excited electronic state. Both these restrictions can be relaxed, but only at the expense of increased notational complexity.[117]

The mixed crystal Hamiltonian for one guest at site p is

$$\mathcal{H} = \sum_{s(\neq p)} H_p + \tfrac{1}{2} \sum_{\substack{s \neq s' \\ (\neq p)}} \sum V_{ss'} + H_p' + \sum_{s(\neq p)} \mathcal{V}_{ps} \tag{48}$$

where $\mathcal{V}_{ps}$ is the Coulombic interaction between the guest p and the host molecule s. The ground state energy of the mixed crystal is

$$E_G' = E_G + (\epsilon_g{}^0 - \epsilon^0) + \sum_{s(\neq p)} [(\psi_p \varphi_s | \mathcal{V}_{ps} | \psi_p \varphi_s) - (\varphi_p \varphi_s | V_{ps} | \varphi_p \varphi_s)] \tag{49}$$

where the ground state of the isolated guest has been denoted ψ_p for emphasis. In the localized basis of single-exciton functions [see (14)], the representatives of the mixed crystal Hamiltonian relative to E_G' are

$$\mathcal{H}_{ss'} = \mathcal{H}_{0,ss'} + \delta_{ss'}[\delta_{ps}\epsilon' + (1 - \delta_{ps})\, d_{ps}'] + (1 - \delta_{ss'})(\delta_{ps}i'_{ps'} + \delta_{ps'}i_{ps}') \qquad (50)$$

where

$$\epsilon' = [(\epsilon_g{}^u - \epsilon_g{}^0) + D_g{}^u] - [(\epsilon^u - \epsilon^0) + D^u] \qquad (51)$$

$$i_{ps}' = (\psi_p{}^u\varphi_s|\mathcal{V}_{ps}|\psi_p\varphi_s{}^u) - (\varphi_p{}^u\varphi_s|V_{ps}|\varphi_p\varphi_s{}^u) \qquad (52)$$

$$d_{ps}' = [(\psi_p{}^u\varphi_s|\mathcal{V}_{ps}|\psi_p{}^u\varphi_s) - (\psi_p\varphi_s|\mathcal{V}_{ps}|\psi_p\varphi_s)] - [(\varphi_p\varphi_s{}^u|V_{ps}|\varphi_p\varphi_s{}^u) - (\varphi_p\varphi_s|V_{ps}|\varphi_p\varphi_s)] \qquad (53)$$

The first term is just the pure crystal matrix element, ϵ' a site shifted primary trap-depth, i_{ps}' the difference in guest–host and host–host exciton interactions, and d_{ps}' the secondary trap interactions.[117] In the dipole approximation the secondary traps are zero for hydrocarbons of symmetry D_{2h} or higher. They are, therefore, short-range in character and likely to be small even for nearest neighbors. Consequently, since a minimum interaction of approximately one quarter of the exciton band width is necessary before a discrete state splits from the band, it is unlikely that secondary traps will cause the formation of new bound states. Inside the host band the secondary traps promote the localization of exciton amplitude in the vicinity of the guest, which may partly explain why chemical guests are such efficient quenchers of host fluorescence. A host exciton once localized in the proximity of the guest site could degrade easily by phonon emission into increasingly tighter states and then drop into a primary traplevel. Further discussion of secondary traps may be found in the papers by Philpott and Craig[117] and Osad'ko.[120] The latter author introduces a separable interaction to simulate the secondary interactions and has investigated the polarization properties of the secondary trap state.

The energy of the primary-trap state (due to ϵ') is easily found once the secondary traps are neglected. The secular determinant reduces to the following form[122]

$$E = \epsilon_g{}^u + G_2' - \frac{(G_1')^2}{G_0} \qquad (54)$$

where

$$G_n' = N^{-1} \sum_k [I_g(\mathbf{k})]^n\, [E(\mathbf{k}) - E]^{-1} \qquad (55)$$

and

$$I_g(\mathbf{k}) = \sum_n{}' (\psi_p{}^u\varphi_n|\mathcal{V}_{pn}|\psi_p\varphi_n{}^u) \exp[i\mathbf{k}\cdot(\mathbf{r}_p - \mathbf{r}_n)] \qquad (56)$$

For a guest with a stack of vibronic levels that are separable according to the crude Born-Oppenheimer approximation[123,124] the secular determinant is

$$1 + \left[G_2' - \frac{(G_1')^2}{G_0} \right] \sum_u \frac{\xi_u^2}{\epsilon_g^u - E} = 0 \tag{57}$$

where ξ_u^2 is the Franck-Condon factor[125] for the 0–u vibronic transition. The dipole transition amplitude is

$$\mathbf{M}_\mathbf{q}(E) = \left[\boldsymbol{\mu}_g^u + \boldsymbol{\mu}^u \frac{I_g(\mathbf{q}) - (G_1'/G_0)}{E - E(\mathbf{q})} \right] a_p \tag{58}$$

where a_p is the guest site coefficient determined by normalizing the wave function to unity. The transition dipole of the hosts is $\boldsymbol{\mu}^u$. Observe that in the deep-trap limit the term G_1'/G_0 is very small and (59) becomes identical with the expressions obtained from perturbation theory by Craig and Thirunamachandran.[126]

The relations derived above can be simplified in a number of situations. One way, due to Merrifield,[119] assumes that the guest–host and corresponding host–host interactions are proportional. In the isotopic impurity limit the interactions are identical and the previously described isotopic equations are recovered.

4. Crystals Heavily Doped with Isotopic Guests

One of the most interesting developments in recent years has been the application of the theory of excitations in random alloys[127–135] to heavily doped isotopic mixed crystals.[136–139] The magnitude of the problem can be seen by examining the secular determinant for crystals with one molecule per unit cell and guest molecules occupying sites $p, p', \ldots$.

$$\det \left| \delta_{pp'} + \frac{\delta}{N} \sum_\mathbf{k} \frac{e^{i\mathbf{k}\cdot(\mathbf{r}_p - \mathbf{r}_{p'})}}{E(\mathbf{k}) - E} \right| = 0 \tag{59}$$

In an early paper Lifschitz tackled the problem by a clever expansion of this determinant and was able to draw a number of interesting conclusions concerning the erosion of the host band edges and the simultaneous build-up in impurity levels as the guest concentration increased. The tack of the latest work is different, instead of manipulating the secular determinant, the averaged matrix of the Greens' operator in **k**-space is found. The theory for crystals with one molecule per unit cell contains all the essential physics of the problem. For the treatment of crystals with two molecules per unit cell, a necessary complication since the persistence of the Davydov splitting in mixed crystals is observed, the reader is referred to the review by Robinson[26] and the papers by Hong and Kopelman[137] and Hoshen and Jortner.[138,139]

In the minds of most theoretical chemists the secular determinant occupies a position of hallowed reverence. It is pertinent, therefore, to show why the Greens' operator is the more useful quantity to calculate for heavily doped crystals. Let the exact eigenstates of the mixed crystal be denoted $|i\rangle$ then

$$\mathcal{H}|i\rangle = \mathcal{E}_i|i\rangle \tag{60}$$

is the Schrödinger equation for the mixed crystal. The probability amplitude for a one-photon transition to state $|i\rangle$ is proportional to

$$\mathbf{M}_{\mathbf{q}} = \langle 0|\sum_n \boldsymbol{\mu}_n \exp(-i\mathbf{q}\cdot\mathbf{r}_n)|i\rangle \tag{61}$$

where $|0\rangle$ is the ground state. If the exact eigenstates $|i\rangle$ are expanded in the "delocalized" crystal basis $|\mathbf{k}\rangle$[140]

$$|\mathbf{k}\rangle = N^{-1/2}\sum_n \exp(i\mathbf{k}\cdot\mathbf{r}_n)|\phi_n{}^u\rangle \tag{62}$$

then

$$\mathbf{M}_{\mathbf{q}} = N^{1/2}\boldsymbol{\mu}^u\langle\mathbf{q}|i\rangle \tag{63}$$

and so a normalized absorption profile may be defined by

$$\mathcal{F}_{\mathbf{q}}(E) = \sum_i |\langle\mathbf{q}|i\rangle|^2\delta(E - \mathcal{E}_i) \tag{64}$$

Now the spectrally resolved Greens' operator $(E - \mathcal{H})^{-1}$ is

$$\mathcal{G}(E) = \sum_i \frac{|i\rangle\ \langle i|}{E - \mathcal{E}_i} \tag{65}$$

and the imaginary part of the **q**th matrix element is given by

$$Im\langle\mathbf{q}|\mathcal{G}(E^-)|\mathbf{q}\rangle = \pi\sum_i |\langle\mathbf{q}|i\rangle|^2\delta(E - \mathcal{E}_i) \tag{66}$$

where E^- has an infinitesimal negative imaginary part. Comparison of (64) and (66) shows that

$$\mathcal{F}_{\mathbf{q}}(E) = \frac{1}{\pi}\, Im\langle\mathbf{q}|\mathcal{G}(E^-)|\mathbf{q}\rangle \tag{67}$$

The hard part of the calculation is finding $\langle\mathbf{q}|\mathcal{G}|\mathbf{q}\rangle$ for a random lattice. The matrix elements of $\mathcal{G}^{-1}$ in the $|\mathbf{k}\rangle$ basis are

$$\langle\mathbf{k}|[\mathcal{G}(E)]^{-1}|\mathbf{k}'\rangle = [E - E(\mathbf{k})]\delta_{\mathbf{k}\mathbf{k}'} - \frac{\delta}{N}\rho(\mathbf{k} - \mathbf{k}') \tag{68}$$

where

$$\rho(\mathbf{k} - \mathbf{k}') = \sum_p \exp\,[i(\mathbf{k}' - \mathbf{k})\cdot\mathbf{r}_p] \tag{69}$$

consists of a sum over guest sites only. The operator $\mathcal{G}$ can be expanded using the identity

$$(A - B)^{-1} = A^{-1} + A^{-1}B(A - B)^{-1} \tag{70}$$

and we find

$$\mathcal{G} = \mathcal{G}_0 + \sum_{m=1}^{\infty} \left(\frac{\delta}{N}\right)^m \mathcal{G}_0(\rho\mathcal{G}_0)^m \tag{71}$$

where $\mathcal{G}_0$ is the unperturbed Greens' operator with matrix elements $[E - E(\mathbf{k})]^{-1}\delta_{\mathbf{kk}'}$. Since the actual impurity sites are of no interest random averages are taken in series using replacement

$$\sum_p (\cdots) = C \sum_n (\cdots) \tag{72}$$

where on the right-hand side C is the concentration of guests and the sum runs over all sites n of the crystal. Upon applying this averaging procedure to $\langle \mathbf{q} | \mathcal{G} | \mathbf{q} \rangle$ we see that averages of the type

$$\langle \rho(\mathbf{k}_1)\rho(\mathbf{k}_2)\rho(\mathbf{k}_3)\cdots\rho(\mathbf{k}_m)\rangle_{\text{Av}} \tag{73}$$

occur. Yonezawa and Matsubara express these moment averages in terms of cumulant averages; once this is done the structure of the expansion becomes crystal clear and each term can be represented by a diagram. The solution of the diagrammatic equation has the Dyson form

$$(\langle \mathbf{q} | \mathcal{G}(E) | \mathbf{q} \rangle)_{\text{Av}} = [E - E(\mathbf{q}) - \Sigma(\mathbf{q},E)]^{-1} \tag{74}$$

where $\Sigma(\mathbf{q},E)$ is a complex self energy. Returning to the normalized absorption function we have

$$\mathfrak{F}_{\mathbf{q}}(E) = \frac{1}{\pi} \frac{\Sigma_2(\mathbf{q},E)}{[E - E(\mathbf{q}) - \Sigma_1(\mathbf{q},E)]^2 + [\Sigma_2(\mathbf{q},E)]^2} \tag{75}$$

where Σ_1 and Σ_2 are the real and imaginary parts of the self-energy. Approximate formulas for Σ have been derived by Onodera and Toyazawa[130] and by Hong and Robinson[136] for crystals containing one and two molecules per unit cell, respectively. Thus far, for molecular crystals, only $\mathbf{q}$ independent self-energies have been obtained for use in the absorption calculations. This is not expected to result in intensities severely in error since the energy gap is always of the same order as the exciton band width. In fact, considering the difficulties in performing the experiments and the approximations made in calculations, theory and experiment are in very good agreement.

It should be added, by way of qualification, that the exciton energies used to calculate the state densities required by the theory are experimentally determined using the restricted Frenkel and noninteracting band approximations.

B. Classical Dipole Theory

One can understand a great deal about the spectra of aggregates using the dipole approximation for the exciton interaction between molecules. The predictions of the dipole theory are only semi-quantitative for strong transitions and fail completely for the weak ones. Due to their non-analyticity dipole interactions can be used to probe the $\mathbf{k} = 0$ region of the exciton bands of medium- and high-intensity transitions. Another unique feature of dipole theory is the simple form of the intermolecular interaction, namely, a second rank tensor that is the same for all molecular polarizations and orientations. Because the interaction is simple a number of quite complex problems have been solved exactly. This means that a dipole theory can serve as a useful guide when attempting to develop more general theories that include the nondipolar part of the exciton interaction.

Before proceeding with particular problems it is necessary to briefly derive the classical equations of exciton theory for the dipole approximation. The starting point is the equation of motion of an oscillating electric dipole, and Maxwell's equations which describe the field causing the oscillation. This approach has an advantage in revealing the origin and nature of retarded interactions, a subject that has received much attention in molecular exciton theory.

To each molecular transition from the ground state an oscillating electric dipole $\mathbf{d}_{su}(t)$ is assigned. The molecular site and orientation are specified by s and the excited state by u, just as was done in Section II.A; t is the time variable. With the neglect of magnetic field effects the equation of motion of the classical dipole is

$$\left(\frac{\partial^2}{\partial t^2} + \omega_{su}{}^2\right) \mathbf{d}_{su}(t) = \frac{e^2 f_{su}}{m} \hat{\mathbf{d}}_{su}\hat{\mathbf{d}}_{su} \cdot \mathbf{E}'(\mathbf{r}_s,t) \tag{76}$$

where ω_{su} and f_{su} are the frequency and oscillator strength of the quantum transition $0 \rightarrow u$ in molecule s, and $\hat{\mathbf{d}}_{su}$ is a unit vector pointing along the direction of this transition. The exciting electric field $\mathbf{E}'$ consists of a free electromagnetic field term $\mathbf{E}_0$ and the combined fields of all the oscillating dipoles at the sites $s'(\neq s)$:

$$\mathbf{E}'(\mathbf{r}_s,t) = \mathbf{E}_0(\mathbf{r}_s,t) + \mathbf{E}_a{}'(\mathbf{r}_s,t) \tag{77}$$

To solve the equation of motion (76) the free field $\mathbf{E}_0$ must first be specified and then the aggregate field $\mathbf{E}_a$ calculated. The behavior of electromagnetic field intensities $\mathbf{E}(\mathbf{x},t)$ and $\mathbf{B}(\mathbf{x},t)$ for general points $\mathbf{x}$ is governed by Maxwell's equations[141]:

$$\nabla \times \mathbf{B} - \frac{1}{c}\frac{\partial}{\partial t}\mathbf{E} = \frac{4\pi}{c}\tilde{\mathbf{j}} \tag{78}$$

$$\nabla \times \mathbf{E} + \frac{1}{c}\frac{\partial}{\partial t}\mathbf{B} = 0 \tag{79}$$

$$\nabla \cdot \mathbf{E} = 4\pi\tilde{\rho} \tag{80}$$

$$\nabla \cdot \mathbf{B} = 0 \tag{81}$$

The sources of the field, $\tilde{\mathbf{j}}$ and $\tilde{\rho}$, are given by

$$\tilde{\mathbf{j}} = \frac{\partial}{\partial t}\mathbf{P} + c\nabla \times \mathbf{M} \tag{82}$$

and

$$\tilde{\rho} = -\nabla \cdot \mathbf{P} \tag{83}$$

where **P** and **M** are the electric and magnetic polarization vectors of the medium. In terms of oscillating electric and magnetic dipoles $\mathbf{d}_{su}(t)$ and $\mathbf{m}_{su}(t)$ on each molecule

$$\mathbf{P}(\mathbf{x},t) = \sum_{su} \mathbf{d}_{su}(t)\delta(\mathbf{x} - \mathbf{r}_s) \tag{84}$$

$$\mathbf{M}(\mathbf{x},t) = \sum_{su} \mathbf{m}_{su}(t)\delta(\mathbf{x} - \mathbf{r}_s) \tag{85}$$

The electric displacement **D** and the magnetic induction **B** can be eliminated from Maxwell's equations using the "additive relations":

$$\mathbf{D} = \mathbf{E} + 4\pi\mathbf{P} \tag{86}$$

$$\mathbf{B} = \mathbf{H} + 4\pi\mathbf{M} \tag{87}$$

It is not always convenient to work with the field intensities $\mathbf{E}(\mathbf{x},t)$ and $\mathbf{H}(\mathbf{x},t)$, and so vector and scalar potential $\tilde{\mathbf{A}}$ and $\tilde{\phi}$ are introduced by

$$\mathbf{E} = -\frac{1}{c}\frac{\partial}{\partial t}\tilde{\mathbf{A}} - \nabla\tilde{\phi} \tag{88}$$

$$\mathbf{B} = \nabla \times \tilde{\mathbf{A}} \tag{89}$$

The potentials are not uniquely defined by these last two equations and may be subjected to a number of restrictions. This is called choosing the gauge. To facilitate easy comparison with the second quantized theory we choose to work mostly in Coulomb gauge.

In Coulomb gauge

$$\nabla \cdot \tilde{\mathbf{A}}(\mathbf{x},t) = 0 \tag{90}$$

and Maxwell's equations become

$$\left(\nabla^2 - \frac{1}{c^2}\frac{\partial^2}{\partial t^2}\right)\tilde{\mathbf{A}} = -\frac{4\pi}{c}\tilde{\mathbf{j}} + \frac{1}{c}\frac{\partial}{\partial t}\nabla\tilde{\phi} \tag{91}$$

$$\nabla^2\tilde{\phi} = -4\pi\tilde{\rho} \tag{92}$$

From (90) the vector potential $\tilde{\mathbf{A}}$ is transverse; therefore the longitudinal part of the right-hand side of (91) is zero:

$$4\pi\tilde{\mathbf{j}}^{\parallel}(\mathbf{x},t) + \frac{\partial}{\partial t}\nabla\tilde{\phi}(\mathbf{x},t) = 0 \tag{93}$$

Thus in Coulomb gauge $\tilde{\mathbf{A}}$ and $\tilde{\phi}$ are uncoupled and satisfy

$$\left(\nabla^2 - \frac{1}{c^2}\frac{\partial^2}{\partial t^2}\right)\tilde{\mathbf{A}}(\mathbf{x},t) = -\frac{4\pi}{c}\tilde{\mathbf{j}}^{\perp}(\mathbf{x},t) \tag{94}$$

$$\nabla^2\tilde{\phi}(\mathbf{x},t) = -4\pi\tilde{\rho}(\mathbf{x},t) \tag{95}$$

Now $\tilde{\phi}$ is determined, because of (83), entirely by electric dipoles, so that

$$\mathbf{E}^{\parallel}(\mathbf{x},t) = -\nabla\tilde{\phi}(\mathbf{x},t) \tag{96}$$

is the longitudinal electric field at $\mathbf{x},t$ due to dipoles $\mathbf{d}_{su}(t)$. The time dependence of $\mathbf{E}^{\parallel}$ is precisely the same as $\mathbf{d}_{su}(t)$. There is no retardation of the signal so that

$$\mathbf{E}^{\parallel}(\mathbf{x},t) = -\sum_{su} R^{-3}(\mathbf{1} - 3\hat{\mathbf{R}}\hat{\mathbf{R}}) \cdot \mathbf{d}_{su}(t) \tag{97}$$

where $\mathbf{R} = \mathbf{x} - \mathbf{r}_s$.

At $\mathbf{x} = \mathbf{r}_s$ the field due to the dipoles $\mathbf{d}_{su}(t)$ must be omitted so that

$$\mathbf{E}'^{\parallel}(\mathbf{r}_s,t) = -\sum_{s'u}(1 - \delta_{ss'})R^{-3}(\mathbf{1} - 3\hat{\mathbf{R}}\hat{\mathbf{R}}) \cdot \mathbf{d}_{s'u}(t) \tag{98}$$

where $\mathbf{R} = \mathbf{r}_s - \mathbf{r}_{s'}$. The transverse electric field

$$\mathbf{E}^{\perp}(\mathbf{x},t) = -\frac{1}{c}\frac{\partial}{\partial t}\mathbf{A}(\mathbf{x},t) \tag{99}$$

is the sum of any free (or driving) transverse field and a retarded transverse field due to the oscillating dipoles.

For monochromatic driving fields with frequency ω all dynamical quantities have the time dependence $\exp(-i\omega t)$. This is a convenient point at which to drop the polarization **M** since magnetic effects are of no interest here. After carrying out the time differentiations and canceling through by the factor $\exp(-i\omega t)$ we obtain

$$\sum_{s'u'} \left[(\omega_{su}{}^2 - \omega^2)\delta_{ss'}\delta_{uu'}\mathbf{1} + \frac{e^2 f_{su}}{m} \hat{\mathbf{d}}_{su}\hat{\mathbf{d}}_{su} \cdot R^{-3}(\mathbf{1} - 3\hat{\mathbf{R}}\hat{\mathbf{R}}) \right] \cdot \mathbf{d}_{s'u'} = \left(\frac{e^2 f_{su}}{m}\right)\left(\frac{i\omega}{c}\right) \hat{\mathbf{d}}_{su}\hat{\mathbf{d}}_{su} \cdot \mathbf{A}(\mathbf{r}_s) \quad (100)$$

$$\left[\nabla^2 + \frac{\omega^2}{c^2}\right]\mathbf{A}(\mathbf{x}) = i\,\frac{4\pi\omega}{c}\,\mathbf{P}^{\perp}(\mathbf{x}) \quad (101)$$

where $\mathbf{P}^{\perp}(\mathbf{x})$ is the transverse part[142] of the vector field

$$\mathbf{P}(\mathbf{x}) = \sum_{su} \mathbf{d}_{su}\delta(\mathbf{x} - \mathbf{r}_s) \quad (102)$$

The equations can be simplified further with the help of the following definitions of dipole polarizability (103), point dipole interaction tensor (104), and unit-cell polarization (105):

$$4\pi\boldsymbol{\alpha}_s(\omega) = \sum_u \frac{\omega_0{}^2 f_{su}}{\omega_{su}{}^2 - \omega^2}\,\hat{\mathbf{d}}_{su}\hat{\mathbf{d}}_{su} \quad (103)$$

$$\mathbf{T}_{ss'} = \frac{v_0}{4\pi}(R_{ss'})^{-3}[\mathbf{1} - 3\hat{\mathbf{R}}_{ss'}\hat{\mathbf{R}}_{ss'}](1 - \delta_{ss'}) \quad (104)$$

$$\mathbf{P}_s = \frac{1}{v_0}\sum_u \mathbf{d}_{su} \quad (105)$$

The quantity ω_0 is the "plasma frequency" defined by

$$\omega_0{}^2 = \frac{4\pi e^2}{(m v_0)} \quad (106)$$

and for crystals v_0 is the unit cell volume. Equations (100) and (101) can be rearranged to

$$\sum_{s'} [\delta_{ss'}\mathbf{1} + 4\pi\boldsymbol{\alpha}_s(\omega) \cdot \mathbf{T}_{ss'}] \cdot \mathbf{P}_{s'} = \boldsymbol{\alpha}_s(\omega) \cdot \mathbf{E}^{\perp}(\mathbf{r}_s) \quad (107)$$

$$\left[\nabla^2 + \frac{\omega^2}{c^2}\right]\mathbf{A}(\mathbf{x}) = i\,\frac{4\pi\omega}{c}\,\mathbf{P}^{\perp}(\mathbf{x}) \quad (108)$$

which are valid for any aggregate in a monochromatic driving field.

When the transverse field

$$\mathbf{E}^{\perp}(\mathbf{r}_s) = \frac{i\omega}{c}\mathbf{A}(\mathbf{r}_s) \tag{109}$$

is ignored the equation of motion for the dipoles becomes

$$\sum_{s'} [\delta_{ss'}\mathbf{1} + 4\pi\boldsymbol{\alpha}_s(\omega) \cdot \mathbf{T}_{ss'}] \cdot \mathbf{P}_{s'}^{(0)} = 0 \tag{110}$$

The allowed frequencies of the undriven motion are found by solving

$$\det|\delta_{ss'}\mathbf{1} + 4\pi\boldsymbol{\alpha}_s(\omega) \cdot \mathbf{T}_{ss'}| = 0 \tag{111}$$

and these are the Coulombic exciton frequencies of the aggregate. If the system possesses no symmetry the determinant is $3N \times 3N$ where N is the number of molecules. Note that all the molecular transitions are included in the theory through the polarizability tensors $\boldsymbol{\alpha}_s(\omega)$.

The simplest problems to analyze are those with translational symmetry, in other words, excitons in infinite crystals and polymers, and polaritons in crystals. Next in order of increasing dificulty are infinite crystals with impurities, and then the problem of semi-infinite crystals.

1. *Infinite Pure Crystals with One Molecule per Unit Cell*

In this case all the molecules have the same polarizability and the normal mode solutions of (110) have the form

$$\mathbf{P}_n^{(0)} = \mathbf{P}_0^{(0)} \exp(i\mathbf{k} \cdot \mathbf{r}_n) \tag{112}$$

where $\mathbf{k}$ lies in the first Brillouin zone. The secular determinant factorized because it is diagonal in $\mathbf{k}$, each block having the form

$$\det|\mathbf{1} + 4\pi\boldsymbol{\alpha}(\omega) \cdot \mathbf{T}(\mathbf{k})| = 0 \tag{113}$$

where

$$\mathbf{T}(\mathbf{k}) = \sum_n \mathbf{T}_{nn'} \exp[i\mathbf{k} \cdot (\mathbf{r}_n - \mathbf{r}_{n'})] \tag{114}$$

is a lattice sum of point dipole–dipole interactions.

Specializing further, if all transitions are polarized along the direction $\hat{\mathbf{d}}$ the determinant reduces to

$$1 + 4\pi\alpha(\omega)T(\mathbf{k}) = 0 \tag{115}$$

where

$$\alpha(\omega) = \hat{\mathbf{d}} \cdot \boldsymbol{\alpha}(\omega) \cdot \hat{\mathbf{d}} \tag{116}$$

and

$$T(\mathbf{k}) = \hat{\mathbf{d}} \cdot \mathbf{T}(\mathbf{k}) \cdot \hat{\mathbf{d}} \tag{117}$$

In the case of molecules with an isolated transition u, not coupled to any other excited state, (115) can be solved

$$\omega^2 = \omega_u^{\,2} + \omega_0^{\,2} f_u T(\mathbf{k}) \tag{118}$$

Expressed in terms of energy rather than frequency this result is

$$E^2 = (\epsilon^u)^2 + 2\epsilon^u I(\mathbf{k}) \tag{119}$$

where

$$I(\mathbf{k}) = \frac{4\pi}{v_0} \,|\boldsymbol{\mu}^u|^2 T(\mathbf{k}) \tag{120}$$

Now (118) and (119) have the "oscillator" form, that is, they are quadratic in ω and E, respectively. After taking the square root and expanding we have

$$E = \epsilon^u + I(\mathbf{k}) - \frac{[I(\mathbf{k})]^2}{\epsilon^u} + \cdots$$

This expansion is almost always valid since ϵ^u is greater than 3 eV and $I(\mathbf{k})$ an order of magnitude less than 1 eV for most crystals.[143–145] Hence the series will converge very rapidly for most systems.

As an example of a case where the expansion does not converge consider two transitions polarized along $\hat{\mathbf{d}}$ with frequencies ω_1 and ω_2. The secular determinant can be rearranged to

$$\omega^2 = \omega_1^{\,2} + \omega_0^{\,2} f_1 T(\mathbf{k}) \left[1 + \frac{\omega_0^{\,2} f_2 T(\mathbf{k})}{\omega_2^{\,2} - \omega^2} \right]^{-1} \tag{121}$$

which cannot be solved by expansion of the round bracket and iteration if

$$\left| \frac{\omega_0^{\,2} f_2 T(\mathbf{k})}{\omega_2^{\,2} - \omega_1^{\,2}} \right| > 1 \tag{122}$$

since the series will not converge. Inequality (122) simply states that the interlevel coupling is greater than the unperturbed energy separation.

In the harmonic (crude Bŏrn-Oppenheimer) approximation all the vibrational sublevels of an isolated electronic state have the same polarization. The exciton frequencies are therefore found by solving (115). According to this equation a stack of vibronic levles is collectively coupled; that is, as $|T(\mathbf{k})|$ increases, the lowest ($T < 0$) or highest ($T > 0$) level breaks away from the stack. It is shown at the end of this section that as this collective level moves away from the rest of the stack it steals all the intensity in the system of transitions (a sort of Matthew effect).[146] For very large $T(\mathbf{k})$, (115) predicts one (intense) level far removed from the rest (the latter remain fairly close to their free molecule positions). This can be seen by plotting $-[4\pi\alpha(\omega)]^{-1}$ as a function of ω as shown in Fig. 1. An alternative method is to consider the form of the function

$$L(\omega) = 1 + 4\pi\alpha(\omega)T(\mathbf{k}) \qquad (122')$$

which is entirely determined by its limiting values as ω tends to zero, infinity, and as ω passes through the poles at $\omega = \omega_u$. See Fig. 2. The zeros of $L(\omega)$ correspond to the one-exciton frequencies. Suppose $T(\mathbf{k}) < 0$, as shown in Fig. 2; then as $|T(\mathbf{k})|$ increases all the roots move to lower frequencies. However, the lowest root is unbounded from below and continues to move to lower frequencies as the dipole sum increases.

At this point it must be mentioned that this description of collective coupling for $T(\mathbf{k}) < 0$ is only approximate, being the description afforded by single-particle theory. For large interactions $T(\mathbf{k})$ (strong vibronic coupling limit[147]) the two-particle states play an important role in stealing intensity away from the collective level. It is likely that the two-particle levels completely dominate the spectral properties of the aggregate starting at their threshold, one vibrational quantum in energy above the collective level.[148]

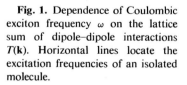

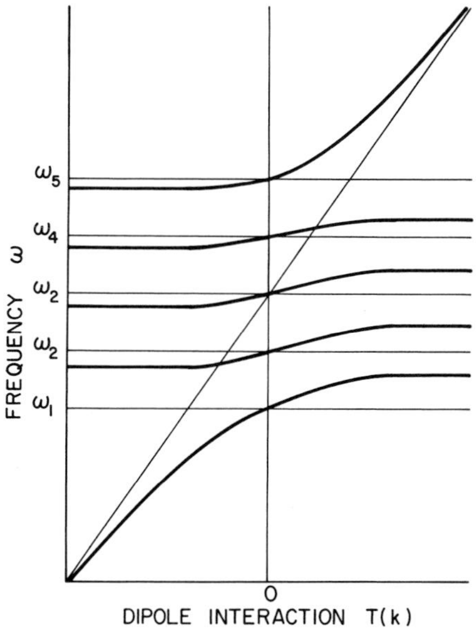

Fig. 1. Dependence of Coulombic exciton frequency ω on the lattice sum of dipole–dipole interactions $T(\mathbf{k})$. Horizontal lines locate the excitation frequencies of an isolated molecule.

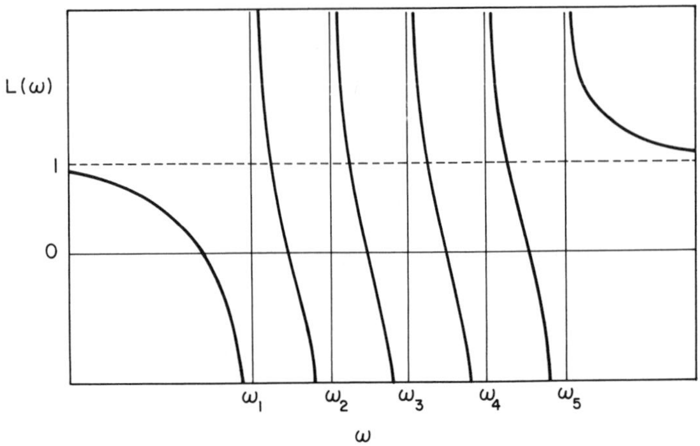

Fig. 2. Schematic representation of the form of the function $L(\omega)$ for $T(\mathbf{k}) < 0$. Vertical asymptotes locate the molecular excition frequencies ω_u.

A number of the cationic dyes aggregate in concentrated solutions to form long polymers. Upon aggregation the absorption spectra change completely. For example the aggregate of the dye psuedoisocyanine displays a very narrow (about 50 cm^{-1}) intense absorption band roughly 1500 cm^{-1} below the onset of absorption in the monomer. This could be an example of a collective vibronic state with a large negative $T(\mathbf{k})$.

The concept of the polariton arose from the work of Pekar,[149] Fano,[150] Hopfield,[151] and Agranovich.[152] In an infinite crystal an exciton with wave vector $\mathbf{k}$ is coupled to photons with wave vectors $\mathbf{k} + \mathbf{K}$, where $\mathbf{K}$ is any reciprocal lattice vector, including the null vector. This coupled excitation is called a polariton. The basic theory of polaritons, as it applies to molecular crystals, is outlined next. For an infinite crystal with one molecule per unit cell we assume that the vector potential has the form

$$\mathbf{A}(\mathbf{x}) = \mathbf{A}_0 \exp(i\mathbf{k} \cdot \mathbf{x}) \tag{123}$$

where $\mathbf{A}_0$ is perpendicular to $\mathbf{k}$. The wave vector $\mathbf{k}$ is assumed to lie in the first zone. It must be stressed that (123) is an approximation that holds extremely well for the optical and ultraviolet region of the electromagnetic spectrum. The coupling of excitons to photons with wave vectors $\mathbf{k} + \mathbf{K}$ (with $\mathbf{K} \neq 0$) has been ignored. This is entirely reasonable since these photons have energies in the range 10^6 cm^{-1}. However, it does mean that part of the retardation correction to the instantaneous dipole interaction

$\mathbf{T}(\mathbf{k})$ is lost. We shall consider this question separately in Section II.B.7. The dipoles $\mathbf{d}_{su}$ have the normal mode form

$$\mathbf{d}_{su} = \mathbf{d}_u \exp(i\mathbf{k} \cdot \mathbf{r}_s) \tag{124}$$

so that (108) may be solved for $\mathbf{A}_0$

$$\mathbf{A}_0 = i\left(\frac{4\pi\omega}{c}\right)\left[\left(\frac{\omega}{c}\right)^2 - \mathbf{k}^2\right]^{-1}(\mathbf{1} - \hat{\mathbf{k}}\hat{\mathbf{k}}) \cdot \mathbf{P}_0 \tag{125}$$

where

$$\mathbf{P}_0 = (v_0)^{-1} \sum_u \mathbf{d}_u \tag{126}$$

Using (109), (124), and (125) the equation of motion (107) can now be solved for the polariton dispersion relation. The result is

$$\det|(1 - n^2)\mathbf{1} + [\mathbf{1} + 4\pi\boldsymbol{\alpha}(\omega) \cdot \mathbf{T}(\mathbf{k})]^{-1} \cdot 4\pi\boldsymbol{\alpha}(\omega) \cdot (\mathbf{1} - \hat{\mathbf{k}}\hat{\mathbf{k}})| = 0 \tag{127}$$

where

$$n = \frac{ck}{\omega} \tag{128}$$

is the refractive index. Alternatively one may construct a dielectric tensor using the relations.

$$\mathbf{P}_0 = \mathbf{X}(\omega,\mathbf{k})\mathbf{E}_0 \tag{129}$$

$$\boldsymbol{\varepsilon}(\omega,\mathbf{k}) = \mathbf{1} + 4\pi\mathbf{X}(\omega,\mathbf{k}) \tag{130}$$

The susceptibility $\mathbf{X}(\omega,\mathbf{k})$ as obtained from (107) is

$$\mathbf{X}(\omega,\mathbf{k}) = [\mathbf{1} + 4\pi\boldsymbol{\alpha}(\omega) \cdot \mathbf{T}(\mathbf{k})]^{-1} \cdot \boldsymbol{\alpha}(\omega) \cdot (\mathbf{1} - \hat{\mathbf{k}}\hat{\mathbf{k}}) \tag{131}$$

Suppose all the transitions have the same polarization $\hat{\mathbf{d}}$, then

$$[\mathbf{1} + 4\pi\boldsymbol{\alpha}(\omega) \cdot \mathbf{T}(\mathbf{k})]^{-1} \cdot \boldsymbol{\alpha}(\omega) = [1 + 4\pi\alpha(\omega)T(\mathbf{k})]^{-1}\alpha(\omega)\hat{\mathbf{d}}\hat{\mathbf{d}} \tag{132}$$

and (127) simplifies accordingly. For photons linearly polarized along $\mathbf{e}_\lambda$ the refractive index is given by

$$[n_\lambda(\omega,\mathbf{k})]^2 = 1 + \frac{4\pi\alpha(\omega)(\hat{\mathbf{e}}_\lambda \cdot \hat{\mathbf{d}})^2}{1 + 4\pi\alpha(\omega)T(\mathbf{k})} \tag{133}$$

The dipole sums $T(\mathbf{k})$ are only very slowly varying functions of $|\mathbf{k}|$ for a fixed direction; that is, there is very little spatial dispersion. At one time the presence of additional electromagnetic waves resulting from spatial dispersion was actively sought after in crystalline anthracene but none were found. This is not surprising now, since exciton band structure calculations show that there is very little change in exciton energy with wave vector $\mathbf{k}$. Spatial dispersion requires the small effective masses that arise primarily from the large electron exchange effects that are common in inorganic insulators but rare in crystals of organic molecules. Therefore in the absence of damping and spatial dispersion the refractive index has a pole at each allowed Coulombic exciton frequency satisfying

$$1 + 4\pi\alpha(\omega)T(\mathbf{k}) = 0 \tag{134}$$

and a zero at each frequency satisfying

$$1 + 4\pi\alpha(\omega)[T(\mathbf{k}) + (\mathbf{e}_\lambda \cdot \hat{\mathbf{d}})^2] = 0 \tag{135}$$

The solutions of (135) also correspond to frequencies of Coulombic excitons. This can be demonstrated by writing the macroscopic (non-analytic) part of the lattice sum separately:

$$T(\hat{\mathbf{k}}) = t(0) + (\hat{\mathbf{k}} \cdot \hat{\mathbf{d}})^2 - \tfrac{1}{3} \tag{136}$$

where, since spatial dispersion is small for the $|\mathbf{k}|$ range at optical wavelengths, the analytic part $t(\mathbf{k})$ has been set equal to its value at $\mathbf{k} = 0$. The range of $T(\mathbf{k})$ is therefore

$$t(0) - \tfrac{1}{3} \leqslant T(\mathbf{k}) \leqslant t(0) + \tfrac{2}{3} \tag{137}$$

and since

$$0 < [(\mathbf{e}_\lambda \cdot \hat{\mathbf{d}})^2 + (\hat{\mathbf{k}} \cdot \hat{\mathbf{d}})^2] \leqslant 1 \tag{138}$$

we see that $[T(\mathbf{k}) + (\mathbf{e}_\lambda \cdot \hat{\mathbf{d}})^2]$ falls in the allowed range. The wave vectors $\mathbf{k}'$ of the excitons satisfying (135) are found from

$$(\hat{\mathbf{k}}' \cdot \hat{\mathbf{d}})^2 = (\hat{\mathbf{k}} \cdot \mathbf{e}_\lambda)^2 + (\hat{\mathbf{k}} \cdot \hat{\mathbf{d}})^2 \tag{139}$$

By arranging for the three vectors $\mathbf{e}_\lambda$, $\hat{\mathbf{d}}$, and $\hat{\mathbf{k}}$ to be coplanar we have

$$\hat{\mathbf{k}}' = \hat{\mathbf{d}}$$

(sign of $\hat{\mathbf{k}}'$ is unimportant because of time reversal) and the excitons whose frequencies coincide with the zeros of $n_\lambda^2(\omega,\mathbf{k})$ have wave vectors $\mathbf{k}'$ parallel to the transition dipole of the molecule. The frequencies, obtained by solving

$$1 + 4\pi\alpha(\omega)[t(0) + 1] = 0 \tag{141}$$

are effectively independent of $\mathbf{k}$, the wave vector of the incident light. We shall refer to these excitons as *longitudinal*. The reflection spectra of several faces of the crystalline dye BDP[153] have been partially analyzed using this assumption.[154]

The dependence of ω on $\mathbf{k}$, obtained from (133), is shown schematically in Fig. 3 for molecular transitions that are of weak, intermediate, and high intensity. In each case $T(\mathbf{k})$ is assumed to be a constant, independent of $|\mathbf{k}|$ for a given direction $\hat{\mathbf{k}}$. Note that the gap between polariton branches is large for the intense transition. In the absence of damping processes there is no wave in the crystal with a frequency in the gap and propagating in direction $\mathbf{k}$. Therefore light at normal incidence is completely reflected from the crystal surface perpendicular to $\mathbf{k}$.

The reflectivity R of a crystal for light with normal incidence is given by

$$R = \left| \frac{n - 1}{n + 1} \right|^2 \tag{142}$$

where $n = n(\omega,\mathbf{k})$ is the refractive index for frequency ω and direction $\mathbf{k}$. In Figs. 4–6 the reflectivity is shown for a weak, medium, and intense transition. A modest amount of phenomenological damping has been included by making $\alpha(\omega)$ complex:

$$4\pi\alpha(\omega) = \sum_u \frac{\omega_0{}^2 f_u}{\omega_u{}^2 - \omega^2 - i\omega\gamma_u} \tag{143}$$

The refractive index is now complex and its real part remains finite as the real part of $[1 + 4\pi\alpha(\omega)T(\mathbf{k})]$ goes through zero. For a weak transition the reflection spectrum consists of one narrow reflection band per vibronic level. At high intensity the reflectivity consists of a massive block with small dips marking the positions of the vibronic levels. The low energy

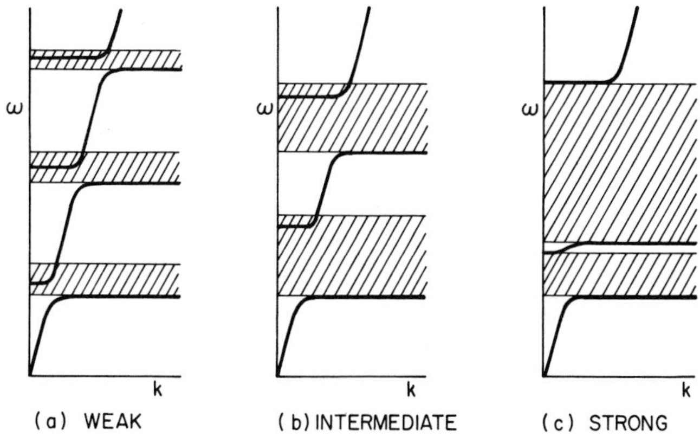

Fig. 3. Schematic representation of the polariton dispersion curve relative to single-particle (exciton) bands for a weak (a), intermediate (*b*), and strong (*c*) molecular transition.

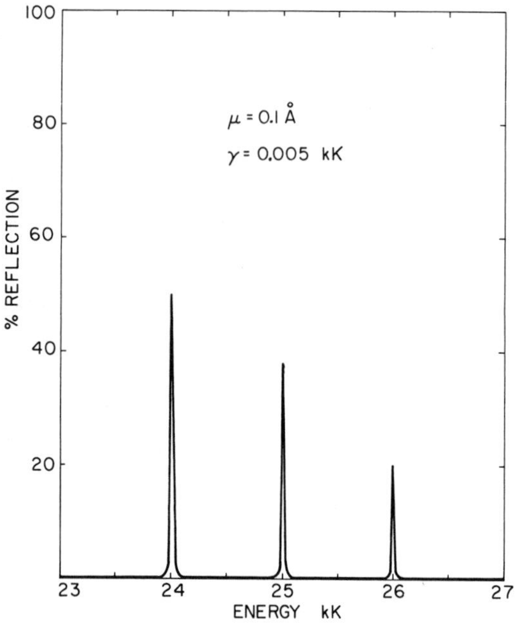

Fig. 4. Classical reflection spectrum for a weak transition.

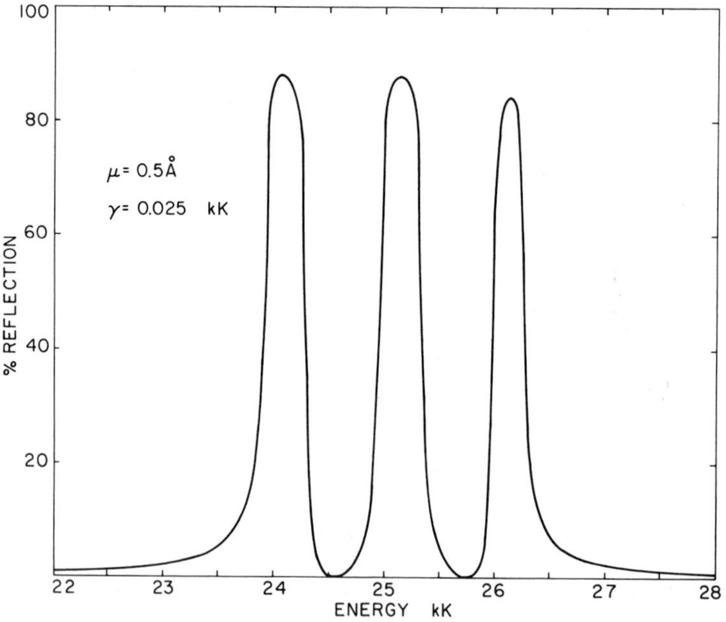

Fig. 5. Classical reflection spectrum for a transition of medium intensity.

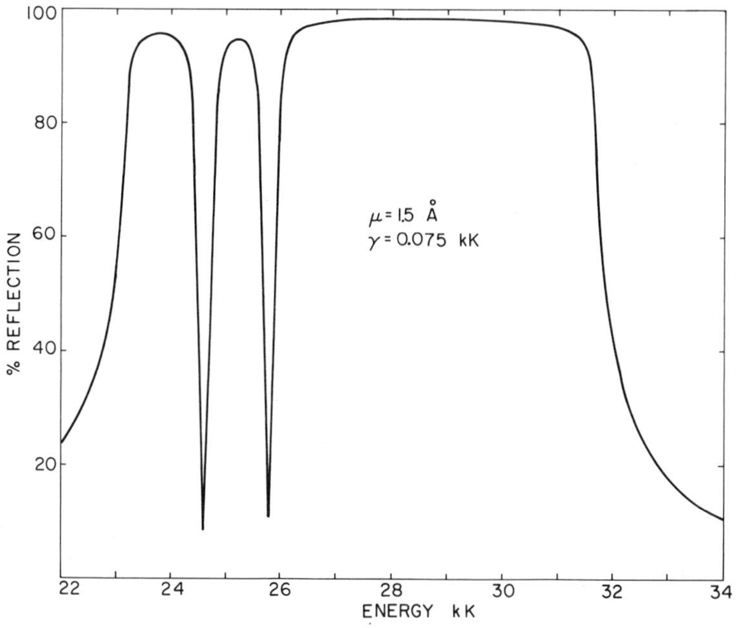

Fig. 6. Classical reflection spectrum for a strong molecular transition.

edge of each reflection band locates the Coulombic exciton **k** (a singularity of n) while the longitudinal excitons **k**′ are located at the high energy edges where the reflectivity drops off rapidly (a zero of n).

For very strong molecular transitions ($f \simeq 3$) reflection bands over an electron volt wide have been observed for certain crystalline cationic dyes[144,145] and for the second singlet transition of crystalline anthracene.[155] This phenomenon is referred to as *metallic reflection*. Single-particle theory cannot explain the vibrational fine structure of these metallic reflectors; only a multi-particle theory can do that. For example the single-particle theory predicts dips near the frequency of the vibronic transitions in the isolated molecule, whereas all the observed structure is located close to the low energy edge. The reason for this is that in the limit of strong coupling the electronic part of the excitation moves independently of any intramolecular vibration. The vibrational structure seen is due to the excitation of two-particle states containing an electronic and a vibrational exciton. However, since the two-particle band starts one vibrational quantum above the lowest single-particle level, the properties of a band of electronic exciton states roughly one vibrational quantum wide can be treated using single-particle theory.

If the vibrational levels of the excited state are ignored and all the intensity concentrated in one level with frequency ω_1 and oscillator strength f_1, a simple relation connects the width W of the metallic reflection with molecular and crystal parameters. Let ω_+ and ω_- be the frequencies of the upper and lower edges of the reflection band. Then since

$$4\pi\alpha(\omega) = \frac{\omega_0{}^2 f_1}{\omega_1{}^2 - \omega^2} \tag{144}$$

we have

$$\omega_+{}^2 = \omega_1{}^2 + \omega_0{}^2 f_1[T(\mathbf{k}) + (\mathbf{e}_\lambda \cdot \hat{\mathbf{d}})^2] \tag{145}$$

$$\omega_-{}^2 = \omega_1{}^2 + \omega_0{}^2 f_1 T(\mathbf{k}) \tag{146}$$

and

$$W = \omega_+ - \omega_- \simeq \left(\frac{4\pi}{\hbar\nu_0}\right) |\boldsymbol{\mu}|^2(\mathbf{e}_\lambda \cdot \hat{\mathbf{d}})^2 \tag{147}$$

Even if the crystal structure is unknown this relation is still useful since ν_0 can be obtained from a density measurement and $|\boldsymbol{\mu}|^2$ from $\epsilon_{\max}$. In deriving (147) it was assumed that background dielectric effects (due to higher transitions) were absent. If estimates of W using (147) greatly exceed the experimentally observed width this is evidence for the existence of an appreciable background polarization.

Next we derive the oscillator strength for any member of a stack of similarly polarized crystal transitions. Equation (122′) is rearranged to[156]

$$L(\omega) = \frac{\{[\Omega_u(\omega)]^2 - \omega^2\}[1 + T(\mathbf{k})\Sigma_u(\omega)]}{\omega_u{}^2 - \omega^2} \tag{148}$$

where

$$[\Omega_u(\omega)]^2 = \omega_u{}^2 + \omega_0{}^2 f_u T(\mathbf{k})[1 + T(\mathbf{k})\Sigma_u(\omega)]^{-1} \tag{149}$$

and

$$\Sigma_u(\omega) = \sum_{v(\neq u)} \frac{\omega_0{}^2 f_v}{\omega_v{}^2 - \omega^2} \tag{150}$$

The secular determinant for the uth Coulombic exciton is

$$\omega^2 = [\Omega_u(\omega)]^2 \tag{151}$$

which can be solved self-consistently. Therefore, for ω close to $\Omega_u(\omega)$

$$(\omega_u{}^2 - \omega^2)4\pi\alpha(\omega) = \omega_0{}^2 f_u[1 + T(\mathbf{k})\Sigma_u(\omega)]^{-1} \tag{152}$$

and the refractive index equation reduces to a single "resonance" formula

$$[n_\lambda(\omega,\mathbf{k})]^2 = 1 + \frac{\omega_0{}^2 F_{u\lambda}}{[\Omega_u(\omega)]^2 - \omega^2} \tag{153}$$

where $F_{u\lambda}$ is the oscillator strength of the uth transition

$$F_{u\lambda} \simeq \frac{f_u(\hat{\mathbf{d}} \cdot \mathbf{e}_\lambda)^2}{[1 + T(\mathbf{k})\Sigma_u(\omega)]^2} \tag{154}$$

For weak transitions $|T(\mathbf{k})\Sigma_u(\omega)| \ll 1$ and

$$F_{u\lambda} \simeq f_u(\hat{\mathbf{d}} \cdot \mathbf{e}_\lambda)^2 \tag{155}$$

On the other hand, for strong transitions there is one collective state with frequency given approximately by

$$[\Omega(\omega)]^2 \simeq \omega_u{}^2 + \omega_0{}^2 T(\mathbf{k}) \left(\sum_u f_u\right) \tag{156}$$

and an oscillator strength

$$F_\lambda \simeq (\hat{\mathbf{d}} \cdot \mathbf{e}_\lambda)^2 \sum_u f_u$$

The oscillator strengths of the levels in the depleted stack die away to zero as the collective level moves away. This of course is the prediction of the single-particle theory and, as has been discussed, the inclusion of two-particle states may seriously modify this view.

The hyper-or hypochromicity of a transition is the ratio of the crystal to oriented gas oscillator strengths. For a stack of similarly polarized transitions

$$h_u(\omega) = [1 + T(\mathbf{k})\Sigma_u(\omega)]^{-2} \tag{158}$$

The behavior of this expression is readily seen for two transitions. For $T(\mathbf{k}) < 0$ the lowest state is hyperchromic and the other hypochromic and for $T(\mathbf{k}) > 0$ the upper state is hyperchromic and the lower hypochromic.

2. Infinite Crystals with Several Molecules per Unit Cell

Most aromatic hydrocarbons crystallize with an even number of molecules per unit cell. Notable exceptions are hexamethylbenzene which has one per cell and dibenzyl which has three (and is optically active). Interactions between translationally inequivalent molecules give rise to branches in the exciton bands that have different polarization properties. In the crystal absorption spectra each molecular term gives rise to a multiplet. The energy separations of components within the same multiplet are called the Davydov or factor group splittings. The number of components of a multiplet is always less than or equal to the number of molecules in the unit cell because of the operation of the factor group selection rules.

In this section the formal results of dipole theory applied to complex molecular crystals are described. Few details of the derivations are presented since the methods are essentially the same as described in the previous section (II.B.1).

In complex crystals with many sites per cell the molecules are labeled $s = n\beta$ where n denotes the unit cell and $\beta = 1, 2, \ldots, h$ refers to the sites within the cell. The equation determining the frequencies of Coulombic excitons is

$$\det|\delta_{\beta\gamma}\mathbf{1} + 4\pi\boldsymbol{\alpha}_\beta(\omega) \cdot \mathbf{T}_{\beta\gamma}(\mathbf{k})| = 0 \tag{159}$$

with

$$\mathbf{T}_{\beta\gamma}(\mathbf{k}) = \frac{v_0}{4\pi}\sum_m{}' R^{-3}(\mathbf{1} - 3\hat{\mathbf{R}}\hat{\mathbf{R}}) \exp[i\mathbf{k}\cdot(\mathbf{r}_{n\beta} - \mathbf{r}_{m\gamma})] \tag{160}$$

and

$$4\pi\boldsymbol{\alpha}_\beta(\omega) = \sum_u \frac{\omega_0{}^2 f_u \hat{\mathbf{d}}_{\beta u}\hat{\mathbf{d}}_{\beta u}}{\omega_u{}^2 - \omega^2} \tag{161}$$

The primed sum in (160) means that the origin site is not to be summed when $\beta = \gamma$.

For molecules with three principal axes defined by symmetry the secular determinant can also be written

$$\det|\delta_{\beta\gamma}\delta_{XY} + 4\pi\alpha_X(\omega)T_{\beta X,\gamma Y}(\mathbf{k})| = 0 \tag{162}$$

where

$$4\pi\alpha_X(\omega) = \sum_x \frac{\omega_0{}^2 f_{Xx}}{\omega_{Xx}{}^2 - \omega^2} \tag{163}$$

is the polarizability component from all transitions polarized X and

$$\mathbf{T}_{\beta X,\gamma Y}(\mathbf{k}) = \hat{\mathbf{d}}_{\beta X}\cdot\mathbf{T}_{\beta\gamma}(\mathbf{k})\cdot\hat{\mathbf{d}}_{\gamma Y} \tag{164}$$

As described before, the secular equations can be simplified for certain special directions in **k**-space by using the group of the wave vector. For definiteness anthracene is considered here. There are two molecules in the unit cell related by a screw axis parallel to the b-direction. For the states with **k** pointing along the b-axis or lying in the ac-plane the secular determinant simplifies to

$$\mathfrak{L}(\omega) = \det|\delta_{XY} + 4\pi\alpha_X(\omega)T_{XY}(\mathbf{k})| = 0 \tag{165}$$

Here the factor group dipole sums are defined by

$$T_{XY}(\mathbf{k}) = T_{1X,1Y}(\mathbf{k}) \pm T_{1X,2Y}(\mathbf{k}) \tag{166}$$

The plus sign refers to states with A and the minus sign to states with B symmetry. There are separate secular determinants (165) for the A and B states.

If only π–π^* transitions are considered the secular matrix is 2×2 and can be rearranged to

$$\omega^2 = [\Omega_{Mm}(\omega)]^2 \tag{167}$$

where

$$[\Omega_{Mm}(\omega)]^2 = (\omega_{Mm})^2 + \omega_0{}^2 f_{Mm}\Lambda_M[1 + \Lambda_M \Sigma_{Mm}]^{-1} \tag{168}$$

with

$$\Sigma_{Mm}(\omega) = \sum_{m'(\neq m)} \frac{\omega_0{}^2 f_{Mm'}}{(\omega_{Mm'})^2 - \omega^2} \tag{169}$$

and

$$\Lambda_M = T_{MM} - \frac{T_{ML} 4\pi\alpha_L(\omega) T_{LM}}{1 + 4\pi\alpha_L(\omega) T_{LL}} \tag{170}$$

For clarity the explicit **k** dependence of the factor group sums $T_{XY}(\mathbf{k})$ has been omitted.

Equation (167) is solved separately for the A and B factor group sums to give the frequencies $\Omega_{Mm}(A)$ and $\Omega_{Mm}(B)$ of the components of the multiplet. The Davydov splitting is

$$\Delta_{Mm} = \Omega_{Mm}(B) - \Omega_{Mm}(A) \tag{171}$$

or, after using (168),

$$\Delta_{Mm} \simeq \frac{\omega_0{}^2 f_{Mm}}{2\omega_{Mm}} \left[\frac{\Lambda_M(A)}{1 + \Lambda_M(A)\Sigma_{Mm}(A)} - \frac{\Lambda_M(B)}{1 + \Lambda_M(B)\Sigma_{Mm}(B)} \right] \tag{172}$$

If power series expansions of Λ_M and $(1 + \Lambda_M \Sigma_{Mm})^{-1}$ exist then the result of perturbation theory is recovered as the leading term

$$\Delta_{Mm} \simeq \frac{\omega_0{}^2 f_{Mm}}{2\omega_{Mm}} [T_{MM}(A) - T_{MM}(B)] = \frac{8\pi}{\hbar\nu_0} |\mu_{Mm}|^2 T_{1M,2M}(\mathbf{k}) \tag{173}$$

When the power series expansions are only slowly convergent (or do not exist) the full expression for the splitting must be used.

Calculations, using the formulas just described, were first performed by Mahan[156] for anthracene. Later Philpott[157a] extended the theory and calculated the splittings and polarization ratios for a number of aromatic hydrocarbons. Taken all together, the calculations indicate that the dipole theory can give qualitatively correct and quantitatively reasonable results while revealing that nondipolar interactions, background polarizations, and possibly collective coupling effects need careful consideration in any deeper analysis. These conclusions parallel some of those drawn from the MO method of calculating the splittings.

The refractive index and dielectric constant of a complex molecular crystal are found by a straightforward extension of the elementary methods described in Section II.A for the one per cell case. For completeness the molecules are allowed transitions polarized along all three principal axes *L*, *M*, and *N*. Only wave vectors **k** parallel to the *b* axis or lying in the *ac* crystal plane are considered (crystals with monoclinic or higher symmetry.) The Coulombic exciton frequencies are found by solving equations which are the same as (167), (168), and (169) except that Λ_M is now defined by

$$\Lambda_M(\omega) = \mathfrak{J}_{MM} - \frac{\mathfrak{J}_{ML}[4\pi\alpha_L(\omega)]\mathfrak{J}_{LM}}{1 + 4\pi\alpha_L(\omega)\mathfrak{J}_{LL}} \tag{174}$$

with

$$\mathfrak{J}_{XY} = T_{XY} - \frac{T_{XN}[4\pi\alpha_N(\omega)]T_{NY}}{1 + 4\pi\alpha_N(\omega)T_{NN}} \tag{175}$$

Again T_{XY} are the factor group dipole sums defined by (166) for crystals with two molecules per unit cell.

The formula for the oscillator strength is

$$F_{Mm,\lambda} \simeq \frac{2f_{Mm}}{[1 + \Lambda_M\Sigma_{Mm}]^2}\left[(\mathbf{e}_\lambda \cdot \hat{\mathbf{d}}_{1M}) - \frac{4\pi\alpha_L\mathfrak{J}_{LM}(\mathbf{e}_\lambda \cdot \hat{\mathbf{d}}_{1L})}{1 + 4\pi\alpha_L\mathfrak{J}_{LL}}\right.$$
$$\left. - \frac{4\pi\alpha_N}{1 + 4\pi\alpha_N T_{NN}}\left(T_{NM} - \frac{4\pi\alpha_L\mathfrak{J}_{LM}T_{NL}}{1 + 4\pi\alpha_L\mathfrak{J}_{LL}}\right)(\mathbf{e}_\lambda \cdot \hat{\mathbf{d}}_{1N})\right]^2 \tag{176}$$

for crystals with two molecules per unit cell. If **k** lies in the *ac* plane then $\mathbf{e}_\lambda = \hat{\mathbf{b}}$ for the *B* state and $\mathbf{e}_\lambda = \hat{\mathbf{b}} \times \hat{\mathbf{k}}$ for the *A* state. For **k** parallel to the *b*-axis only the *A* state is observable and $\mathbf{e}_\lambda$ lies in the *ac* plane.

From (176) formally exact expressions for the hyperchromism and polarization ratio can be obtained. The hyperchromism of level *Mm* is

$$h_{Mm}(\omega) = \frac{F_{Mm,\lambda}}{2f_{Mm}(\mathbf{e}_\lambda \cdot \hat{\mathbf{d}}_{1M})^2} \tag{177}$$

The polarization ratio is

$$\mathcal{P}\left(\frac{B}{A}\right) = \mathcal{P}_{og}\frac{h_{Mm}(B)}{h_{Mm}(A)} \tag{178}$$

where $\mathcal{P}_{og}$ is the oriented gas value.

The equation for the refractive index is too long to be written in full. In matrix form

$$n_\lambda^2 = 1 + \sum_X (\mathbf{A}^{-1}\mathbf{B})_{XX} \tag{179}$$

where the summation is over $X = L, M, N$, and the matrices **A** and **B** are defined by

$$\mathbf{A}_{XY} = \delta_{XY} + 4\pi\alpha_X(\omega)T_{XY} \tag{180}$$

$$\mathbf{B}_{XY} = 2\pi\alpha_X(\omega)(\mathbf{e}_\lambda \cdot \mathbf{d}_X)(\mathbf{d}_Y \cdot \mathbf{e}_\lambda) \tag{181}$$

This last equation contains vectors $\mathbf{d}_X$ defined by

$$\mathbf{d}_X = \hat{\mathbf{d}}_{1X} \pm \hat{\mathbf{d}}_{2X} \tag{182}$$

where plus and minus signs hold for the A and B polarized states, respectively.

The foregoing equations summarized the current status of dipole theory for molecular crystals. Since the mean polarizability and anisotropy of the polarizability are known for a number of aromatic hydrocarbons, there is some hope that these experimental quantities can be incorporated into the theory, thereby reducing the number of parameters. Efforts to carry out such a program have so far met with only limited success.[157b] One reason for this may be the large strength of the interaction coupling the lowest transitions with the many unobserved higher ones. The total oscillator strength in these higher transitions is considerable, and in a dipole theory, where there are no compensating nondipolar interactions, the high-energy transitions exert an unrealistically large influence on the lower ones.

3. Coulombic Excitons in Helical Polymers

A large class of biopolymers assume a helical conformation in certain ranges of temperature, pH, and ionic strength. The helical regions can vary from a few to thousands of angstroms in length. For the longer polymers of this sort the model of an infinite single-stranded helix of monomers is often used to calculate optical properties. Now a helix is an asymmetric structure without an inversion center and some dipole sums can be pure imaginary, unlike their counterparts in crystals (which are generally real).

In a helix of identical monomers the screw operations map any monomer onto all others. Choose a right-handed coordinate system fixed on the helix with unit vectors $\mathbf{e}_x$, $\mathbf{e}_y$, $\mathbf{e}_z$ such that $\mathbf{e}_z$ is parallel to the helix axis. Let $\mathbf{R}(\phi)$ be the rotational part of the basic screw operation; then in matrix form

$$\mathbf{R}(\varphi) = \begin{pmatrix} \cos\varphi & -\sin\varphi & 0 \\ \sin\varphi & \cos\varphi & 0 \\ 0 & 0 & 1 \end{pmatrix} \tag{183}$$

Next cyclic boundary conditions are imposed on the polymer and solutions sought of (110)

$$\mathbf{P}_n^{(0)} = [\mathbf{R}(\varphi)]^n \cdot \mathbf{P}_0^{(0)} \exp(i\mathbf{k} \cdot \mathbf{t}_n) \tag{184}$$

Here $\mathbf{t}_n$ are the positions of the monomers n measured along the helix axis. The equations of motion reduce to

$$[\mathbf{1} + 4\pi\boldsymbol{\alpha}_0(\omega) \cdot \mathbf{S}(\mathbf{k})] \cdot \mathbf{P}_0^{(0)} = 0 \tag{185}$$

where $\boldsymbol{\alpha}_0(\omega)$ is the polarizability of the "zeroth" monomer and

$$\mathbf{S}(\mathbf{k}) = \sum_n{}' \mathbf{T}_{0n} \cdot [\mathbf{R}(\varphi)]^n \exp(i\mathbf{k} \cdot \mathbf{t}_n) \tag{186}$$

The polarizability of the nth monomer is related to $\boldsymbol{\alpha}_0$ by

$$\boldsymbol{\alpha}_n(\omega) = [\mathbf{R}(\varphi)]^n \cdot \boldsymbol{\alpha}_0(\omega) \cdot [\mathbf{R}(\varphi)]^{-n} \tag{187}$$

The secular determinant for exciton frequencies is

$$\det |\mathbf{1} + 4\pi\boldsymbol{\alpha}_0(\omega) \cdot \mathbf{S}(\mathbf{k})| = 0 \tag{188}$$

which is solved by specifying the directions of $\mathbf{e}_x$, $\mathbf{e}_y$ relative to the zeroth monomer and then multiplying out the 3 × 3 determinant. If all the transitions are polarized parallel to the helix axis then (188) becomes

$$1 + 4\pi\alpha_z(\omega)S_z(\mathbf{k}) = 0 \tag{189}$$

where

$$\alpha_z(\omega) = \mathbf{e}_z \cdot \boldsymbol{\alpha}_0(\omega) \cdot \mathbf{e}_z \tag{190}$$

$$S_z(\mathbf{k}) = \mathbf{e}_z \cdot \mathbf{S}(\mathbf{k}) \cdot \mathbf{e}_z \tag{191}$$

When all the transitions are polarized perpendicular to the axis the determinant becomes

$$[1 + 4\pi\alpha_x(\omega)S_x][1 + 4\pi\alpha_y(\omega)S_y] - [4\pi\alpha_x(\omega)][4\pi\alpha_y(\omega)]|S_{xy}|^2 = 0 \tag{192}$$

The sums S_{xy} are pure imaginary and $S_{xy} = -S_{yx}$. Equation (192) may be solved after the methods of Section II.B.1. If the x transitions are of interest then the secular determinant can be written

$$1 + 4\pi\alpha_x(\omega)\left[S_x - \frac{4\pi\alpha_y(\omega)|S_{xy}|^2}{1 + 4\pi\alpha_y(\omega)S_y}\right] = 0 \tag{193}$$

4. *Interstitial Impurities in Infinite Crystals*

A number of impurity problems can be solved exactly in the dipole limit, the simplest being the hypothetical case of interstitial impurities that do not disturb the lattice of the host crystal. It is of considerable theoretical interest to determine how the host modifies the long-range (dipole–dipole) exciton interaction between the impurities.[158–161] In truth it must be admitted that interstitial impurities are, as yet, unknown in aromatic hydrocarbon crystals. However, the theory described here serves as the prototype for a class of systems which have important biological functions. It is well known that many biopolymers, especially the polynucleotides like DNA, bind ligands containing chromophores. Also, DNA forms complexes with proteins, like the histones, that appear to have a role in the process of transcription.

For simplicity the host is assumed to be a crystal with one molecule per unit cell. The impurity sites, which do not coincide with any lattice site, are denoted by x, x', ... From (110) the secular equations for Coulombic excitations are

$$\sum_{x'} [\delta_{xx'}\mathbf{1} + 4\pi\boldsymbol{\alpha}_x(\omega) \cdot \mathbf{T}_{xx'}] \cdot \mathbf{P}_{x'}^{(0)} + 4\pi\boldsymbol{\alpha}_x(\omega) \cdot \sum_{n} \mathbf{T}_{xn} \cdot \mathbf{P}_{n}^{(0)} = 0 \quad (194)$$

and

$$\sum_{n'} [\delta_{nn'}\mathbf{1} + 4\pi\boldsymbol{\alpha}(\omega) \cdot \mathbf{T}_{nn'}] \cdot \mathbf{P}_{n'}^{(0)} + 4\pi\boldsymbol{\alpha}(\omega) \cdot \sum_{x} \mathbf{T}_{nx} \cdot \mathbf{P}_{x}^{(0)} = 0 \quad (195)$$

All the host molecules are identical and so have the same polarizability $\boldsymbol{\alpha}(\omega)$. The interstitial impurities are not assumed to be identical, and the site labels x, x', . . . are retained on the impurity polarizabilities $\boldsymbol{\alpha}_x(\omega)$, $\boldsymbol{\alpha}_{x'}(\omega)$, ...

The equations are solved by eliminating the host polarizations $\mathbf{P}_n^{(0)}$. This is achieved by defining

$$\mathbf{T}(\mathbf{k}) = \sum_{n'}{}' \mathbf{T}_{nn'} \exp [i\mathbf{k} \cdot (\mathbf{r}_n - \mathbf{r}_{n'})] \quad (196)$$

$$\mathbf{T}_x(\mathbf{k}) = \sum_{n} \mathbf{T}_{xn} \exp (i\mathbf{k} \cdot \mathbf{r}_n) \quad (197)$$

$$\mathbf{P}_{\mathbf{k}} = N^{-1} \sum_{n} \mathbf{P}_n^{(0)} \exp (i\mathbf{k} \cdot \mathbf{r}_n) \quad (198)$$

and transforming to $\mathbf{P}_{\mathbf{k}}$ variables in (194) and (195). The results are

$$\sum_{x'} [\delta_{xx'}\mathbf{1} + 4\pi\boldsymbol{\alpha}_x(\omega) \cdot \mathbf{T}_{xx'}] \cdot \mathbf{P}_{x'}^{(0)} + 4\pi\boldsymbol{\alpha}_x(\omega) \cdot \sum_{\mathbf{k}} \mathbf{T}_x(\mathbf{k}) \cdot \mathbf{P}_{\mathbf{k}} = 0 \quad (199)$$

$$[\mathbf{1} + 4\pi\boldsymbol{\alpha}(\omega) \cdot \mathbf{T}(\mathbf{k})] \cdot \mathbf{P}_{\mathbf{k}} + 4\pi\boldsymbol{\alpha}(\omega) \cdot \sum_{x} \mathbf{T}_x^*(\mathbf{k}) \cdot \mathbf{P}_x^{(0)} = 0 \quad (200)$$

Equation (200) gives a relation for $\mathbf{P}_{\mathbf{k}}$ which is then substituted into (199) to give

$$\sum_{x'} [\delta_{xx'}\mathbf{1} + 4\pi\boldsymbol{\alpha}_x(\omega) \cdot \boldsymbol{\Phi}_{xx'}(\omega)] \cdot \mathbf{P}_{x'}^{(0)} = 0 \quad (201)$$

where $\boldsymbol{\Phi}_{xx'}(\omega)$ is the effective exciton interaction between impurities x and x' when $x \neq x'$:

$$\boldsymbol{\Phi}_{xx'}(\omega) = \mathbf{T}_{xx'} + N^{-1} \sum_{\mathbf{k}} \mathbf{T}_x(\mathbf{k}) \cdot [\mathbf{1} + 4\pi\boldsymbol{\alpha}(\omega) \cdot \mathbf{T}(\mathbf{k})]^{-1} \cdot 4\pi\boldsymbol{\alpha}(\omega) \cdot \mathbf{T}_{x'}{}^*(\mathbf{k}) \tag{202}$$

If there is only one impurity the frequencies of exciton states localized on or near site x are found by solving

$$\det |\mathbf{1} + 4\pi\boldsymbol{\alpha}_x(\omega) \cdot \boldsymbol{\Phi}_{xx}(\omega)| = 0 \tag{203}$$

so that $\boldsymbol{\Phi}_{xx}$ is the effective field at the impurity site due to host–impurity interactions. The meaning of this equation becomes clear from examining special cases. Suppose, for instance, that all impurity states are polarized along $\hat{\mathbf{d}}'$ and all host states along $\hat{\mathbf{d}}$. Then the secular determinant straightway reduces to

$$1 = 4\pi\alpha_x(\omega) \frac{1}{N} \sum_{\mathbf{k}} \frac{4\pi\alpha(\omega)|T_x(\mathbf{k})|^2}{1 + 4\pi\alpha(\omega)T(\mathbf{k})} \tag{204}$$

where

$$T_x(\mathbf{k}) = \hat{\mathbf{d}}' \cdot \mathbf{T}_x(\mathbf{k}) \cdot \hat{\mathbf{d}}' \tag{205}$$

$$T(\mathbf{k}) = \hat{\mathbf{d}} \cdot \mathbf{T}(\mathbf{k}) \cdot \hat{\mathbf{d}} \tag{206}$$

$$\alpha(\omega) = \hat{\mathbf{d}} \cdot \boldsymbol{\alpha}(\omega) \cdot \hat{\mathbf{d}} \tag{207}$$

$$\alpha_x(\omega) = \hat{\mathbf{d}}' \cdot \boldsymbol{\alpha}_x(\omega) \cdot \hat{\mathbf{d}}' \tag{208}$$

If the impurity and host have only one transition, each with frequencies ω_x and ω_1, respectively, then (204) gives

$$\omega^2 = \omega_x{}^2 - \omega_0{}^2 f_x \frac{1}{N} \sum_{\mathbf{k}} \frac{\omega_0{}^2 f_1 |T_x(\mathbf{k})|^2}{[\Omega_1(\mathbf{k})]^2 - \omega^2} \tag{209}$$

where

$$[\Omega_1(\mathbf{k})]^2 = \omega_1{}^2 + \omega_0{}^2 f_1 T(\mathbf{k}) \tag{210}$$

Note that for impurity levels lying below the host band of frequencies $\Omega_1(\mathbf{k})$, the shift is always negative.

The long-range interaction between impurities in their ground states was investigated by Mahan[158] using a model that can also be applied to the exciton problem. The host molecules are assumed to be isotropic and the crystal cubic:

$$\boldsymbol{\alpha}(\omega) = \alpha(\omega)\mathbf{1} \tag{211}$$

$$\hat{\mathbf{e}}_i \cdot \mathbf{T}(\mathbf{k}) \cdot \hat{\mathbf{e}}_j = \delta_{ij} T_i(\mathbf{k}) \tag{212}$$

where $\{\mathbf{e}_i\}$ are a set of orthogonal unit vectors. Therefore

$$[1 + 4\pi\boldsymbol{\alpha}(\omega) \cdot \mathbf{T}(\mathbf{k})]^{-1} = \sum_i [1 + 4\pi\alpha(\omega)T_i(\mathbf{k})]^{-1}\hat{\mathbf{e}}_i\hat{\mathbf{e}}_i \qquad (213)$$

The lattice sums $T_i(\mathbf{k})$ are approximated by their long wavelength ($\mathbf{k} \to 0$) values:

$$T_1(\mathbf{k}) = T_2(\mathbf{k}) = -\tfrac{1}{3} \qquad (214)$$

$$T_3(\mathbf{k}) = \tfrac{2}{3} \qquad (215)$$

where $\hat{\mathbf{e}}_3 = \hat{\mathbf{k}}$. Next the lattice sums $\mathbf{T}_x(\mathbf{k})$ times $\exp(i\mathbf{k} \cdot \mathbf{x})$ are approximated in the same way as the host sums $\mathbf{T}(\mathbf{k})$ and the effective interaction, (202), becomes

$$\boldsymbol{\Phi}_{xx'}(\omega) = \mathbf{T}_{xx'} - \frac{(4\pi\alpha/3)\,[1 - (8\pi\alpha/3)]}{[1 - (4\pi\alpha/3)]\,[1 + (8\pi\alpha/3)]} \cdot \frac{1}{N}\sum_{\mathbf{k}} \hat{\mathbf{k}}\hat{\mathbf{k}} \exp[i\mathbf{k} \cdot (\mathbf{x}' - \mathbf{x})] \qquad (216)$$

Now it can be shown (see Section II.B.7) that

$$\mathbf{T}_{xx'} = N^{-1}\sum_{\mathbf{k}} \hat{\mathbf{k}}\hat{\mathbf{k}} \exp[i\mathbf{k} \cdot (\mathbf{x}' - \mathbf{x})] \qquad (217)$$

so that

$$\boldsymbol{\Phi}_{xx'}(\omega) = \mathbf{T}_{xx'}\left[\left(1 + \frac{8\pi\alpha}{3}\right)\left(1 - \frac{4\pi\alpha}{3}\right)\right]^{-1} \qquad (218)$$

The dielectric constant of the host crystal is defined by

$$\epsilon = \frac{1 + (8\pi\alpha/3)}{1 - (8\pi\alpha/3)} \qquad (219)$$

and upon eliminating the host polarizability from (218) we find

$$\boldsymbol{\Phi}_{xx'}(\omega) = \frac{1}{\epsilon}\left(\frac{\epsilon + 2}{3}\right)^2 \mathbf{T}_{xx'} \qquad (210)$$

This is the result for the long-range exciton interaction between two interstitial impurities modified by a host crystal. The host exerts two effects that act in opposite directions. First, the impurity interaction is shielded by a dielectric constant ϵ; and second, it is amplified by a Lorentz factor $[(\epsilon + 2)/3]^2$. The latter can be interpreted as an increase in the impurity oscillator strength due to coupling with the host states.

The frequency-dependent dielectric constants ϵ of most molecular crystals are fairly small except in the neighborhood of absorption bands.

TABLE I

Change in Interaction of Interstitial Impurities with Host Dielectric

ϵ	1	2	3	4	6	8	10
$\frac{1}{\epsilon}\left(\frac{\epsilon+2}{3}\right)^2$	1.000	0.889	0.926	1.000	1.185	1.389	1.600

Table I indicates how the factor $\epsilon^{-1}[(\epsilon+2)/3]^2$ varies with ϵ. It can be seen that the change is less than 20% for ϵ smaller than 6 (or refractive index $\sqrt{\epsilon} = 2.45$).

5. *Substitutional Impurities in Infinite Crystals*[158,162]

For simplicity only host crystals containing one molecule per unit cell are considered here. It is assumed that the host lattice is not disturbed when guests substitutionally replace host molecules at sites $p, p', p'', \ldots$, etc. Sites occupied by host molecules are denoted $n, n', \ldots$ and when all sites, guest and host, are to be summed the indices s, s' are used. In this section it is understood that $\mathbf{T}_{ss'}$ vanishes if $s = s'$.

The polarizability tensors of hosts and guests are denoted $\boldsymbol{\alpha}(\omega)$ and $\boldsymbol{\alpha}'(\omega)$, respectively. From (110) the secular equations written separately for guest and host sites are

$$\sum_{p'} [\delta_{pp'}\mathbf{1} + 4\pi\boldsymbol{\alpha}'(\omega) \cdot \mathbf{T}_{pp'}] \cdot \mathbf{P}_{p'}^{(0)} + 4\pi\boldsymbol{\alpha}'(\omega) \cdot \sum_{n} \mathbf{T}_{pn} \cdot \mathbf{P}_{n}^{(0)} = 0 \quad (211)$$

$$\sum_{n'} [\delta_{nn'}\mathbf{1} + 4\pi\boldsymbol{\alpha}(\omega) \cdot \mathbf{T}_{nn'}] \cdot \mathbf{P}_{n'}^{(0)} + 4\pi\boldsymbol{\alpha}(\omega) \cdot \sum_{p} \mathbf{T}_{np} \cdot \mathbf{P}_{p}^{(0)} = 0 \quad (212)$$

Now the tensor $\mathbf{T}_{ss'}$ from (104), is independent of whether s and s' are guest or host sites, and so the two sets of (211) and (212) may be combined into the single set

$$\sum_{s'} [\delta_{ss'}\mathbf{1} + 4\pi\boldsymbol{\alpha}(\omega) \cdot \mathbf{T}_{ss'}] \cdot \mathbf{P}_{s'}^{(0)} + \sum_{s'}\sum_{p'} \delta_{sp'} 4\pi[\boldsymbol{\alpha}'(\omega) - \boldsymbol{\alpha}(\omega)] \cdot \mathbf{T}_{p's'} \cdot \mathbf{P}_{s'}^{(0)} = 0 \quad (213)$$

Amplitudes $\mathbf{P}_{\mathbf{k}}$ are introduced in the same way as for interstitial impurities by defining

$$\mathbf{P}_s = N^{-1} \sum_{\mathbf{k}} \mathbf{P}_{\mathbf{k}} \exp(-i\mathbf{k} \cdot \mathbf{r}_s) \quad (214)$$

After substituting (214) into (213), followed by some rearranging, we find

$$\mathbf{P}_{\mathbf{k}} + \sum_{p} [\mathbf{1} + 4\pi\boldsymbol{\alpha}(\omega) \cdot \mathbf{T}(\mathbf{k})]^{-1} \exp(-i\mathbf{k} \cdot \mathbf{r}_p) \cdot 4\pi[\boldsymbol{\alpha}'(\omega) - \boldsymbol{\alpha}(\omega)] \cdot \mathbf{U}_p = 0 \quad (215)$$

$$\mathbf{U}_p = N^{-1} \sum_{\mathbf{k}} \mathbf{T}(\mathbf{k}) \cdot \mathbf{P}_{\mathbf{k}} \exp(i\mathbf{k} \cdot \mathbf{r}_p) \tag{216}$$

Equation (215) is converted into one for $\mathbf{U}_p$ by multiplying by N^{-1} exp $(i\mathbf{k} \cdot \mathbf{r}_{p'})\mathbf{T}(k)$ and summing over all $\mathbf{k}$:

$$\sum_{p'} \{\delta_{pp'}\mathbf{1} + N^{-1} \sum_{\mathbf{k}} \exp[i\mathbf{k} \cdot (\mathbf{r}_p - \mathbf{r}_{p'})]\mathbf{T}(\mathbf{k}) \cdot \mathbf{G}(\mathbf{k}) \cdot 4\pi[\boldsymbol{\alpha}'(\omega) - \boldsymbol{\alpha}(\omega)]\} \mathbf{U}_{p'} = 0 \tag{217}$$

where

$$\mathbf{G}(\mathbf{k}) = [\mathbf{1} + 4\pi\boldsymbol{\alpha}(\omega) \cdot \mathbf{T}(\mathbf{k})]^{-1} \tag{218}$$

Since the quantities $\mathbf{U}_p$ are nonzero the determinant of their coefficients obtained from (217) vanishes. After postmultiplying by $[4\pi(\boldsymbol{\alpha}' - \boldsymbol{\alpha})]^{-1}$ and premultiplying by $4\pi(\boldsymbol{\alpha}' - \boldsymbol{\alpha})$ the secular determinant for exciton frequencies becomes

$$\det |\delta_{pp'}\mathbf{1} + 4\pi[\boldsymbol{\alpha}'(\omega) - \boldsymbol{\alpha}(\omega)]N^{-1} \sum_{\mathbf{k}} \mathbf{T}(\mathbf{k}) \cdot \mathbf{G}(\mathbf{k}) \exp[i\mathbf{k} \cdot (\mathbf{r}_p - \mathbf{r}_{p'})]| = 0 \tag{219}$$

Another form taken by this determinant is

$$\det |N^{-1} \sum_{\mathbf{k}} [\mathbf{1} + 4\pi\boldsymbol{\alpha}'(\omega) \cdot \mathbf{T}(\mathbf{k})] \cdot \mathbf{G}(\mathbf{k}) \exp[i\mathbf{k} \cdot (\mathbf{r}_p - \mathbf{r}_{p'})]| = 0 \tag{220}$$

These determinants have order $3M \times 3M$ where M is the number of guests.

Equations (219) and (220) are the formally exact solutions of a complicated problem; the off-diagonal elements are host-modified guest–guest interactions. For isotropic hosts, cubic crystals, and the long-wavelength approximation, the off-diagonal elements in determinant (219) are

$$4\pi[\boldsymbol{\alpha}'(\omega) - \alpha(\omega)\mathbf{1}] \frac{1}{\epsilon}\left(\frac{\epsilon + 2}{3}\right)^2 \cdot \mathbf{T}_{pp'} \tag{221}$$

which resembles the case of interacting interstitial impurities. There is, however, an essential difference: a crystal with substitutional impurities has a different dielectric response function. This introduces a subtle difference into the long-range interaction between substitutional impurities compared with interstitial ones. Consider two widely separated guests molecules at $\mathbf{r}_p$ and $\mathbf{r}_{p'}$ with polarizabilities given by

$$4\pi\boldsymbol{\alpha}'(\omega) = \frac{\omega_0{}^2 f_g}{\omega_g{}^2 - \omega^2} \hat{\mathbf{d}}'\hat{\mathbf{d}}' \tag{222}$$

The solutions of the secular determinant are

$$[\Omega_\pm]^2 = \omega_g^2 + \omega_0^2 f_g \hat{\mathbf{d}}' \cdot N^{-1} \sum_{\mathbf{k}} \mathbf{T}(\mathbf{k}) \cdot \mathbf{G}(\mathbf{k})\{1 \pm \exp[i\mathbf{k} \cdot (\mathbf{r}_p - \mathbf{r}_{p'})]\}$$

$$\cdot (\mathbf{1} - 4\pi\alpha(\omega)N^{-1} \sum_{\mathbf{k}} \mathbf{T}(\mathbf{k}) \cdot \mathbf{G}(\mathbf{k})\{1 \pm \exp[i\mathbf{k} \cdot (\mathbf{r}_p - \mathbf{r}_p')]\})^{-1} \cdot \hat{\mathbf{d}}' \quad (223)$$

In the denominator the rapidly oscillating exponential contributes little and is neglected. The effective interaction between the guests is then

$$\nu_{\text{eff}} = \tfrac{1}{2}(\Omega_+ - \Omega_-) \quad (224)$$

and, evaluated for the model adopted earlier, is

$$\nu_{\text{eff}} = \frac{1}{\epsilon}\left(\frac{\epsilon + 2}{3}\right)^2 |\mu_g|^2 \left(\frac{4\pi}{\hbar\nu_0}\right) \hat{\mathbf{d}}' \cdot \mathbf{T}_{pp'} \cdot [N^{-1} \sum_{\mathbf{k}} \mathbf{G}(\mathbf{k})]^{-1} \cdot \hat{\mathbf{d}}' \quad (225)$$

where $\mathbf{G}(k)$ is evaluated for $\omega = \omega_g$. This effective interaction contains a factor, not present for the interstitial impurities, which cannot be approximated satisfactorily any further using the model. Mahan[158] found a similar factor for the interaction between substitutional impurities in their ground electronic states. In the deep-trap limit

$$N^{-1} \sum_{\mathbf{k}} \mathbf{G}(\mathbf{k}) \simeq \mathbf{1} + [4\pi\alpha(\omega)]^2 \hat{\mathbf{d}} \cdot \sum_{n'} \mathbf{T}_{nn'} \cdot \hat{\mathbf{d}}\hat{\mathbf{d}} \cdot \mathbf{T}_{n'n} \quad (226)$$

and the second term becomes negligible, and so the effective interactions between substitutional and interstitial impurities are the same.

6. *Excitons and Polaritons in Semi-Infinite Crystals*[163,164]

In practice all crystals are finite and therefore exciton properties are subject to boundary conditions not met with in infinite periodic systems. Absorption spectra are generally measured using thin slablike crystals, and reflection spectra using thick crystals that have surfaces of good quality. In an idealized reflection experiment an infinite plane wave photon, at normal incidence, is reflected from a semi-infinite crystal. This is the problem analyzed next. It is assumed that the crystal surface is perfect, with the same arrangement of molecules found for similar planes deep inside the crystal. In a way it is remarkable that a formally exact theory of the reflectivity of a semi-infinite crystal can be derived in view of the long range of dipole–dipole interactions. The problem can be solved because dipole sums done in planewise fashion fall off very rapidly.[164,165]

For simplicity only crystals with one molecule per unit cell are considered. It is also assumed that all the transitions are polarized in the same direction $\hat{\mathbf{d}}$. The derivation given here is based on the work of Deutsche and Mead,[163] and Mahan and Obermair.[164]

First we shall make some brief comments on Coulombic excitons in thin[168,169] and semi-infinite crystals without assuming any particular polarization for the molecular transitions. Since the crystal is infinite in extent in the two lateral directions, periodic boundary conditions are assumed for these directions. The positions of the unit cells $\mathbf{r}_n$ may be written

$$\mathbf{r}_n = \mathbf{r}_l + \mathbf{r}_\tau \qquad l = 1, 2, 3, \ldots \tag{227}$$

where $\mathbf{r}_l$ is parallel to the z axis (perpendicular to crystal surface) and $\mathbf{r}_\tau$ is the plane parallel to the surface. Let $\boldsymbol{\kappa}$ be a wave vector belonging to the two-dimensional Brillouin zone; then the polarization vectors have the form

$$\mathbf{P}_n^{(0)} = \mathbf{P}_l^{(0)} \exp\,(i\boldsymbol{\kappa}\cdot\mathbf{r}_\tau) \tag{228}$$

and the secular determinant for Coulombic exciton frequencies is

$$\det\left|\delta_{ll'}\mathbf{1} + 4\pi\boldsymbol{\alpha}(\omega)\cdot\mathbf{v}_{l-l'}(\boldsymbol{\kappa})\right| = 0 \tag{229}$$

where

$$\mathbf{v}_{l-l'}(\boldsymbol{\kappa}) = \sum_{\tau'} \mathbf{T}_{l\tau,l'\tau'} \exp\,[i\boldsymbol{\kappa}\cdot(\mathbf{r}_\tau - \mathbf{r}_{\tau'})] \tag{230}$$

In this two-dimensional sum the origin dipole is at $n = (l, \tau)$ and all interactions with dipoles (l', τ') in the l'th plane are summed. It is known that for $\boldsymbol{\kappa} = 0$ the planewise sums $\mathbf{v}_{l-l'}(0)$ fall off very rapidly with increasing $|l - l'|$. Presumably the additional oscillations present when $\boldsymbol{\kappa} \neq 0$ make the sums $\mathbf{v}_{l-l'}(\boldsymbol{\kappa})$ die away even more rapidly with $|l - l'|$. In actual calculations it has been found that the sums $\mathbf{v}_{l-l'}$ are effectively zero for $|l - l'| > 2$. Thus the calculation of polarizations $\mathbf{P}_l$ and exciton frequencies reduces to an exercise in solving finite difference equations. For a crystal N layers thick with nonzero interactions between neighboring planes the secular determinant (229) separates into N factors

$$\det\left|\mathbf{1} + 4\pi\boldsymbol{\alpha}(\omega)\cdot\left[\mathbf{v}_0(\boldsymbol{\kappa}) + 2\mathbf{v}_1(\boldsymbol{\kappa})\cos\left(\frac{l\pi}{N+1}\right)\right]\right| = 0 \tag{231}$$

with $l = 1, 2, \ldots, N$.

Next we turn to the polariton problem for semi-infinite crystals. For plane-polarized light propagating along the z axis only the $\boldsymbol{\kappa} = 0$ exciton modes are coupled to the electromagnetic field. The equations of motion of the dipoles are

$$\sum_{l'} [\delta_{ll'} + 4\pi\alpha(\omega)v_{l-l'}]P_{l'} = (\hat{\mathbf{d}}\cdot\hat{\mathbf{e}})\alpha(\omega)E^{\perp}(\mathbf{r}_l) \tag{232}$$

Henceforth, as in (232), we work with scalar quantities. The field equation is

$$\left(\frac{d^2}{dz^2} + \frac{\omega^2}{c^2}\right) E^{\perp}(z) = -\left(\frac{4\pi\omega^2 a}{c^2}\right) (\hat{\mathbf{d}} \cdot \hat{\mathbf{e}}) \sum P_l \delta(z - la) \tag{233}$$

where a is the interplanar distance. The field equation is solved with the aid of the retarded Green function

$$G^{\text{ret}}(z,z') = \frac{1}{2\pi} \int_{-\infty}^{+\infty} dk \frac{\exp [ik(z - z')]}{k^2 - (\omega^+/c)^2} \tag{234}$$

$$\omega^+ = \omega + i\sigma \tag{235}$$

which satisfies

$$\left(\frac{d^2}{dz^2} + \frac{\omega^2}{c^2}\right) G^{\text{ret}}(z,z') = -\delta(z - z') \tag{236}$$

The solution of (233) is

$$E^{\perp}(z) = E_{\text{in}}(z) + \left(\frac{4\pi\omega^2 a}{c^2}\right) (\hat{\mathbf{e}} \cdot \hat{\mathbf{d}}) \sum_{l'=1}^{\infty} P_l \frac{1}{2\pi} \int_{-\infty}^{+\infty} dk \frac{\exp [ik(z - l'a)]}{k^2 - (\omega^+/c)^2} \tag{237}$$

where the incident field is given by

$$E_{\text{in}}{}^{\perp}(z) = E_{\text{in}}{}^{\perp} \exp \left(\frac{i\omega z}{c}\right) \tag{238}$$

Suppose the lattice sums $v_{l-l'}(0)$ are nonzero for L values of $|l - l'|$. The surface layers of the crystal have $l = 1, 2, \ldots, L$ and the bulk layers have $l \geqslant L + 1$. Inside the crystal these are $L + 1$ polariton modes and each is characterized by a refractive index $n_j(\omega)$, a polarization amplitude $P_j(\omega)$, and a transverse electric field $E_j(\omega)$. Expressed in terms of these polariton amplitudes

$$E^{\perp}(la) = \sum_j E_j \exp (in_j\Omega l) \tag{239}$$

$$P_l = \sum_j P_j \exp (in_j\Omega l) \tag{240}$$

where

$$\Omega = \frac{\omega a}{c} \tag{241}$$

is a dimensionless frequency. Next (239) and (240) are substituted into

(237) for the point $z = la$. The one-dimensional integral over k is readily evaluated

$$\frac{1}{2\pi}\int_{-\infty}^{+\infty}\frac{dk\,\exp\,[ik(l-l')a]}{k^2-(\omega^+/c)^2} = \frac{1}{2}\,ia\,\frac{\exp\,(i\Omega|l-l'|)}{\Omega} \tag{242}$$

and (237) becomes

$$\sum_j\left[E_j - \frac{2\pi\Omega\sin\Omega}{\cos\Omega-\cos(\Omega n_j)}(\hat{\mathbf{e}}\cdot\hat{\mathbf{d}})P_j\right]\exp\,(in_j\Omega l)$$
$$= \left\{E_{\text{in}}^{\perp} - i2\pi\sum_j\frac{P_j}{1-\exp\,[i\Omega(1-n_j)]}\right\}\exp\,(i\Omega l) \tag{243}$$

Each side of this equation is independent of the other; therefore

$$E_j = \frac{2\pi\Omega\sin\Omega}{\cos\Omega-\cos(\Omega n_j)}(\hat{\mathbf{e}}\cdot\hat{\mathbf{d}})P_j \tag{244}$$

$$E_{\text{in}}^{\perp} = i(2\pi\Omega)(\hat{\mathbf{e}}\cdot\hat{\mathbf{d}})\sum_j\frac{P_j}{1-\exp\,[i\Omega(1-n_j)]} \tag{245}$$

These equations were first obtained by Deutsche and Mead.[163] The electric field can now be eliminated from the equations of motion (232) and a set of simultaneous equations for the modes P_j obtained

$$\sum P_j\exp\,(in_j\Omega l)\left\{1-4\pi\alpha(\omega)\left[\frac{1}{2}(\hat{\mathbf{e}}\cdot\hat{\mathbf{d}})^2\frac{\Omega\sin\Omega}{\cos\Omega-\cos(\Omega n_j)}\right.\right.$$
$$\left.\left.-\sum_{l'=1-l}^{L}v_{l'}\exp\,(in_j\Omega l')\right]\right\} = 0 \tag{246}$$

In the bulk region of the crystal $l \geq L + 1$, the summation over l' is independent of l, and so the coefficients of P_j must vanish giving

$$1 + 4\pi\alpha(\omega)\left\{v(k_j) - \frac{1}{2}(\hat{\mathbf{e}}\cdot\hat{\mathbf{d}})^2\frac{\Omega\sin\Omega}{\cos\Omega-\cos(\Omega n_j)}\right\} = 0 \tag{247}$$

where

$$k_j = \frac{n_j\omega}{c} \tag{248}$$

and

$$v(k_j) = v_0 + 2\sum_{l'=1}^{L}v_{l'}\cos\,(alk_j) \tag{249}$$

Equation (247) was first derived by Mahan and Obermair[164]; it is the dispersion relation for polaritons of mode j. It assumes a more familiar form if the trigonometric functions are expanded assuming $\Omega \ll 1$:

$$1 + 4\pi\alpha(\omega) \left\{ v(k_j) - \frac{(\hat{\mathbf{e}} \cdot \hat{\mathbf{d}})^2}{n_j^2 - 1} \right\} = 0 \tag{250}$$

which is identical with (133).

If fictitious polarizations P_l with $l = 0, -1, -2, \ldots, 1 - L$ are defined by means of (240) it can be shown that[164]

$$P_l = 0 \qquad l = 0, -1, -2, \ldots, 1 - L \tag{251}$$

These are the boundary conditions of the problem.

The electric field at the first surface layer ($l = 1$) consists of incident and reflected parts

$$E^{\perp}(a) = E_{\text{in}}^{\perp}\, e^{i\Omega} + R^{\perp} e^{-i\Omega} \tag{252}$$

where from (243)

$$R^{\perp} = -i(2\pi\Omega)(\hat{\mathbf{e}} \cdot \hat{\mathbf{d}}) \sum_j \frac{P_j}{1 - \exp\left[-i\Omega(1 + n_j)\right]} \tag{253}$$

The reflectivity for normal incidence is

$$R(\omega) = \left| \frac{R^{\perp}}{E_{\text{in}}^{\perp}} \right|^2 \tag{254}$$

Mahan and Obermair[164] have shown that substitution of (245) and (253) into this expression leads, after some lengthy algebraic manipulations, to the remarkable result

$$R(\omega) = \prod_{j=1}^{L+1} \left| \frac{n_j(\omega) - 1}{n_j(\omega) + 1} \right|^2 \tag{255}$$

This expression, which generalizes an earlier one due to Hopfield,[166,167] shows that the reflectivity is the product of separate reflectivity factors for each polariton mode. In practice most of the refractive indices $n_j(\omega)$ calculated from (247) are complex and therefore contribute little to $R(\omega)$, and those that are pure imaginary contribute nothing at all.

On the basis of dipole calculations Mahan and Obermair[164] concluded that local field variations near the surface of perfect cubic crystals contribute very little to the optical reflectivity at normal incidence. It is very likely that the same holds true for aromatic hydrocarbon crystals, for preliminary calculations[165] indicate that for several faces of anthracene the planar components $v_{l-l'}(0)$ of dipole sums also fall off rapidly with increasing $|l - l'|$. This observation appears to open up a new avenue of research by making calculations of the effects of surface disorder on crystal spectra a practical possibility. It also suggests that Davydov splittings of the stronger transitions are *not* dependent on thickness for crystals more than a few layers thick. The dipole theory has, of course, oversimplified the local field at the surface; site shifts and nondipolar exciton interactions must be included in any refinement of this theory. One way of including site shifts in the dipole theory is to alter the excitation energies of molecules in the surface layers that is their polarizabilities $\alpha(\omega)$.

7. *Lattice Sums of Instantaneous and Retarded Dipole Interactions*[170–175]

No subject has been more hotly debated among theoreticians interested in molecular excitons than the method of performing lattice sums of point dipole-dipole interactions. The origin of much of the confusion is a thoughtless application of the approximate $\Delta\mathbf{k} = 0$ selection rule. A dilemma arose in the following way. According to the $\Delta\mathbf{k} = 0$ rule a transition from the ground state must end on a $\mathbf{k} = 0$ upper state, and from perturbation theory the factor group splitting is proportional to

$$I_{\beta\gamma}(\mathbf{k} = 0) = \boldsymbol{\mu}_\beta \cdot \sum_m R^{-3}(\mathbf{1} - 3\hat{\mathbf{R}}\hat{\mathbf{R}}) \cdot \boldsymbol{\mu}_\gamma \tag{256}$$

where $\mathbf{R} = \mathbf{r}_{\beta n} - \mathbf{r}_{\gamma m}$. This dipole sum contains no phase modulation and is conditionally convergent; that is, the result depends on the order in which the individual terms are summed. In particular the sum depends on the shape of the volume (of the crystal) containing the lattice points included in the sum. This volume was equated to that occupied by the exciton created by photon absorption and thus to the boundary conditions on the exciton wave function. The dipole sums depend markedly on the shape assumed, being quite different for spheres, cylinders, and slabs. At this stage it was suggested that because of retardation (propagation of all signals with velocity of light c) the interactions between widely separated dipoles would be weakened in some way and the dipole sum would actually converge. To change from an instantaneous sum to a retarded one in (256) is *incorrect*, amounting to an unjustifiable change of gauge in only part of the total Hamiltonian (crystal and field). This does not deny that exciton theory can be formulated in terms of retarded interactions, only that it must be done consistently with proper consideration of the gauge of the electromagnetic field. It should be mentioned, however, that retarded dipole sums are singular due to a term proportional to $[k^2 - \omega^2/c^2]^{-1}$ which causes the sum to diverge when the magnitude of the wave vector $k(= |\mathbf{k}|)$ get close to the frequency ω divided by c. Computer calculations of retarded dipole interactions must be based on formulas from which this divergent term has been removed.

The dilemma described above for instantaneous dipole sums can be avoided in several ways. The first way is to consider an infinite crystal without damping processes. For transitions from the crystal ground state the exact selection rule is $\mathbf{k} = \mathbf{q}$ where $\mathbf{q}$ is the photon's wave vector. Thus the dipole sum has a small, but nevertheless finite, phase modulation. According to classical electrodynamics an infinite, continuous medium with polarization $(\mathbf{d}/v_0) \exp(i\mathbf{k} \cdot \mathbf{r})$ has a macroscopic electric field equal to

$$\mathbf{E}^{\parallel}_{\text{cont}}(\mathbf{r}) = -\left(\frac{4\pi}{v_0}\right) \hat{\mathbf{k}}(\hat{\mathbf{k}} \cdot \mathbf{d}) \exp(i\mathbf{k} \cdot \mathbf{r}) \tag{257}$$

Therefore, to be consistent with classical theory the dipole sum must be computed in a manner that does not alter this macroscopic field. One such way is Ewald's method[170] which leads to a formula of the type

$$\mathbf{T}_{\beta\gamma}(\mathbf{k}) = \mathbf{t}_{\beta\gamma}(\mathbf{k}) + [\hat{\mathbf{k}}\hat{\mathbf{k}} - \tfrac{1}{3}\mathbf{1}] \tag{258}$$

The exciton energy calculated in this way contains the following direction-dependent term

$$\frac{4\pi}{v_0} (\boldsymbol{\mu}_\beta \cdot \hat{\mathbf{k}}) (\hat{\mathbf{k}} \cdot \boldsymbol{\mu}_\gamma) \tag{259}$$

where $\boldsymbol{\mu}_\beta$ and $\boldsymbol{\mu}_\gamma$ are the oriented transition dipoles.

The method of dipole summation just described may be criticized as irrelevant since real crystals are not infinite and in practice are often very thin, being in some cases a thousand angstroms thick. This is countered by the fact that Ewald's method gives results identical with those of de Wette-Schacher's planewise method of summation.[171] We have already mentioned that the planar sums $\mathbf{v}_{l-l'}$ (computed parallel to the crystal surface and therefore perpendicular to $\mathbf{k}$) fall off so rapidly that all molecules except those in the first few layers see the same dipole field.[172] In other words, the Ewald sums are good for thin crystals.

A complete resolution of the dipole sum problem will come only with the development of a practical theory of molecular excitons and polaritons in semi-infinite and thin crystals. The treatment described in the last section is an encouraging beginning.

Retarded dipole interactions arise naturally in an exciton theory formulated in Lorentz gauge. In Lorentz gauge the vector and scalar potentials are connected together by the invariant relation

$$\boldsymbol{\nabla} \cdot \tilde{\mathbf{A}} + \frac{1}{c}\frac{\partial}{\partial t}\tilde{\phi} = 0 \tag{260}$$

The equation of motion of the dipoles in this gauge reduces to

$$\sum_{s'} [\delta_{ss'}\mathbf{1} + 4\pi\boldsymbol{\alpha}_s(\omega) \cdot \boldsymbol{\Psi}_{ss'}(\omega)] \cdot \mathbf{P}_{s'} = \boldsymbol{\alpha}_s(\omega) \cdot \boldsymbol{\varepsilon}_0(\mathbf{r}_s) \tag{261}$$

where $\boldsymbol{\varepsilon}_0$ is the externally applied field. $\boldsymbol{\Psi}_{ss'}$ is the dimensionless dyadic for retarded dipole interactions

$$\boldsymbol{\Psi}_{ss'}(\omega) = -\frac{v_0}{4\pi}\left(\boldsymbol{\nabla}_s\boldsymbol{\nabla}_s + \frac{\omega^2}{c^2}\mathbf{1}\right) R^{-1} \exp\left(\frac{iR\omega}{c}\right) \tag{262}$$

where $\mathbf{R} = \mathbf{r}_s - \mathbf{r}_{s'}$. For an infinite crystal the secular determinant is

$$\det |\mathbf{1} + 4\pi\boldsymbol{\alpha}(\omega) \cdot \boldsymbol{\Psi}(\mathbf{k},\omega)| = 0 \tag{263}$$

where

$$\boldsymbol{\Psi}(\mathbf{k},\omega) = \sum_{s'} \boldsymbol{\Psi}_{ss'}(\omega) \exp [i\mathbf{k} \cdot (\mathbf{r}_s - \mathbf{r}_{s'})] \tag{264}$$

Unlike the previous cases in which Coulomb gauge was used, (263) *is the polariton dispersion relation.* In other words, (263) is *not* a secular determinant for exciton frequencies.

The connection between instantaneous and retarded dipole sums becomes transparent if the summations are carried out over the reciprocal lattice instead of the more usual real lattice. For simplicity only inequivalent dipole sums are considered here. The instantaneous dipole sum for a dipole $\boldsymbol{\mu}_\beta$ at $\mathbf{x}$ interacting with a Bravais lattice of dipoles $\boldsymbol{\mu}_\gamma \exp (i\mathbf{k} \cdot \mathbf{r}_n)$ at sites $\mathbf{r}_n$ is

$$I_{\beta\gamma}(\mathbf{k}) = -\boldsymbol{\mu}_\beta \cdot \mathbf{E}_\gamma^{\parallel}(\mathbf{x}) \exp (-i\mathbf{k} \cdot \mathbf{x}) \tag{265}$$

where

$$\mathbf{E}_\gamma^{\parallel}(\mathbf{x}) = -\sum_n R^{-3}(\mathbf{1} - 3\hat{\mathbf{R}}\hat{\mathbf{R}}) \cdot \boldsymbol{\mu}_\gamma \exp (i\mathbf{k} \cdot \mathbf{r}_n) \tag{266}$$

with $\mathbf{R} = \mathbf{x} - \mathbf{r}_n$. The electric field can be written more simply as

$$\mathbf{E}_\gamma^{\parallel}(\mathbf{x}) = \boldsymbol{\nabla\nabla} \cdot \sum_n R^{-1}\boldsymbol{\mu}_\gamma \exp (i\mathbf{k} \cdot \mathbf{r}_n) \tag{267}$$

The lattice sum in (267) is a function periodic in the lattice and as such may be expressed as a sum over the reciprocal lattice

$$\mathbf{E}_\gamma^{\parallel}(\mathbf{x}) = \boldsymbol{\nabla\nabla} \cdot \left(\frac{4\pi}{v_0}\right) \sum_{\mathbf{K}} |\mathbf{k} + \mathbf{K}|^{-2} \exp [i(\mathbf{k} + \mathbf{K}) \cdot \mathbf{x}]\boldsymbol{\mu}_\gamma \tag{268}$$

where $\mathbf{K}$ are the reciprocal lattice vectors. Therefore

$$\mathbf{E}_\gamma^{\parallel}(\mathbf{x}) = -\left(\frac{4\pi}{v_0}\right) \sum_{\mathbf{K}} \frac{(\mathbf{k} + \mathbf{K})(\mathbf{k} + \mathbf{K}) \cdot \boldsymbol{\mu}_\gamma}{|\mathbf{k} + \mathbf{K}|^2} \exp [i(\mathbf{k} + \mathbf{K}) \cdot \mathbf{x}] \tag{269}$$

The retarded dipole sum for the same Bravais lattice is

$$I_{\beta\gamma}^{\text{ret}}(\mathbf{k},\omega) = -\boldsymbol{\mu}_\beta \cdot \mathbf{E}_\gamma^{\text{ret}}(\mathbf{x}) \exp(-i\mathbf{k}\cdot\mathbf{x}) \tag{270}$$

where

$$\mathbf{E}_\gamma^{\text{ret}}(\mathbf{x}) = \sum_n \exp\left(\frac{i\omega R}{c}\right)\left[R^{-1}(\mathbf{1} - \hat{\mathbf{R}}\hat{\mathbf{R}})\frac{\omega^2}{c^2} - R^{-3}(\mathbf{1} - 3\hat{\mathbf{R}}\hat{\mathbf{R}})\left(1 - \frac{i\omega R}{c}\right)\right]\cdot \boldsymbol{\mu}_\gamma \exp(i\mathbf{k}\cdot\mathbf{r}_n) \tag{271}$$

This electric field can also be written in a simpler compact fashion

$$\mathbf{E}_\gamma^{\text{ret}}(\mathbf{x}) = \left[\boldsymbol{\nabla\nabla} + \frac{\omega^2}{c^2}\mathbf{1}\right]\cdot \sum_n R^{-1}\exp\left[i\left(\frac{\omega R}{c} + \mathbf{k}\cdot\mathbf{r}_n\right)\right]\boldsymbol{\mu}_\gamma \tag{272}$$

The sum in the right-hand side (272) is another function periodic in the lattice and after transforming to a **K**-sum we find

$$\mathbf{E}_\gamma^{\text{ret}}(\mathbf{x}) = -\left[\boldsymbol{\nabla\nabla} + \frac{\omega^2}{c^2}\mathbf{1}\right]\cdot\left(\frac{4\pi}{v_0}\right)\sum_{\mathbf{K}}\frac{\exp[i(\mathbf{k}+\mathbf{K})\cdot\mathbf{x}]\boldsymbol{\mu}_\gamma}{(\omega^2/c^2) - |\mathbf{k}+\mathbf{K}|^2} \tag{273}$$

or

$$\mathbf{E}_\gamma^{\text{ret}}(\mathbf{x}) = -\left(\frac{4\pi}{v_0}\right)\sum_{\mathbf{K}}\frac{[(\omega^2/c^2)\mathbf{1} - (\mathbf{k}+\mathbf{K})(\mathbf{k}+\mathbf{K})]\cdot\boldsymbol{\mu}_\gamma}{(\omega^2/c^2) - |\mathbf{k}+\mathbf{K}|^2}\exp[i(\mathbf{k}+\mathbf{K})\cdot\mathbf{x}] \tag{274}$$

This retarded field may now be divided into longitudinal and transverse parts. The longitudinal part equals $\mathbf{E}_\gamma^{\parallel}(\mathbf{x})$, the field from the instantaneous dipole interactions

$$\mathbf{E}_\gamma^{\parallel\,\text{ret}}(\mathbf{x}) = \mathbf{E}_\gamma^{\parallel}(\mathbf{x}) \tag{275}$$

and the transverse part is

$$\mathbf{E}_\gamma^{\perp\text{ret}}(\mathbf{x}) = -\left(\frac{4\pi}{v_0}\right)\frac{\omega^2}{c^2}\sum_{\mathbf{K}}\frac{[\mathbf{1} - (\mathbf{k}+\mathbf{K})(\mathbf{k}+\mathbf{K})]\cdot\boldsymbol{\mu}_\gamma}{(\omega^2/c^2) - |\mathbf{k}+\mathbf{K}|^2}\exp[i(\mathbf{k}+\mathbf{K})\cdot\mathbf{x}] \tag{276}$$

All reciprocal lattice vectors, including $\mathbf{K} = 0$, are included in this sum. For frequencies in the optical region the most important term in (276) is the $\mathbf{K} = 0$ term, namely

$$-\left(\frac{4\pi}{v_0}\right)\frac{\omega^2}{c^2}\frac{[\mathbf{1} - \hat{\mathbf{k}}\hat{\mathbf{k}}]\cdot\boldsymbol{\mu}_\gamma}{(\omega^2/c^2) - k^2}\exp(i\mathbf{k}\cdot\mathbf{x}) \tag{277}$$

This particular term is singular at $\omega^2 = c^2k^2$ so that the whole retarded dipole sum, (271), "blows up" when the magnitude of the exciton wave vector $|\mathbf{k}|$ equals ω/c. This justifies our comment made earlier that machine calculations of retarded dipole interactions, like (271), must in some fashion subtract out the singular part (277). To our knowledge no calculation of retarded dipole interactions using equations like (271) have subtracted out the singular part. An Ewald method of performing the retarded sums with the singular part subtracted has been described by Mahan.[173]

The derivation of the polariton dispersion relation described in Section II.B.1 neglected the $\mathbf{K} \neq 0$ terms of $\mathbf{E}_\gamma$. The $\mathbf{K} = 0$ term of (276) is recognizable as the exciton–photon coupling constant divided by the vacuum dispersion relation for the photon $[(\omega^2/c^2) - k^2 = 0]$. This proves the assertion, made earlier, that (263) is the polariton dispersion relation. The crucial question yet to be answered is how important are the $\mathbf{K} \neq 0$ terms. Rough estimates suggest they are very small compared to $\mathbf{E}_\gamma{}^{\parallel}$.[173,174] The question could be finally settled by some careful machine calculations.

C. Second Quantized Theory of Single Exciton States

The treatment of molecular excitons using second quantization techniques was pioneered by Agranovich.[152] The theoretical framework is now sufficiently well developed to be able to account for all single-particle properties of polymers and molecular crystals. In principle there are few restrictions on the nature of the exciton interactions which may be of the dipole-dipole, higher multipole, electron-exchange, etc., type. Generality, however, is synonymous with complexity in notation, and the equations sometimes allow little in the way of physical insight.

Before considering specific problems by the method of second quantization the basic theory is developed for Coulombic excitons in an aggregate with an arbitrary arrangement of monomers (no spatial symmetry). Coulomb gauge is used for the electromagnetic potentials so that the exciton part of the total Hamiltonian contains instantaneous intermolecular interactions.

The total Hamiltonian of a molecular aggregate and radiation field can be written:

$$\mathcal{H} = \mathcal{H}_{\text{agg}} + \mathcal{H}_{\text{rad}} + \mathcal{H}_{\text{int}} \tag{278}$$

The aggregate Hamiltonian has been defined already [see Eq. (12)]. In terms of creation and annihilation operators $a_{\mathbf{q}\lambda}{}^{+}$ and $a_{\mathbf{q}\lambda}$ for photon, the radiation Hamiltonian is

$$\mathcal{H}_{\text{rad}} = \sum_{\mathbf{q}\lambda} \hbar c q a_{\mathbf{q}\lambda}{}^{+} a_{\mathbf{q}\lambda} \tag{279}$$

where $\lambda = 1,2$ is the linear polarization label. The interaction operator is

$$\mathcal{H}_{\text{int}} = -\sum_{sa} \frac{e_a}{m_a c} \mathbf{p}_{sa} \cdot \mathbf{A}(\mathbf{r}_{sa}) + \sum_{sa} \frac{e_a^2}{2m_a c^2} [\mathbf{A}(\mathbf{r}_{sa})]^2 \tag{280}$$

where $\mathbf{p}_{sa}$ is the momentum of the ath charge (position $\mathbf{r}_{sa}$) on molecule s. The vector potential is given by

$$\mathbf{A}(\mathbf{r}) = \sum_{\mathbf{q}\lambda} \left(\frac{2\pi\hbar c}{qV}\right)^{1/2} \mathbf{e}_{\mathbf{q}\lambda}(a_{\mathbf{q}\lambda}{}^{+}e^{-i\mathbf{q}\cdot\mathbf{r}} + a_{\mathbf{q}\lambda}e^{i\mathbf{q}\cdot\mathbf{r}}) \tag{281}$$

where V is the quantization volume of the electromagnetic field.

In the second quantization representation the Hamiltonian of the aggregate is usually taken as

$$\mathcal{H}_{\text{ex}} = E_G + \sum_{su} \Delta_{su} B_{su}{}^{+} B_{su} + \tfrac{1}{2} \sum_{su} \sum_{s'u'}{}' V_{ss'}(0u' \mid u0)(B_{su} + B_{su}{}^{+})(B_{s'u'} + B_{s'u'}{}^{+}) \tag{282}$$

where $B_{su}{}^{+}$, and B_{su} are creation and annihilation operators for excitons satisfying Boson commutation rules, and

$$\Delta_{su} = \epsilon_s{}^u - \epsilon_s{}^0 + D_s{}^u \tag{283}$$

$$V_{ss'}(0u' \mid u0) = (\varphi_s{}^0 \varphi_{s'}{}^{u'} \mid V_{ss'} \mid \varphi_s{}^u \varphi_{s'}{}^0) \tag{284}$$

The notation introduced in Section II,A has been used in (283) and (284).

The exciton Hamiltonian is derived via a number of approximations worth explaining. First the exciton operators actually satisfy the Pauli commutation rules.[176–178] However, at the very low exciton concentrations encountered in practice the Pauli commutators are quite accurately approximated by Bose commutators. Pauli operators can be expanded in a series of Bose or Fermi operators. If there is only one excited state the expansion is given by the Holstein-Primakoff transformation.[176] the expansions become more complicated when there are several excited states.[178] Second, if electron exchange between molecules has important energetic consequences then the basis functions (13)–(16) should be properly normalized before going over to the second quantized representation. If, on the other hand, electron exchange is not of major importance, it is probably sufficient to slip the antisymmetrizer into matrix elements (284). Multiparticle terms describing three or more exciton processes have been ignored since these are of secondary importance in single-exciton theory. Finally, the site shift matrix $D_s^{uu'}$, (30), is assumed to be diagonal in the excited state indices.

In the exciton approximation the total Hamiltonian is quadratic in creation and annihilation operators for photons and excitons, and can be formally diagonalized by a canonical transformation. The diagonalization procedure is fully described in the book by Tyablikov.[179] The total Hamiltonian may be diagonalized in one step, in which case equations similar to those of the dipole theory of polaritons are obtained, or in two steps. In the latter method, $\mathcal{H}_{ex}$ is diagonalized first, the interaction $\mathcal{H}_{int}$ expressed in terms of the new exciton operators, and then the complete Hamiltonian treated. This two-step route opens a different perspective compared to classical dipole theory.

According to Tyablikov's method the formal diagonalization of $\mathcal{H}_{ex}$ is achieved by finding a set of Bose operators ξ_ν^+, ξ_ν such that

$$[\mathcal{H}_{ex}, \xi_\nu^+] = E_\nu \xi_\nu^+ \tag{285}$$

The new operators are given by the transformation

$$\xi_\nu = \sum_{su} [u_{su}^*(\nu) B_{su} - v_{su}(\nu) B_{su}^+] \tag{286}$$

The inverse transformation is

$$B_{su} = \sum_{\nu} [u_{su}(\nu)\xi_\nu + v_{su}^*(\nu)\xi_\nu^+] \tag{287}$$

In terms of the quasiparticle operators ξ_ν^+, ξ_ν

$$\mathcal{H}_{ex} = \mathcal{E}_G + \sum_{\nu} E_\nu \xi_\nu^+ \xi_\nu \tag{288}$$

where

$$\mathcal{E}_G = E_G - \sum_{\nu} \sum_{su} E_\nu |v_{su}(\nu)|^2 \tag{289}$$

is the "exact" ground-state energy of the aggregate. The Tyablikov parameters u and v satisfy sets of simultaneous equations obtained by substituting (282), (287), and complex conjugate into (285), and then comparing coefficients. We find

$$(E - \Delta_{su})u_{su} = \sum_{s'u'} V_{ss'}(0u'|u0)\tilde{u}_{s'u'} \tag{290}$$

$$-(E + \Delta_{su})v_{su} = \sum_{s'u'} V_{ss'}(0u'|u0)\tilde{u}_{s'u'} \tag{291}$$

where

$$\tilde{u}_{su} = u_{su} + v_{su} \tag{292}$$

There is also a normalization condition

$$\sum_{su} [u_{su}(\nu)u_{su}^*(\nu') - v_{su}(\nu)v_{su}^*(\nu')] = \delta_{\nu\nu'} \tag{293}$$

The simultaneous equations (290) and (291) are readily combined into a single set for the coefficients $\tilde{u}$:

$$\sum_{s'u'} [(\Delta_{su}^2 - E^2)\delta_{ss'}\delta_{uu'} + 2\Delta_{su}V_{ss'}(0u' \mid u0)]\tilde{u}_{s'u'} = 0 \tag{294}$$

In the dipole approximation these equations are completely equivalent to the secular equations for Coulombic excitons in the classical theory of Section II.A [see Eq. (110)]. Finally the normalization equation for the coefficients $\tilde{u}$ is

$$\sum_{su} (\Delta_{su})^{-1}\tilde{u}_{su}(\nu)\tilde{u}_{su}^*(\nu') = (E_\nu)^{-1}\delta_{\nu\nu'} \tag{295}$$

1. Coulombic Excitons in Infinite Crystals and Polymers

It is assumed that the crystal or helical polymer has periodic boundary conditions with N unit cells (or monomers in the case of a helix) in the periodic volume. The quasiparticle label ν now becomes $\mathbf{k}\nu$ where $\mathbf{k}$ is the exciton wave vector and ν corresponds to branch and excited state. For a crystal with sites $n\beta$ the Tyablikov coefficients are

$$\tilde{u}_{n\beta u}(\mathbf{k}\nu) = N^{-1/2}\tilde{u}_{\beta u}(\mathbf{k}\nu) \exp(i\mathbf{k} \cdot \mathbf{r}_{n\beta}) \tag{296}$$

and the secular equations are diagonal in $\mathbf{k}$:

$$\sum_{\gamma u'} \{[\Delta_u^2 - E^2(\mathbf{k}\nu)]\delta_{\beta\gamma}\delta_{uu'} + 2\Delta_u I_{\beta\gamma}^{uu'}(\mathbf{k})\}\tilde{u}_{\gamma u'}(\mathbf{k}\nu) = 0 \tag{297}$$

These equations resemble those of the classical dipole theory in structure except for the appearance of the full exciton interaction $I_{\beta\gamma}(\mathbf{k})$ which we last saw in the finite basis theories. Further simplification of (297) depends on the direction of $\mathbf{k}$ and the conditions are essentially those discussed in Section II.A.

Equations (297) represent a generalization of the dipole secular equations to include nondipole exciton interactions. These are the equations from which exciton energies should be calculated for crystals and polymers with several coupled intense transitions.

For those directions and points of $\mathbf{k}$-space of sufficiently high symmetry that different exciton branches do not interact, the secular determinant is

$$\det |(\Delta_u^2 - E^2)\delta_{uu'} + 2\Delta_u I^{uu'}(\mathbf{k}\zeta)| = 0 \tag{298}$$

where ζ is the branch label and $I^{uu'}(\mathbf{k}\zeta)$ a factor group lattice sum. The order of determinant (298) is just the number of excited molecular states included in the calculation. By some formal manipulations (298) can be written in the following form

$$E^2 = \Delta_u^2 + 2\Delta_u \mathcal{G}^u(\mathbf{k}\zeta,E) \tag{299}$$

where

$$\mathcal{G}^u(\mathbf{k}\zeta,E) = I^{uu}(\mathbf{k}\zeta) - \sum_{v(\neq u)} \sum_{v'(\neq u)} I^{uv}(\mathbf{k}\zeta)(\mathfrak{M}^{-1})_{vv'} 2\Delta_{v'} I^{v'u}(\mathbf{k}\zeta) \quad (300)$$

with matrix $\mathfrak{M}$ defined by

$$\mathfrak{M}_{vv'} = (\Delta_v{}^2 - E^2)\delta_{vv'} + 2\Delta_v I^{vv'}(\mathbf{k}\zeta) \quad (301)$$

If the state u is fairly isolated, (299) should prove a convenient starting point for an iterative solution. The factor group splittings are obtained by solving (299) for the various branches ζ.

Collective coupling of vibronic levels was discussed earlier (Section II.B.1) in the dipole approximation. The problem is briefly reconsidered here.[175] Suppose the first electronic state is isolated (a common occurrence for cationic dye molecules) and let the vibronic levels be denoted $1(u)$. Assuming the vibronic wave functions to be separable, then for crystals with one molecule per unit cell, the secular determinant reduces to

$$1 + I(\mathbf{k}) \sum_u \frac{2\Delta_{1(u)}\xi^2_{1(u)}}{\Delta^2_{1(u)} - E^2} = 0 \quad (302)$$

where $\xi_{1u}{}^2$ is the Franck-Condon factor for the transition $0(0) \rightarrow 1(u)$ in the free molecule, and $I(\mathbf{k})$ is the purely electronic part of the lattice sum $I^{1(u),1(u)}(\mathbf{k})$. The source of the collective coupling is $I(\mathbf{k})$, the full interaction, not just the dipole component.

The result of the formal diagonalization of $\mathcal{H}_{\text{ex}}$ is written

$$\mathcal{H}_{\text{ex}} = \mathcal{E}_G + \sum_{\mathbf{k}v} E(\mathbf{k}\nu)\xi_{k\nu}{}^+\xi_{k\nu} \quad (303)$$

where the quasiparticle operators are defined by

$$\xi_{\mathbf{k}\nu} = \sum_{\beta u} [u_{\beta u}{}^*(\mathbf{k}\nu)B_{\mathbf{k}\beta u} - v_{\beta u}(\mathbf{k}\nu)B^+_{\mathbf{k}\beta u}] \quad (304)$$

and the complex conjugate equation. Here

$$B_{\mathbf{k}\beta u} = N^{-1/2} \sum_n B_{n\beta u} \exp(-i\mathbf{k} \cdot \mathbf{r}_{n\beta}) \quad (305)$$

2. Polaritons in Infinite Molecular Crystals

In this section the dispersion relation for polaritons in infinite crystals with periodic boundaries is derived, following mainly the work of Agranovich.[152,180] For simplicity only crystals with a center of inversion and one molecule per unit are considered. This excludes crystals that are optically active due either to asymmetric molecules or to lack of an inversion center. The lattice sums $I^{uu}(\mathbf{k})$ are real, as are the Tyablikov coefficients $\tilde{u}$.

The interaction between crystal and photons is

$$\mathcal{H}_{\text{int}} = \mathcal{H}'_{\text{int}} + \mathcal{H}''_{\text{int}} \tag{306}$$

where

$$\mathcal{H}'_{\text{int}} = -\sum_n \sum_{\mathbf{q}\lambda} \left(\frac{2\pi\hbar c}{qV}\right)^{1/2} \left(\frac{e}{mc}\right) (a_{\mathbf{q}\lambda} e^{i\mathbf{q}\cdot\mathbf{r}_n} + a_{\mathbf{q}\lambda}{}^+ e^{-i\mathbf{q}\cdot\mathbf{r}_n}) (\mathbf{e}_{\mathbf{q}\lambda} \cdot \mathbf{J}_n) \tag{307}$$

and

$$\mathcal{H}''_{\text{int}} = \frac{1}{4} n_e \sum_{\mathbf{q}\lambda} \left(\frac{\hbar\omega_0{}^2}{qc}\right) (a_{\mathbf{q}\lambda} a_{-\mathbf{q}\lambda} + a_{\mathbf{q}\lambda}{}^+ a_{\mathbf{q}\lambda} + a_{\mathbf{q}\lambda} a_{\mathbf{q}\lambda}{}^+ + a_{\mathbf{q}\lambda}{}^+ a_{-\mathbf{q}\lambda}) \tag{308}$$

The operator $\mathbf{J}_n$ is given by

$$\mathbf{J}_n = \sum_a \mathbf{p}_{na} \tag{309}$$

and the number of electrons per molecule is n_e. In deriving (307) and (308) the vector potential has been evaluated at the center of the molecule by setting

$$\exp(i\mathbf{q} \cdot \mathbf{r}_{na}) = \exp(i\mathbf{q} \cdot \mathbf{r}_n) \tag{310}$$

for all electrons a. This approximation means that electric quadrupole and magnetic dipole coupling of molecules to the electromagnetic field, important for optical activity, has been neglected.

The explicit sum over unit cells n in $\mathcal{H}'_{\text{int}}$ is removed by defining

$$\mathbf{J}_{\pm\mathbf{q}} = \sum_n \mathbf{J}_n \exp(\pm i\mathbf{q} \cdot \mathbf{r}_n) \tag{311}$$

which is related to $\mathbf{D}_{\mathbf{q}}$, (22), by

$$\mathbf{J}_{\mp q} \simeq \frac{im}{\hbar e} [\mathcal{H}_{\text{ex}}, \mathbf{D}_{\pm q}] \tag{312}$$

This result is approximate because the full Hamiltonian $\mathcal{H}$ has been replaced by $\mathcal{H}_{\text{ex}}$. After expressing the dipole operator $\mathbf{D}_{\mathbf{q}}$ in terms of the quasiparticle operators $\xi_{\mathbf{q}\nu}$, $\xi_{\mathbf{q}\nu}{}^+$

$$\mathcal{H}'_{\text{int}} = \sum_\nu \sum_{\mathbf{q}\lambda} iC(\mathbf{q}\lambda\nu)[a_{\mathbf{q}\lambda}(\xi_{-\mathbf{q}\nu} - \xi_{\mathbf{q}\nu}{}^+) + a_{\mathbf{q}\lambda}{}^+(\xi_{\mathbf{q}\nu} - \xi_{-\mathbf{q}\nu}{}^+)] \tag{313}$$

where

$$C(\mathbf{q}\lambda\nu) = \sum_u \left(\frac{2\pi N}{\hbar c q V}\right)^{1/2} (\mathbf{e}_{q\lambda} \cdot \boldsymbol{\mu}^u) E(\mathbf{q}\nu) \tilde{u}_u(\mathbf{q}\nu) \tag{314}$$

It should be pointed out here that although the sum over $\mathbf{q}$ is unrestricted, only the projection of $\mathbf{q}$ onto the first Brillouin zone has any meaning for the operators $\xi_{\mathbf{q}\nu}$ because of the relation

$$\exp\,[i(\mathbf{k} + \mathbf{K}) \cdot \mathbf{r}_n] = \exp\,(i\mathbf{k} \cdot \mathbf{r}_u) \tag{315}$$

where $\mathbf{K}$ is any reciprocal lattice vector.

In the second stage of the diagonalization of $\mathcal{H}$, Tyablikov equations are set up for the Hamiltonian $\mathcal{H}$ consisting of a sum of (303), (313), and (308). If the coupling between an exciton with wave vector $\mathbf{k}$ to photons $\mathbf{k} + \mathbf{K}$ lying *outside* the first zone is ignored (sometimes called neglect of photon Umklapp processes), the secular determinant reduces to

$$\det \left| [E^2 - (\hbar ck)^2 - n_e(\hbar\omega_0)^2]\delta_{\lambda\lambda'} + 2\hbar ck \sum_\nu \frac{2E(\mathbf{k}\nu)C(\mathbf{k}\lambda\nu)C(\mathbf{k}\lambda'\nu)}{E^2 - [E(\mathbf{k}\nu)]^2} \right| = 0 \tag{316}$$

Notice that a term $n_e(\hbar\omega_0)^2$ appears in the diagonal part of this 2×2 determinant. This odd term can be eliminated by use of the oscillator strength sum rule

$$\sum_\nu \mathbf{F}(\mathbf{k}\nu) = n_e\mathbf{1} \tag{317}$$

where

$$\mathbf{F}(\mathbf{k}\nu) = \frac{2m}{\hbar^2 e^2} E(\mathbf{k}\nu) \sum_{uv} \tilde{u}_u(\mathbf{k}\nu)\tilde{u}_v(\mathbf{k}\nu)\boldsymbol{\mu}^u\boldsymbol{\mu}^v \tag{318}$$

Equation (316) describes the coupling of exciton and photon with wave vector $\mathbf{k}$; it is therefore the polariton dispersion relation. The refractive index is

$$n = \frac{\hbar ck}{E} \tag{319}$$

and after expanding the determinant we find

$$n_\pm^2 = 1 - \frac{1}{2}\sum_\nu \frac{\omega_0^2 F(\mathbf{k}\nu)[\sin\Theta_{3\nu}]^2}{\omega^2 - [\Omega(\mathbf{k}\nu)]^2} \pm \left[\left(\frac{1}{2}\sum_\nu \frac{\omega_0^2 F(\mathbf{k}\nu)}{\omega^2 - [\Omega(\mathbf{k}\nu)]^2}[\cos^2\Theta_{1\nu} - \cos^2\Theta_{2\nu}]\right)^2 + \left(\sum_\nu \frac{\omega_0^2 F(\mathbf{k}\nu)}{\omega^2 - [\Omega(\mathbf{k}\nu)]^2}\cos\Theta_{1\nu}\cos\Theta_{2\nu}\right)^2\right]^{1/2} \tag{320}$$

where $F(\mathbf{k}\nu)$ is the magnitude of the oscillator strength and $\hbar\Omega(\mathbf{k}\nu) = E(\mathbf{k}\nu)$, is the frequency of a Coulombic exciton obtained in the diagonalization of $\mathcal{H}_{ex}$ in the first step. The $\Theta_{i\nu}$ ($i = 1, 2, 3$) are the angles between

$$\sum_u \tilde{u}_u(\mathbf{k}\nu)\boldsymbol{\mu}^u \tag{321}$$

and the unit vectors $\mathbf{e}_{\mathbf{k}\lambda}$ ($\lambda = 1, 2, 3$) where $\mathbf{e}_{\mathbf{k}3} = \mathbf{k}$.

The refractive index according to classical dipole theory was mild compared to (320). In the present case the exciton frequencies appear explicitly, and this can be used to introduce approximations that are difficult to use in the dipole theory. For example, suppose for an isolated transition ν, $\mathbf{e}_{\mathbf{k}1}$, $\hat{\mathbf{k}}$, and transition dipole (321) are coplanar; then

$$n^2 = \epsilon + \frac{\omega_0{}^2F(\mathbf{k}\nu)}{[\Omega(\mathbf{k}\nu)]^2 - \omega^2}\cos^2\Theta_{1\nu} \tag{322}$$

where ϵ is the background dielectric due to all the other transitions. According to this model the reflection band starts at

$$(\omega_-)^2 = [\Omega(\mathbf{k}\nu)]^2 \tag{323}$$

and ends at

$$(\omega_+)^2 = [\Omega(\mathbf{k}\nu)]^2 + \epsilon^{-1}\omega_0{}^2F(\mathbf{k}\nu)\cos^2\Theta_{1\nu} \tag{324}$$

The width of the reflection band is therefore reduced by a factor equal to the background dielectric constant [see comments following (147)], an important consideration for metallic reflection phenomena.

3. *Interstitial Impurities in Infinite Crystals*[159–161]

As has already been mentioned briefly, a crystal containing interstitial impurities may be regarded as the prototype for the optical properties of a large class of biologically important systems. Some of these systems are complexes formed by two large polymers (e.g., protein–DNA complexes) or consist of one long polymer with many bound ligands (e.g., DNA-dye complexes). Distortions of the host structure almost certainly occur upon complex formation. In this section the exciton problem is treated in a way that formally, at least, allows for host distortion. Secondary trap interactions are neglected here.

Let the host sites be denoted n, n', and the impurity sites x, x', ... From (294) the secular equations for the impurities may be written

$$\sum_{x'v'}\left[\delta_{vv'}\delta_{xx'} + \frac{2\Delta_{xv}}{\Delta_{xv}{}^2 - E^2}V_{xx'}(0v'\,|\,v0)\right]\tilde{u}_{x'v'} = -\frac{2\Delta_{xv}}{\Delta_{xv}{}^2 - E^2}\sum_{nu}V_{xn}(0u\,|\,v0)\tilde{u}_{nu} \tag{325}$$

The equations for the host sites are

$$\sum_{n'u'} [(\Delta_{u'}{}^2 - E^2)\delta_{nn'}\delta_{uu'} + 2\Delta_u V_{nn'}(0u' \mid u0)]\tilde{u}_{n'u'} = -2\Delta_u \sum_{xv} V_{nx}(0v \mid u0)\tilde{u}_{xv} \quad (326)$$

The left-hand side of the host equation can be diagonalized by expanding $\tilde{u}_{nu}$ in terms of the $\tilde{u}_{nu}^{(0)}(\nu)$ of the isolated distorted host.

$$\tilde{u}_{nu} = \sum_{\nu} c_\nu \tilde{u}_{nu}^{(0)}(\nu) \quad (327)$$

The amplitudes c_ν are given by

$$c_\nu = \frac{2E(\nu)}{E^2 - [E(\nu)]^2} \sum_{nu} \tilde{u}_{nu}^{(0)*}(\nu) \sum_{xv} V_{nx}(0v \mid u0)\tilde{u}_{xv} \quad 328)$$

where $E(\nu)$ is the exciton energy of the distorted host. Substitution of (328) into (327) gives $\tilde{u}_{nu}$ in terms of impurity coefficients $\tilde{u}_{xv}$ only, so that the host coefficients may be eliminated entirely from (325). The result is

$$\sum_{x'v'} \left[\delta_{vv'}\delta_{xx'} + \frac{2\Delta_{xv}}{\Delta_{xv}{}^2 - E^2} J_{xv,x'v'}(E) \right] \tilde{u}_{x'v'} = 0 \quad (329)$$

where J is an effective impurity–impurity interaction for $x \neq x'$:

$$J_{xv,x'v'}(E) = V_{xx'}(0v' \mid v0) + G^{(2)}_{xv,x'v'}(E) \quad (330)$$

with

$$G^{(2)}_{xv,x'v'}(E) = \sum_{\nu} \frac{2E(\nu)}{E^2 - [E(\nu)]^2} \tilde{I}_x{}^v(\nu)\tilde{I}^{v'*}(\nu) \quad (331)$$

and

$$\tilde{I}_x{}^v(\nu) = \sum_{nu} V_{xn}(0u \mid v0)\tilde{u}_{nu}^{(0)}(\nu) \quad (332)$$

Equation (329) is the complete generalization of the dipole theory result, (203).

It is easily seen that $J_{xv,x'v'}$ is the whole of the effective interaction by considering the special case of two identical impurities with one excited state v. The splitting of the two impurity levels is $2J_{xv,x'v}$.

Equation (329) would be particularly useful for a pair of close impurities interacting via short-range forces. The dipole theory is best suited for widely separated pairs in crystals where dielectric effects are important.

4. *Substitutional Impurities in Infinite Crystals*

The properties of substitutionally mixed crystals have been discussed at length in some of the earlier sections. In this section Coulombic excitons in very dilute isotopic mixed crystals are considered. Two problems are considered, first the theory for the discrete guest state, and second the scattering by the guest of excitons lying within the host band.

Let $p\beta$ be the site occupied by the guest and assume that the only difference between host and guest is the excitation energy of the vth transition. The trap-depth is therefore

$$\delta_v = \Delta_v' - \Delta_v \tag{333}$$

where Δ_v' is the guest excitation energy. With these assumptions the secular equations are

$$\sum_{n'\alpha'u'} (M_{n\alpha u,n'\alpha'u'} - E^2\delta_{nn'}\delta_{\alpha\alpha'}\delta_{uu'})\tilde{u}_{n'\alpha'u'} + \delta_{np}\delta_{\alpha\beta}\delta_{uv} \times \{[(\Delta_v')^2 - \Delta_v^2]\tilde{u}_{p\beta v} + 2\delta_v \sum_{n'\alpha'u'} V_{p\beta,n'\alpha'}(0u'\,|\,v0)\tilde{u}_{n'\alpha'u'}\} = 0 \tag{334}$$

where $M_{n\alpha u,n'\alpha'u'}$ is the pure crystal matrix. The second term in the brace may be neglected to an error of less than 1%.[181] Next $\tilde{u}_{n\alpha u}$ is expanded in the corresponding pure crystal functions

$$\tilde{u}_{n\alpha u} = \sum_{\mathbf{k}\nu} c_{\mathbf{k}\nu}\tilde{u}_{n\alpha u}^{(0)}(\mathbf{k}\nu) \tag{335}$$

with the result

$$\sum_{\mathbf{k}\nu} \{[E(\mathbf{k}\nu)]^2 - E^2\}\tilde{u}_{n\alpha u}^{(0)}(\mathbf{k}\nu)c_{k\nu} + \delta_{np}\delta_{\alpha\beta}\delta_{uv}[(\Delta_v')^2 - \Delta_v^2]\tilde{u}_{p\beta v} = 0 \tag{336}$$

The normalization condition (295) enables this last equation to be solved for $c_{\mathbf{k}\nu}$ and thus from (335) we obtain

$$\tilde{u}_{n\alpha u} + \frac{[(\Delta_u')^2 - \Delta_v^2]}{2\Delta_v} \sum_{\mathbf{k}\nu} \frac{2E(\mathbf{k}\nu)}{[E(\mathbf{k}\nu)]^2 - E^2} [\tilde{u}_{p\beta v}^{(0)*}(\mathbf{k}\nu)\tilde{u}_{n\alpha u}^{(0)}(\mathbf{k}\nu)]\tilde{u}_{p\beta v} = 0 \tag{337}$$

This equation is just a step away from the more familiar result described by (42). To an accuracy of about 1%

$$\frac{\Delta_v' + \Delta_v}{2\Delta_v} = 1 \tag{338}$$

$$\frac{2E(\mathbf{k}\nu)}{E + E(\mathbf{k}\nu)} = 1 \tag{339}$$

and the secular equation simplifies to

$$\tilde{u}_{n\alpha u} + \frac{\delta_v}{N} \sum_{\mathbf{k}\nu} \frac{\tilde{u}_{\beta v}^{(0)*}(\mathbf{k}\nu)\tilde{u}_{\alpha u}^{(0)}(\mathbf{k}\nu)}{E(\mathbf{k}\nu) - E} e^{i\mathbf{k}\cdot[\mathbf{r}_{n\alpha} - \mathbf{r}_{p\beta}]}\,\tilde{u}_{p\beta v} = 0 \tag{340}$$

where (296) has been used for the pure crystal $\tilde{u}$ functions. The secular determinant is therefore

$$1 + \frac{\delta_v}{N} \sum_{\mathbf{k}\nu} \frac{[\tilde{u}_{\beta v}^{(0)}(\mathbf{k}\nu)]^2}{E(\mathbf{k}\nu) - E} = 0 \tag{341}$$

where $E(\mathbf{k}\nu)$ are the exact single-particle states of the pure crystal. Notice that the summation is over *all* pure crystal states, not just those in the host exciton band (v) closest to the guest level; the number of values of the index ν equals the number of molecules per unit cell times the number of excited states. For weakly interacting bands $\tilde{u}_{\beta v}^{(0)}(\mathbf{k}\nu)$ is small unless $\nu = \zeta_v$, the branch label of exciton band v, so (341) becomes

$$1 + \frac{\delta_v}{N} \sum_{\mathbf{k}} \sum_{\zeta_v} \frac{|\tilde{u}_{\beta v}^{(0)}(\mathbf{k}\zeta_v)|^2}{E(\mathbf{k}\zeta_v) - E} = 0 \tag{342}$$

In a crystal with strong interactions between exciton bands belonging to different electronic states the $E(\mathbf{k}\zeta_v)$ should be calculated by diagonalizing $\mathcal{H}_{ex}$ for the pure crystal. For very weak transitions it seems to be sufficient to assume that nearest-neighbor interactions are dominant.

Next the theory of the polarization ratio is briefly examined. Within the approximations made above the Tyablikov coefficients satisfy (340) with ν replaced by ζ_v. In quantized form the dipole operator is

$$\mathbf{D}_{\mathbf{q}} = \sum_{\rho} \sum_{n\alpha u} \boldsymbol{\mu}_{\alpha}{}^{u} \exp(-i\mathbf{q} \cdot \mathbf{r}_{n\alpha})[\tilde{u}_{n\alpha u}(\rho)\xi_{\rho} + \tilde{u}_{n\alpha u}^{*}(\rho)\xi_{\rho}{}^{+}] \tag{343}$$

where ρ is here the quasiparticle label for the mixed crystal. The oscillator strength of a transition from the ground state to $\xi_{\rho}{}^{+}|0\rangle$ is

$$F_{\mathbf{q}\lambda} = \frac{2mE_{\rho}}{\hbar^2 e^2} |\langle 0|(\mathbf{D}_{\mathbf{q}} \cdot \mathbf{e}_{\mathbf{q}\lambda})\xi_{\rho}{}^{+}|0\rangle|^2 \tag{344}$$

A straightforward calculation utilizing (340), (343), and (344) gives

$$F_{\mathbf{q}\lambda} = \frac{2mE_{\rho}}{\hbar^2 e^2} (\delta_v)^2 |\tilde{u}_{\rho\beta v}(E_{\rho})|^2 \left| \sum_{\zeta_v} \hat{\mathbf{e}}_{\mathbf{q}\lambda} \cdot \boldsymbol{\pi}(\mathbf{q}\zeta_v) \tilde{u}_{\beta v}^{(0)*}(\mathbf{q}\zeta_v)[E(\mathbf{q}\zeta_v) - E]^{-1} \right|^2 \tag{345}$$

where

$$\boldsymbol{\pi}(\mathbf{q}\zeta_v) = \sum_{\alpha u} \tilde{u}_{\alpha u}^{(0)}(\mathbf{q}\zeta_v)\boldsymbol{\mu}_{\alpha}{}^{u} \tag{346}$$

The polarization ratio is just the ratio of the oscillator strengths of the transition to the trap-level for light polarized along two orthogonal directions. If in each of the directions chosen there is a unique crystal transition, then the trap polarization ratio $\mathcal{P}'$ is given by

$$\mathcal{P}'(1:2) = \left(\frac{E(\mathbf{q}\zeta_2) - E}{E(\mathbf{q}\zeta_1) - E}\right)^2 \mathcal{P}(1:2) \tag{346'}$$

where $\mathcal{P}$ is the pure crystal polarization ratio.[182]

Equation (346′) for the polarization ratio is a relation between measurable quantities, and as such is rather remarkable. It may be used to determine a host crystal polarization ratio if there are experimental difficulties preventing a direct measurement. Alternatively it may be used to test the validity of the neutral Frenkel exciton picture. Thus far no deviations of any consequence have been reported.

Another point of interest is that for intense transitions where dipole interactions become important the host levels $E(\mathbf{q}\zeta_v)$ will depend on the direction of $\mathbf{q}$. Thus a ratio of trap polarization ratios for different directions $\mathbf{q}$ will not equal the corresponding ratio of ratios for the pure crystal. It would be interesting to look for this effect in crystalline aromatic hydrocarbon with a discrete guest state close to the exciton band.[183]

Shallow traps of magnitude less than the critical value do not lead to the formation of discrete trap-states split away from the band. They are, however, responsible for scattering host excitons and localization of excitation in the vicinity of the guest site. If E lies inside the host band (337) defines $\tilde{u}_{n\alpha u}$ only to within a solution of the pure host with energy E.[118] Let the host solution be $\tilde{u}_{n\alpha u}(\mathbf{k}_0\nu_0)$ then

$$\sum_{n'\alpha'u'} (M_{n\alpha u,n'\alpha'u'} - E^2\delta_{nn'}\delta_{\alpha\alpha'}\delta_{uu'})\tilde{u}^{(0)}_{n'\alpha'u'}(\mathbf{k}_0\nu_0) = 0 \tag{347}$$

In the scattering problem the guest site coefficient $\tilde{u}_{p\beta v}$ is eliminated since the form of the exciton wave function at great distances is known. With an outgoing boundary condition on the scattered waves the solution for $\tilde{u}_{n\alpha u}(E)$ is

$$\tilde{u}_{n\alpha v} = \tilde{u}^{(0)}_{n\alpha v}(\mathbf{k}_0\nu_0) - \delta_v \frac{\mathcal{G}_{n\alpha v,p\beta v}(E)\tilde{u}^{(0)}_{p\beta v}(\mathbf{k}_0\nu_0)}{1 + \delta_v\mathcal{G}_{p\beta v,p\beta v}(E^+)} \tag{348}$$

where

$$\mathcal{G}_{n\alpha u,p\beta v}(E^+) = \frac{1}{N}\sum_k \sum_{\zeta v} \frac{[\tilde{u}^{(0)}_{\beta v}(\mathbf{k}\zeta_v)]^*[\tilde{u}^{(0)}_{\alpha u}(\mathbf{k}\zeta_v)]}{E(\mathbf{k}\zeta_v) - E} \exp[i\mathbf{k}\cdot(\mathbf{r}_{n\alpha} - \mathbf{r}_{p\beta})] \tag{349}$$

There is a "resonance" in the second part of (348) wherever

$$|1 + \delta_v\mathcal{G}(E^+)| \simeq 0 \tag{350}$$

Whether or not such a resonance occurs depends on the density of states in the host band. The development of resonances is described in the next section.

The probability of a transition from the ground to an inband level can be calculated using (348) in (345). For an isolated exciton band in a crystal with one molecule per unit cell the oscillator strength is

$$F_{\mathbf{q}\lambda} = \frac{2mE}{\hbar^2 e^2} \frac{\delta^2(\mathbf{e}_\lambda \cdot \boldsymbol{\mu})^2}{N[E(\mathbf{q}) - E]^2 |1 + \delta \mathcal{G}(E^+)|^2} \tag{351}$$

This quantity vanishes in the limit $N \to 0$ since there is only one guest. For N' noninteracting guests the oscillator strength is N' times (351).

5. *Model Calculations for Dilute Isotopic Mixed Crystals*

In this section some calculations based on model density of states are presented. The idea is to show how the properties of very dilute mixed crystals depend on the density of states function. The equations upon which the calculations are based are summarized here for convenience. The crystal has one molecule per unit cell, N cells total. There is one host exciton band with energies $E(\mathbf{k})$ and one guest on site p.

The energy E' of the discrete guest level (if it exists) is obtained by solving

$$1 - \delta G(E') = 0 \tag{352}$$

where δ is the trap-depth and

$$G(E) = N^{-1} \sum_{\mathbf{k}} [E - E(\mathbf{k})]^{-1} \tag{353}$$

The oscillator strength of the trap-state transition in units of the oriented gas oscillator strength is

$$F' = (\mathbf{e}_{q\lambda} \cdot \boldsymbol{\mu})^2 [E' - E(\mathbf{q})]^{-2} \left(\frac{-dG}{dE}\right)^{-1}_{E=E'} \tag{354}$$

The probability of finding the exciton localized on the guest site p is

$$[\tilde{u}_p(E')]^2 = \left[\delta^2 \left(\frac{-dG}{dE}\right)_{E=E'}\right]^{-1} \tag{355}$$

The oscillator strength (*times* N) of transitions inside the host band is

$$F'_{\text{cont}} = \frac{(\mathbf{e}_\lambda \cdot \boldsymbol{\mu})^2 \delta^2}{[E - E(\mathbf{q})]^2} \{[1 - \delta G'(E)]^2 + [\delta G''(E)]^2\}^{-1} \tag{356}$$

where $G'(E)$ and $G''(E)$ are the real and imaginary parts of $G(E^+)$:

$$G(E^+) = G'(E) + iG''(E) \tag{357}$$

Expressed in terms of a density of states

$$G'(E) = \int d\epsilon \rho(\epsilon)(E - \epsilon)^{-1} \tag{358}$$

$$G''(E) = -\pi \int d\epsilon \rho(\epsilon)\delta(\epsilon - E) \tag{359}$$

where the integration is over the extent of the host band [$\rho(\epsilon) = 0$ outside the host band], and $\int$ stands for principal part.

In the pure crystal the probability of finding the exciton at any site is N^{-1}. In the mixed crystal the probability of finding the guest site cocupied is $|\tilde{u}_p(E)|^2$. The ratio of mixed to pure is therefore $N|\tilde{u}_p(E)|^2$, which for a level lying inside the host band is given by

$$N|\tilde{u}_p(E)|^2 = \{[1 - \delta G'(E)]^2 + [\delta G''(E)]^2\}^{-1} \tag{360}$$

This function measures the localization of the exciton at p relative to the pure crystal.

Model A. In crystals with equal nearest-neighbor interactions a critical point analysis reveals that the density of states is proportional to $|\epsilon \pm \epsilon_0|^{1/2}$ where $\pm\epsilon_0$ are the band edges. A simple density of states with this square-root dependence on ϵ is[184]

$$\rho(\epsilon) = \frac{2\pi}{(\epsilon_0)^2}[(\epsilon_0)^2 - \epsilon^2]^{1/2} \tag{361}$$

where $-\epsilon_0 \leq \epsilon \leq \epsilon_0$. All calculations are performed with $\epsilon_0 = 1.0$. Thus for discrete states $|E| > 1.0$ and

$$G(E) = 2E - 2\,\text{sgn}\,(E)\,(E^2 - 1)^{1/2} \tag{362}$$

$$\frac{-dG}{dE} = 2(1 - E^{-2})^{1/2} - 2$$

For continuum levels $|E| < 1$ and

$$G'(E) = 2E \tag{363}$$

$$G''(E) = -2(1 - E^2)^{1/2} \tag{364}$$

The real and imaginary parts of $G(E)$ are plotted in Figs. 7 and 8.

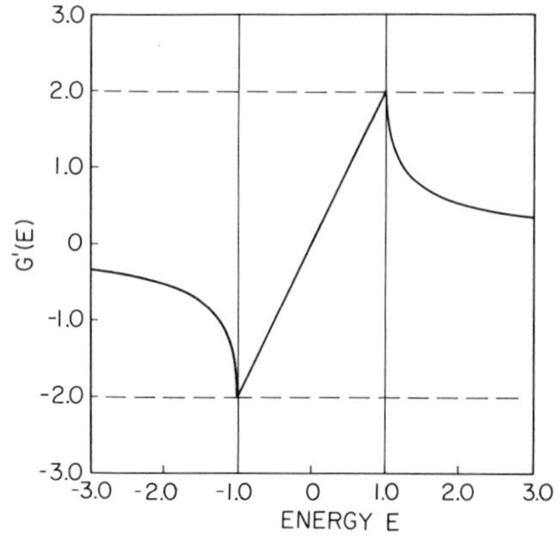

Fig. 7. Model A: plot of $G'(E)$.

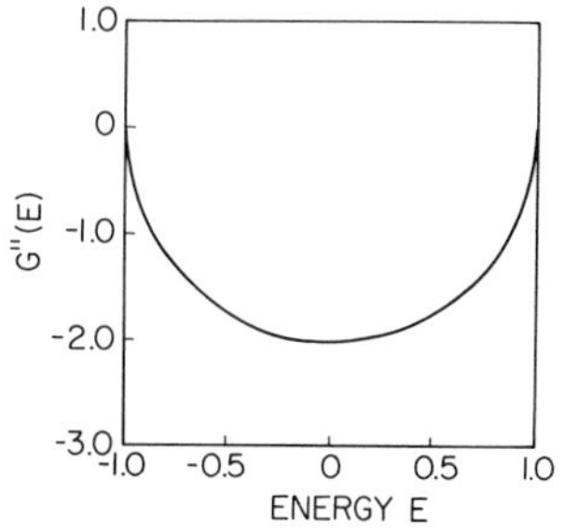

Fig. 8. Model A: plot of $G''(E)$.

For discrete states remarkably simple expressions are found for energy, probability, and intensity:

$$E' = \delta[1 + (4\delta^2)^{-1}] \tag{365}$$

$$[\tilde{u}_p(E')]^2 = 1 - (4\delta^2)^{-1} \tag{366}$$

$$F' = \frac{\delta^2[1 - (4\delta^2)^{-1}]}{[E' - E(\mathbf{q})]^2} (\mathbf{e}_\lambda \cdot \boldsymbol{\mu})^2 \tag{367}$$

The formula for E' has the same form as the result obtained with second-order perturbation theory. In Figs. 9 and 10 the critical value of trap-depth is $\delta_c = \pm 0.5$; that is, this is the minimum value of the perturbation

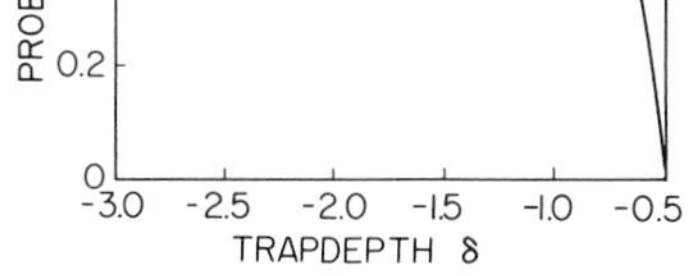

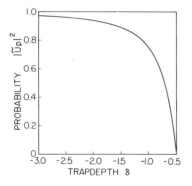

Fig. 9. Model A: probability $(\tilde{u}_p)^2$ that the exciton occupies guest site p for different trap-depths δ.

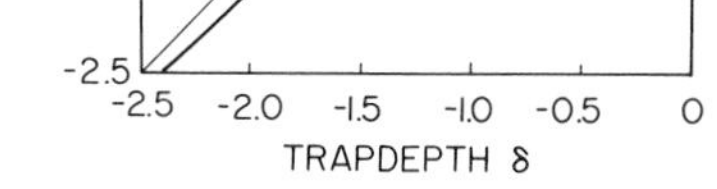

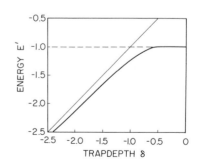

Fig. 10. Model A: change of bound state energy E' with trap-depth δ for model A. Critical value of trap-depth is -0.5.

for the formation of a discrete guest state beyond the edges of the host exciton band.

The δ dependence of the intensity of the discrete state is determined ($\delta < 0$ case) by the ratio of two monotonically decreasing factors. A plot of F' against trap-depth δ is shown in Fig. 11 for $-3 \leq \delta \leq -0.5$ and various values of $E(\mathbf{q})$, the optical level. For convenience $(\boldsymbol{\mu} \cdot \mathbf{e}_\lambda)^2$ has been set equal to unity. For large trap-depths F' tends to a limiting value of 1.0. As δ approaches the critical value -0.5 the intensity F' rises monotonically to a singular value at $\delta = -0.5$ if $E(\mathbf{q}) = -1.0$, or goes through a maximum before dropping to zero if $-1.0 < E(\mathbf{q}) \leq 0$, or drops monotonically to zero if $0 < E(\mathbf{q}) \leq 1.0$.

The rise and fall in the intensity F' is expected to occur in crystals with more than one molecule per unit cell, provided all the optically allowed transitions $E(\mathbf{q}\zeta)$ (ζ labels exciton branch) fall *inside* the band. This effect has not been observed thus far.

Inside the host band

$$N|\tilde{u}_p(E)|^2 = [1 + 4\delta(1 - E)]^{-1} \tag{368}$$

$$F'_{\text{cont}} = \frac{\delta^2}{[E - E(\mathbf{q})]^2}[1 + 4\delta(1 - E)]^{-1} \tag{369}$$

First of all we note that there are no quasibound states (resonances) in the band. This can be seen by inspection of Fig. 7. The continuum intensity

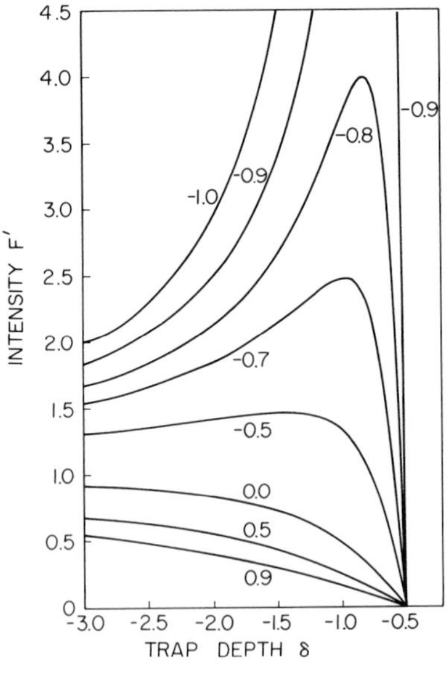

Fig. 11. Model A: change in intensity F' of the guest level with trap-depth δ for different positions $E(\mathbf{q})$ of the allowed host transition.

is dominated by a pole at $E = E(\mathbf{q})$ and has subsidiary spikes at the band edges that develop into singularities as δ approaches the critical values ± 0.5. See Fig. 12 where this is illustrated for the case $E(\mathbf{q}) = 1.0$.

Model B. The main attribute of the second model density is the occurrence of quasibound states ("resonances") within the host band. The normalized density with band edges at $\pm\,\epsilon_0$ is [185]

$$\rho(\epsilon) = \frac{3}{4\epsilon_0}\left[1 - \left(\frac{\epsilon}{\epsilon_0}\right)^2\right] \qquad |\epsilon| \leqslant \epsilon_0 \tag{370}$$

The slope of this density is finite at the band edges unlike the last density, (361). Again we set $\epsilon_0 = 1.0$. The real and imaginary parts of $G(E)$ are

$$G'(E) = \frac{3}{4}\left[2E - (1 - E^2)\ln\left|\frac{1 - E}{1 + E}\right|\right] \tag{371}$$

for all E and

$$G''(E) = -\tfrac{3}{4}\pi(1 - E^2) \tag{372}$$

for $|E| < 1$. Outside the band the derivative of $G(E)$ is given by

$$\frac{-dG}{dE} = -3\left[1 + \tfrac{1}{2}E\ln\left|\frac{1 - E}{1 + E}\right|\right] \tag{373}$$

The real and imaginary parts of $G(E)$ are shown in Figs. 13 and 14. Unlike the last density there are no simple expressions for energy, probability, etc. The critical trap-depths are $\delta = \pm\frac{2}{3}$. Inspection of Fig. 13

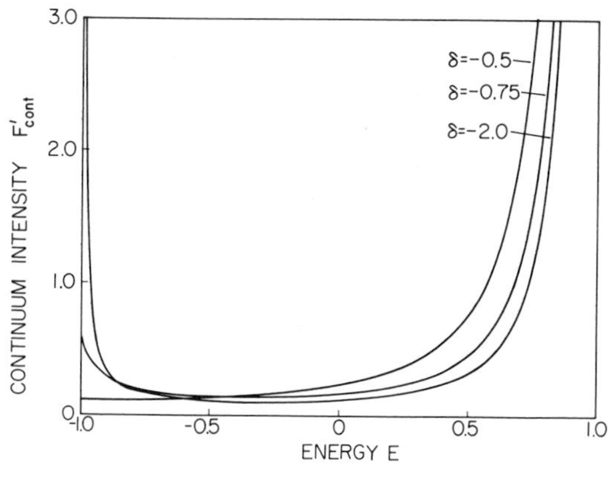

Fig. 12. Model A: continuum intensity F'_{cont} for $E(\mathbf{q}) = 1.0$ and various trap-depths δ. There are no quasibound states in the continuum region.

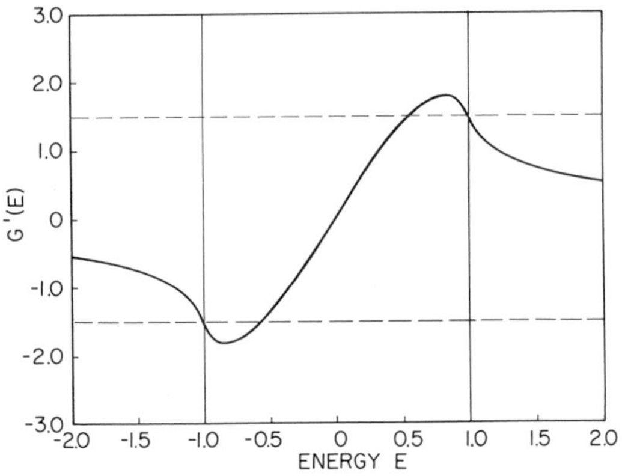

Fig. 13. Model B: plot of $G'(E)$.

shows that $|G'(E)| \geq \delta_c$ for $0.55 \leq |E| \leq 1.0$, and this limits the range of continuum energies where resonances are possible to

$$-1.0 \leq E \leq -0.55 \qquad \text{and} \qquad +0.55 \leq E \leq 1.0$$

For convenience we again set $(\boldsymbol{\mu} \cdot \mathbf{e}_\lambda)^2 = 1.0$. The resonances are clearly visible in Fig. 15 where the continuum intensity is plotted against energy for the case $E(\mathbf{q}) = 1.0$. As δ approaches the critical value -0.6667 the resonance peaks shift toward $E = -1.0$, simultaneously increasing in intensity and decreasing in width (due to the smaller value of $G''(E)$). The question of quasibound states in the host band has been raised previously[113] and a search failed to reveal their presence.[26]

The variation in the intensity F' of the discrete state is qualitatively similar to the first density function. Results of calculations for F' as a function of the energy E are shown in Fig. 16 for different values of the energy $E(\mathbf{q})$ of the allowed transition of the pure host.

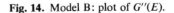

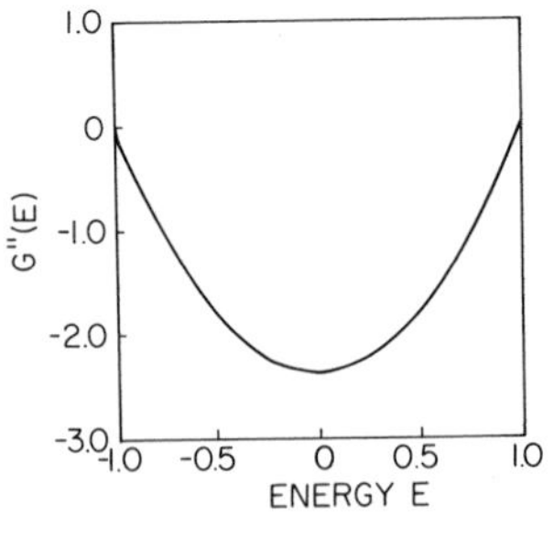

Fig. 14. Model B: plot of $G''(E)$.

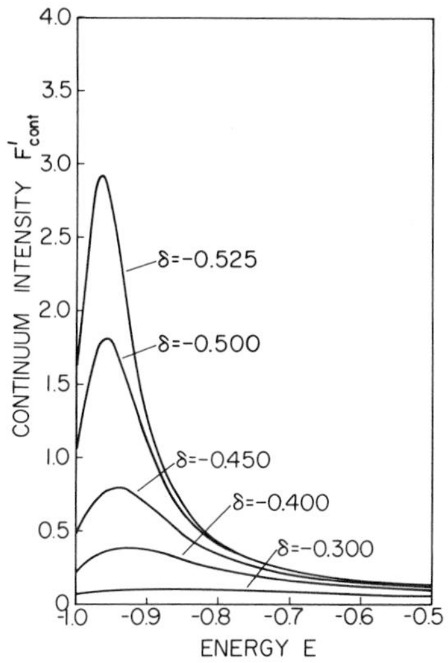

Fig. 15. Model B: continuum intensity F'_{cont} for $E(\mathbf{q} = 1.0)$ and various trap-depths δ. Note the quasibound states.

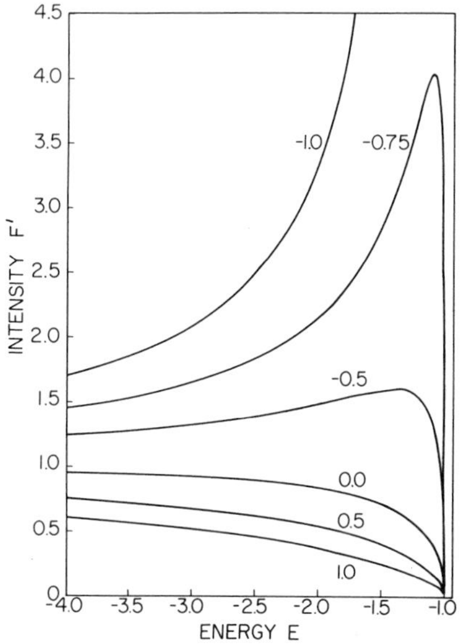

Fig. 16. Model B: intensity F' of discrete guest level for energies E' below the band.

Model C. The previous two models were appropriate for weak transitions where the dipole–dipole interactions were an inconsequential part of $E(\mathbf{k})$. For strong transitions the dipole component dominates and the energy $E(\mathbf{q})$ of the optically allowed level traverses most of the exciton band as the angle θ between $\mathbf{q}$ and $\boldsymbol{\mu}$ changes from 0 to 90°.

In this last model the density of states is derived from the dipole formula for the exciton energies

$$E(\mathbf{k}) = \frac{4\pi}{v_0}\,\mu^2 t(\mathbf{k}) + g(\cos^2\Theta - \tfrac{1}{3}) \qquad (374)$$

Here $t(\mathbf{k})$ is the analytic part, and

$$g = \frac{4\pi\boldsymbol{\mu}^2}{v_0} \qquad (375)$$

where v_0 is the unit cell volume and Θ the angle between the exciton wave vector $\mathbf{k}$ and the transition dipole $\boldsymbol{\mu}$. For the model the analytic part of (374) is neglected leaving

$$E(\mathbf{k}) = g(\cos^2\Theta - \tfrac{1}{3}) \qquad (376)$$

and the sum $N^{-1}\Sigma_{\mathbf{k}}$ is replaced by an integral over a sphere. Thus in place of

$$G(E) = N^{-1} \sum_{\mathbf{k}} [E - g(\cos^2 \Theta - \tfrac{1}{3})]^{-1} \tag{377}$$

we have

$$G(E) = \int_0^1 du[E - g(u^2 - \tfrac{1}{3})]^{-1} \tag{378}$$

where $u = \cos \Theta$. For convenience we set $g = 1.0$.

The density of state function for $\frac{1}{3} < E < \frac{2}{3}$ is equal to $-G''(E)/\pi$ where

$$G''(E) = -\pi \int_0^1 du\delta[E - (u^2 - \tfrac{1}{3})] \tag{379}$$

The integral is evaluated using

$$\delta[F(x)] = \sum_i \delta(x - x_i) \left| \frac{dF}{dx} \right|_{x=x_i}^{-1} \tag{380}$$

where the sum runs over all the simple zeros x_i of $F(x)$.

In terms of the variable

$$\sigma = |E + \tfrac{1}{3}|^{1/2} \tag{381}$$

we obtain

$$G'(E) = -\sigma^{-1} \operatorname{arctg} (\sigma^{-1}) , \qquad E < -\tfrac{1}{3} \tag{382}$$

$$G'(E) = (2\sigma)^{-1} \ln \left| \frac{1 + \sigma}{1 - \sigma} \right| , \qquad E > \tfrac{2}{3} \tag{383}$$

and

$$G''(E) = -\tfrac{1}{2}\pi\sigma^{-1/2} , \qquad -\tfrac{1}{3} < E < \tfrac{2}{3} \tag{384}$$

$$G''(E) = 0 , \qquad E < -\tfrac{1}{3}, E > \tfrac{2}{3} \tag{385}$$

The normalized density of states is

$$\rho(\epsilon) = (2\sigma^{1/2})^{-1} \tag{386}$$

This function is not realistic for real crystals since it is singular at the lower edge $\epsilon = -\frac{1}{3}$. At the upper edge ρ is finite with the value $\frac{1}{2}$. A plot of $\rho(\epsilon)$ is shown in Fig. 17. It is essentially flat except for the steep rise near the singularity. This model density does not permit the existence of quasibound states because $G'(E)$ is singular as the band edge is approached from outside the host band; that is, the critical value of the trap-depth is zero and every value of δ splits a discrete state away from the band.

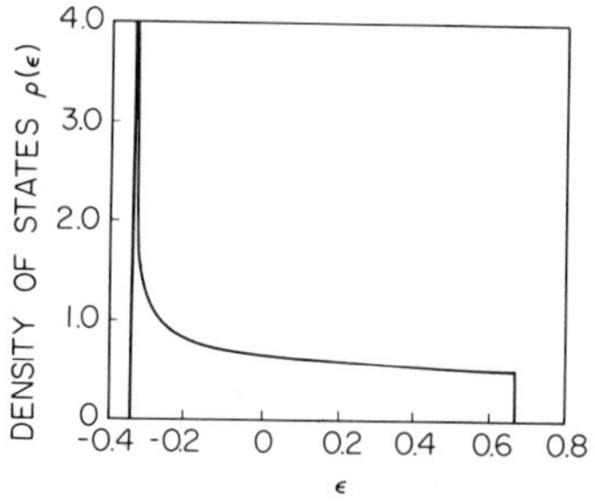

Fig. 17. Model C: plot of density of states function $\rho(\epsilon)$. At the point $\epsilon = -\frac{1}{3}$ the density is singular.

For energies E outside the band the derivative of $G(E)$ is given by

$$\frac{-dG}{dE} = (2\sigma^2)^{-1}\left[(1+\sigma^2)^{-1} + \frac{1}{\sigma}\operatorname{arctg}(\sigma^{-1})\right], \qquad E < -\tfrac{1}{3} \quad (387)$$

$$\frac{-dG}{dE} = (2\sigma^2)^{-1}\left[(\sigma^2-1)^{-1} + (2\sigma)^{-1}\ln\left(\frac{1+\sigma}{1-\sigma}\right)\right], \qquad E > \tfrac{2}{3} \quad (388)$$

The bound state probability $(\tilde{u}_p)^2$ curves for $E < -\frac{1}{3}$ and $E > \frac{2}{3}$ are practically mirror images in spite of the asymmetry in the density of states.

The main interest in this model lies in the explicit dependence of exciton energy on wave vector $\mathbf{q}$. Let ϕ denote the angle between $\mathbf{q}$ and $\boldsymbol{\mu}$. Since the wave vector $\mathbf{q}$ and polarization vector $\mathbf{e}_\lambda$ are orthogonal we set

$$(\boldsymbol{\mu}\cdot\mathbf{e}_\lambda)^2 = \mu^2\sin^2\phi \quad (389)$$

$$(\boldsymbol{\mu}\cdot\mathbf{q})^2 = \mu^2\cos^2\phi \quad (390)$$

so that the intensity of discrete state transitions is given by

$$F'(E') = \frac{\mu^2\sin^2\phi}{[E' - (\cos^2\phi - \frac{1}{3})]^2}\left[\left(\frac{-dG}{dE}\right)_{E=E'}\right]^{-1} \quad (391)$$

The energy of the allowed pure host transition

$$E(\mathbf{q}) = \cos^2\phi - \tfrac{1}{3} \quad (392)$$

lies at the bottom of the band when $\mathbf{q}$ and $\boldsymbol{\mu}$ are perpendicular ($\phi = 90°$) and at the top when $\mathbf{q}$ and $\boldsymbol{\mu}$ are parallel ($\phi = 0°$). The $\sin^2\phi$ factor causes F' to fall rapidly as $\mathbf{q}$ swings from a direction perpendicular to $\boldsymbol{\mu}$ to a parallel one. Intensity calculations are displayed in Figs. 18–20. In Fig. 18 the rise and fall of F' for $\phi < 90°$ occurs close to the lower band edge at $E = -\frac{1}{3}$. Figs. 19 and 20 show the E dependence of F' above the band for various ϕ. Note the order of magnitude difference in the intensity scale. The intensity maxima are spread over a much wider range of energies in Fig. 20 than in Fig. 18.

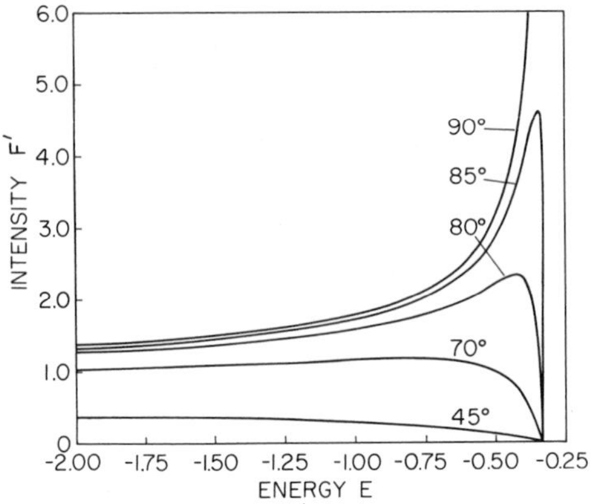

Fig. 18. Model C: intensity F' of discrete guest levels below the host band for various angles ϕ.

6. Optical Rotation and Circular Dichroism of Helical Polymers

In this section the optical rotation and circular dichroism of an infinite single-stranded helix of identical monomers are found using the second quantized formulation of exciton theory. There are several other ways of tackling these problems, for example, linear response theory, which appears to give entirely equivalent results. We shall not discuss any of the other techniques here since it would draw us too far from the theme of this

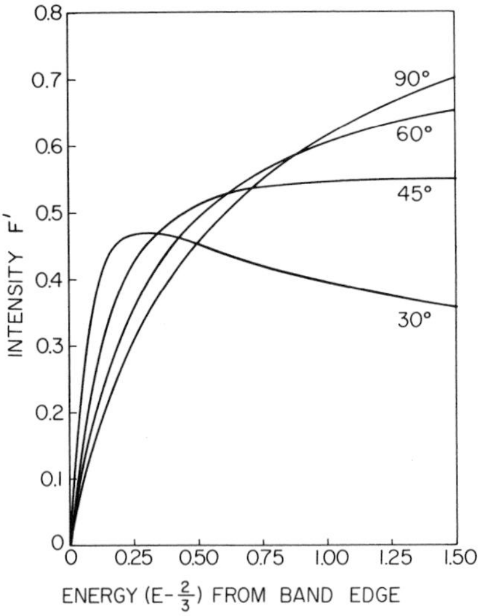

Fig. 19. Model C: intensity F' of discrete guest levels above the host exciton band for various angles ϕ.

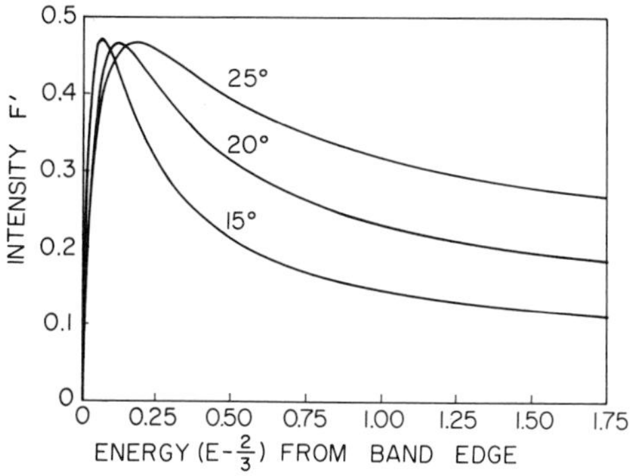

Fig. 20. Model C: intensity F' of discrete states above the host band for various angles ϕ.

review. In the lowest excited states of polypeptides and polynucleotides the promoted electrons belong to chemical units that are not directly linked by covalent bonds. In this respect the biopolymers and molecular crystals are similar. However, the units in a helical polymer, for example, a polynucleotide, are generally much closer together and so electron exchange effects should be more important. In spite of this theoreticians invariably opt for the tight binding approach, and we do not resist the temptation here.

There are two main reasons for analyzing the spectra of biopolymers. The first is to study electronic interactions between the monomer units and the second is to deduce as much as possible about the conformation of the polymer. As probes of structure the CD and ORD spectra, which arise either from the intrinsic asymmetry of monomers or from asymmetric regions (usually helical) of the polymer, are particularly useful because of their known sensitivity to changes in geometry. The absorption and fluorescence spectra of biopolymers have yielded little direct information about the magnitudes of exciton interactions. This is because thermal effects almost always wash out the closely spaced individual vibronic transitions. At low temperatures (80°K) the spectra reveal more stucture, generally shoulders and small peaks on broad absorption bands, but nothing approaching the wealth of detail found in crystal spectra. It appears that still lower temperatures are going to be necessary before any real details emerge from the spectra.

It is only in the last few years that the correct procedure for calculating the optical rotation and circular dichroism spectra of *infinite* helical polymers has been established.[186–193] For about ten years it was generally believed that cyclic boundary conditions could not be used to calculate the ORD, although when used to calculate the energy the correct result was obtained. The error, it turns out, lies not with the periodic boundaries but in the formula used for the magnetic dipole of an array of monomers. A number of papers have appeared in recent years dealing with this and allied problems. In some respects the origins of the error are similar to those of the crystal dipole sum case. In both cases the errors would have been avoided if the approximate selection rules had been used at the end of the calculation instead of in the middle.

After passage of the light beam through an optically active sample the optical rotation can be measured as the angle of rotation $\Delta\Theta$ of the plane of polarization and the circular dichroism as the change in elipticity $\Delta\eta$. These observables are conveniently regarded as the real and imaginary parts of Θ, a complex optical rotation:

$$\Theta = \Delta\Theta - i\Delta\eta \tag{393}$$

For a sample containing N_0 noninteracting polymers, all with the same orientation,

$$\Theta = \frac{1}{i}\left(\frac{N_0 L}{\hbar c}\right) \langle \mathbf{q}\mathbf{e}_{\mathbf{q}2};0 | R(E^+) | \mathbf{q}\mathbf{e}_{\mathbf{q}1};0 \rangle \tag{394}$$

where L is the sample thickness and $|\mathbf{q}e_{\mathbf{q}\lambda};0\rangle$ is the product state for a photon and a ground-level polymer. The reactance operator is defined by

$$R(E^+) = \sum_{m=0}^{\infty} \mathcal{H}'_{\text{int}}[\mathcal{G}_0(E^+)\mathcal{H}'_{\text{int}}]^m \tag{395}$$

with

$$\mathcal{G}_0(E^+) = (\mathcal{H}_0 - E - i\sigma)^{-1} \tag{396}$$

and

$$\mathcal{H}_0 = \mathcal{H}_{\text{ex}} + \mathcal{H}_{\text{rad}} \tag{397}$$

The energy of the states $|\mathbf{q}e_{\mathbf{q}\lambda};0\rangle$ is E, and σ is a positive infinitesimal specifying an outgoing boundary condition for scattered photons.[194] Note that in (395) the interaction $\mathcal{H}''_{\text{int}}$ does not appear; it has been neglected since it contributes nothing to the optical rotation at the level of approximation used. The effect of $\mathcal{H}'_{\text{int}}$ and $\mathcal{H}''_{\text{int}}$ on the exciton energies is small; they are the source of those corrections that turn the instantaneous interactions into fully retarded ones. The real part of this interaction corrects $I^{uu}(\mathbf{k})$ slightly while the imaginary part leads to an increase in the radiative damping of allowed transitions.[195]

Formula (394) applies to polymers at absolute zero in a vacuum. At finite temperatures an ensemble average over thermally accessible levels is necessary and the optical properties arise from transitions between quasiparticle states. In the interest of simplicity and clarity the full problem of interacting exciton states is not described here; it is available elsewhere.[196] We assume that exciton bands are noninteracting, which eliminates hypochromism. The wave functions of the isolated monomers are taken to be real so that the matrix elements of electric dipoles and quadrupoles are real, whereas those of magnetic dipoles are pure imaginary. With these approximations the exciton energies are

$$E(\mathbf{k}u) = \Delta_u \left[1 + \frac{2I^{uu}(\mathbf{k})}{\Delta_u} \right]^{1/2} \tag{398}$$

and the Tyablikov coefficients are

$$\tilde{u}_{nu}(\mathbf{k}\nu) = \delta_{u\nu}\tilde{u}(\mathbf{k}u) \exp (i\mathbf{k} \cdot \mathbf{r}_n) \tag{399}$$

where from the normalization condition (295)

$$|\tilde{u}(\mathbf{k}u)|^2 = \frac{\Delta u}{E(\mathbf{k}u)} \simeq 1 \tag{400}$$

If $\tilde{u}(\mathbf{k}u)$ is normalized to unity, this is equivalent to neglecting the double excitation terms in $\mathcal{H}_{\text{ex}}$, so

$$E(\mathbf{k}u) = \Delta_u + I^{uu}(\mathbf{k}) \tag{401}$$

The interaction Hamiltonian $\mathcal{H}'_{\text{int}}$, (280), contains $\exp (i\mathbf{q} \cdot \mathbf{r}_{na})$, which must be expanded in a power series either about the center of mass of the monomer or about an adjacent center on the helix axis. After making the expansion

$$\mathcal{H}'_{\text{int}} = -\sum_{\mathbf{q}\lambda} \left(\frac{2\pi\hbar}{cqV}\right)^{1/2} \mathbf{e}_{\mathbf{q}\lambda} \cdot \{\mathbf{J}_{\mathbf{q}} a_{\mathbf{q}\lambda} + \mathbf{J}_{-\mathbf{q}} a_{\mathbf{q}\lambda}{}^{+}\} \tag{402}$$

where

$$\mathbf{J}_q = \frac{-i}{\hbar} \sum_n \exp (i\mathbf{q} \cdot \mathbf{t}_n) \cdot \{[\boldsymbol{\mu}_n + i(\mathbf{q} \cdot \boldsymbol{\rho}_n)\boldsymbol{\mu}_n + \tfrac{1}{2} i\mathbf{q} \cdot \mathbf{Q}_n, \mathcal{H}_{\text{ex}}] + \hbar c\, \mathbf{m}_n \times \mathbf{q}\} \tag{403}$$

The commutator with $\mathcal{H}_{\text{ex}}$ arises from approximation (312); $\boldsymbol{\rho}_n$ is the vector ($\perp$ helix axis) to monomer n from the helix axis, that is, $\mathbf{r}_n = \mathbf{t}_n + \boldsymbol{\rho}_n$.

To complete the transformation to the second quantized representation the molecular operators $\boldsymbol{\mu}_n$, $\mathbf{m}_n$, and $\mathbf{Q}_n$ are expressed in terms of the creation and annihilation operators of localized excitons and then these localized excitons are transformed into the quasiparticle operators $\xi_{\mathbf{k}u}{}^{+}$, $\xi_{\mathbf{k}u}$ using

$$B_{nu} = N^{-1/2} \sum_{\mathbf{k}} [u(\mathbf{k}u)\xi_{\mathbf{k}u} e^{i\mathbf{k} \cdot \mathbf{t}_n} + v^*(\mathbf{k}u) e^{-i\mathbf{k} \cdot \mathbf{t}_n} \xi_{\mathbf{k}u}{}^{+}] \tag{404}$$

As has already been mentioned the exciton energies are subject to radiative damping. In real polymers there are numerous nonradiative damping processes too, and it is convenient to lump all these damping terms together as $-\frac{1}{2}i\hbar\gamma_{\mathbf{k}u}$. After a little manipulation the complex optical rotation is found to be

$$\Theta = \frac{1}{i}\left(\frac{N_0 L}{\hbar c}\right)\left(\frac{2\pi\hbar}{cqV}\right)\sum_{\mathbf{k}u} 2E(\mathbf{k}u) \times$$

$$\frac{[\hbar cq - E(\mathbf{k}u) - \frac{1}{2}i\hbar\gamma_{\mathbf{k}u}]\,\langle 0|\mathbf{e}_{q1}\cdot\mathbf{J}_{\mathbf{q}}|\mathbf{k}u\rangle\langle\mathbf{k}u|\mathbf{e}_{\mathbf{q}2}\cdot\mathbf{J}_{-q}|0\rangle}{[\hbar cq + E(\mathbf{k}u)]\{[\hbar cq - E(\mathbf{k}u)]^2 + \frac{1}{4}\hbar^2\gamma_{\mathbf{k}u}{}^2\}} \tag{405}$$

where $|\mathbf{k}u\rangle$ is the quasiparticle state $\xi_{\mathbf{k}u}{}^{+}|0\rangle$.[197]

According to (405) the line shape is Lorentzian. Usually Gaussian curves give a much better fit to experimental curves at room temperature, but this is due to a Boltzmann distribution among initial states plus a high density of final states.

The next stages in calculating Θ are a little tedious. First the matrix elements $\langle 0|\mathbf{e}_{\mathbf{q}\lambda}\cdot\mathbf{J}_{\mathbf{q}}|\mathbf{k}u\rangle$ are determined. This involves relating the dipole and quadrupole operators of monomer n to monomer $n = 0$, using the rotation matrix $\mathbf{R}(\varphi)$, (183). In dyadic notation

$$\mathbf{R}(\varphi) = \exp(iSt_n)\mathbf{e}_{-}\mathbf{e}_{+} + \exp(-iSt_n)\mathbf{e}_{+}\mathbf{e}_{-} + \mathbf{e}_z\mathbf{e}_z \tag{406}$$

where $\mathbf{e}_i$ ($i = x, y, z$) is the helix unit vector set and

$$\mathbf{e}_{\pm} = 2^{-1/2}(\mathbf{e}_x \pm i\mathbf{e}_y) \tag{407}$$

Once the matrix elements of $\mathbf{J}_{\pm\mathbf{q}}$ are known, they are substituted into (405) and the components of Θ determined

$$\Theta = \Theta^{\mu-\mu} + \Theta^{\mu-m} + \Theta^{\mu-p\mu} + \Theta^{\mu-Q} + \cdots \tag{408}$$

The final stage consists of simplifying the components using oscillator and rotational sum rules.

The oscillator strength sum rules for an infinite helix are

$$\sum_{\mathbf{k}}\sum_{u} F_z(\mathbf{k}u)\delta(k + q_z) = n_e \tag{409}$$

$$\sum_{\mathbf{k}}\sum_{u} F_{\pm}(\mathbf{k}u)\delta(k + q_z \pm S) = n_e \tag{410}$$

where n_e is the number of electrons per monomer,

$$F_\sigma(\mathbf{k}u) = \frac{2m}{\hbar^2 e^2}E(\mathbf{k}u)|\mathbf{e}_\sigma\cdot\boldsymbol{\mu}_0{}^u|^2 \tag{411}$$

and

$$\mathbf{e}_\sigma = \mathbf{e}_\pm, \mathbf{e}_z \tag{412}$$

Equation (411) is really nothing more than the isolated molecule sum rule. It is valid here because interband interactions and therefore hypochromism are excluded from the theory.

The term $\Theta^{\mu-\mu}$ was the one lost by Moffit[52] in his original treatment, and of all the components of Θ it is the one requiring most careful consideration. It is the only one examined in detail here. For light propagating parallel to the helix axis with a frequency close to the uth exciton band the angle of rotation is

$$\Delta\Theta^{\mu-\mu} = \Delta\Theta_u^{\mu-\mu} + \Delta\Theta_{\text{back}}^{\mu-\mu} \tag{413}$$

where

$$\Delta\Theta_u^{\mu-\mu} = \left(\frac{2\pi N_0 NL}{\hbar cV}\right)\hbar cq \sum_{\mathbf{k}} \frac{E(\mathbf{k}u)|\mathbf{e}_+ \cdot \boldsymbol{\mu}_0{}^u|^2[\hbar cq - E(\mathbf{k}u)]\,[\delta(k+q+S) - \delta(k+q-S)]}{[\hbar cq + E(\mathbf{k}u)]\{[\hbar cq - E(\mathbf{k}u)]^2 + \frac{1}{4}\hbar^2\gamma_{\mathbf{k}u}{}^2\}} \tag{414}$$

is the "resonance" contribution due to electronic state u, and $\Delta\Theta_{\text{back}}$ is the background arising from all other transitions. The ellipticity $\Delta\eta^{\mu-\mu}$ has no background; it is a "resonance" term given by

$$\Delta\eta^{\mu-\mu} = \left(\frac{2\pi N_0 NL}{\hbar cV}\right)\frac{1}{\hbar cq} \sum_{\mathbf{k}} \frac{[E(\mathbf{k}u)]^3\,\frac{1}{2}\hbar\gamma_{\mathbf{k}u}|\mathbf{e}_+ \cdot \boldsymbol{\mu}_0{}^u|^2[\delta(k+q+S) - \delta(k+q-S)]}{[\hbar cq + E(\mathbf{k}u)]\{[\hbar cq - E(\mathbf{k}u)]^2 + \frac{1}{4}\hbar^2\gamma_{\mathbf{k}u}{}^2\}} \tag{415}$$

The ellipticity drops rapidly to zero for $|\hbar cq - E(\mathbf{k}u)| > \frac{1}{2}\hbar\gamma_{\mathbf{k}u}$ and it is reasonable to approximate $E(\mathbf{k}u) \simeq \hbar cq$ in the numerator of (415) with the result that the "resonance" term

$$\Theta_u^{\mu-\mu} = \Delta\Theta_u^{\mu-\mu} - i\Delta\eta_u^{\mu-\mu} \tag{416}$$

is given by

$$\Theta_u^{\mu-\mu} = \left(\frac{\pi N_0 NL}{\hbar cV}\right)\hbar cq \sum_{\mathbf{k}} \frac{|\mathbf{e}_+ \cdot \boldsymbol{\mu}_0{}^u|^2[\hbar cq - E(\mathbf{k}u) - \frac{1}{2}i\hbar\gamma_{\mathbf{k}u}]\,[\delta(k+q+S) - \delta(k+q-S)]}{\{[\hbar cq - E(\mathbf{k}u)]^2 + \frac{1}{4}\hbar^2\gamma_{\mathbf{k}u}{}^2\}} \tag{417}$$

The optical rotation consists of a couplet, the transitions being polarized perpendicular to the helix axis at energies $E(S \pm q,u)$. Since q is very small compared to S (Eq. (9)], the splitting is also extremely small, being only few cm^{-1} when calculated in the dipole approximation (assuming $\mu \simeq 1$ Å, and helix geometries typical of biopolymers[198]).

If the sample contains polymers with random orientations relative to the direction $\mathbf{q}$ of the light beam it is necessary to average $\Theta^{\mu-\mu}$ over all orientations. The following spatial averages[199] are of help:

$$\langle(\mathbf{e}_{\mathbf{q}\lambda} \cdot \mathbf{a})(\mathbf{b} \cdot \mathbf{e}_{\mathbf{q}\lambda'})\rangle = \tfrac{1}{3}(\mathbf{a} \cdot \mathbf{b})\delta_{\lambda\lambda'} \tag{418}$$

$$\langle(\mathbf{e}_{\mathbf{q}1} \cdot \mathbf{a})(\mathbf{e}_{\mathbf{q}2} \cdot \mathbf{b})(\mathbf{e}_{\mathbf{q}3} \cdot \mathbf{c})\rangle = \tfrac{1}{6}\, \mathbf{a} \cdot \mathbf{b} \times \mathbf{c} \tag{419}$$

Clearly one cannot set $q_z = 0$ for, as hinted by (417), $\Theta^{\mu-\mu}$ becomes zero. The energy denominators must be carefully expanded assuming that the splitting is small compared to $\hbar\gamma$. The leading term in the result is

$$\Theta^{\mu-\mu}_{u,\,\text{random}} = \left(\frac{\pi N_0 NL}{3\hbar c V}\right) \frac{|\boldsymbol{\mu}_0{}^u \cdot \mathbf{e}_+|^2 E(S,u)}{[W^2 + \Gamma^2]^2} \Delta[(W^2 - \Gamma^2) - i\Gamma W] \tag{420}$$

where

$$W = \hbar c q - E(S,u) \tag{421}$$

$$\Gamma = \tfrac{1}{2}\hbar\gamma_{Su} \tag{422}$$

$$\Delta = E(S + q,u) - E(S - q,u) \tag{423}$$

Here the couplet splitting Δ is the maximum attenable which occurs for light propagating parallel to the helix axis. The energy W drops to zero at the center of the couplet.

Under not too restrictive conditions $\Theta^{\mu-\mu}$ is the only important component of the complex optical rotation Θ. If the transition $0 \to u$ is an intense one, with $\boldsymbol{\mu}_0{}^u$ perpendicular to the helix axis, the radial term $\Theta^{\mu-\rho\mu}$ vanishes because of a factor $(\boldsymbol{\mu}_0{}^u \cdot \mathbf{e}_z)$. Almost all molecules with intense transitions have at least a plane of symmetry so that $\Theta^{\mu-m}$ vanishes also. The dipole–quadrupole component $\Theta^{\mu-Q}$ is zero for randomly oriented samples. Therefore suitable systems for experimental study may well be found among the dye aggregates or among the numerous complexes that dyes form with the helical biopolymers.

III. COMPARISON WITH EXPERIMENT

The question to be discussed next is how well the single-particle theory of exciton states agrees with the observed spectra. For the weak transitions the latest measurements of factor group splittings, polarization ratios, etc., may be regarded as pretty well final. The situation for medium- and high-intensity transitions lies at the other extreme. Except for anthracene (which is by no means tied down) very little is known that is reliable about the strong transitions in molecular crystals. In a number of cases, the most important so far being anthracene, reproducible room-temperature, and some low-temperature, reflection spectra are becoming available. Reflection studies avoid all the problems encountered with transmission through thin crystals in the region of strong- or medium-intensity transitions, and at the moment are the best method of probing the exciton bands of stronger transitions in the $\mathbf{k} = 0$ region. There are, however, a number of problems that must be solved and not the least of these is the effect of surface quality and surface and subsurface contaminants on the reflection spectrum.

There follows a series of comparisons of theory and experiment for a number of well-studied molecular crystals, beginning with anthracene. There occur in crystalline anthracene both types of transition of particular interest in this review. The first singlet ($f \simeq 0.3$) at 3800 Å is of medium intensity and the second one ($f \simeq 4.8$) at 2500 Å is very strong. The only other strong transitions that have received as much attention belong to the cationic dyes.

A. Anthracene

The electronic processes known to occur in crystalline anthracene are many and diverse. The Davydov splittings of the first singlets depend on crystal face as do the widths of the reflection bands.[155] Absorption of ruby laser light results in blue fluorescence with a prompt and a delayed component due, respectively, to two-photon transitions and triplet exciton annihilation.[24,200] The quantum yield of fluorescence is greater for the crystal than in solution, owing to a change in the order of the first singlet and second triplet levels.[201] The fluorescence can also be altered by applying a magnetic field to the crystal.[24] These and other phenomena make anthracene one of the most fascinating of molecular crystals.

1. *Electronic States of the Isolated Molecule*

Anthracene is the catacondensed hydrocarbon in which the ^1L_a and ^1L_b states in Platt's scheme coincide,[202] and consequently in contrast to naphthalene and phenanthrene, has only two distinguishable electronic states in the UV region. Some evidence for the missing transition has been noted by Hochstrasser.[203] In the vapor phase the 0–0 transition of the first singlet (^1L_a) occurs at 27,688 cm^{-1} and is polarized along the short in-plane axis M with an oscillator strength of $f \simeq 0.3$.[204] The second observed singlet has a 0–0 vapor transition at 42,270 cm^{-1}, has an oscillator strength $f \simeq 4.8$, and is L-axis polarized.[205] The vapor spectrum shows the presence of Rydberg series near 50,000 cm^{-1} but no unambiguous evidence of the transitions clearly seen in solution spectra (n-heptane solvent) around 45,500 cm^{-1} ($f \simeq 0.6$) and 53,000 cm^{-1} ($f \simeq 1.2$).[202,206] The vapor spectrum also displays a flat continuum absorption spectrum out to 65,000 cm^{-1} and beyond.[205]

The first singlet transition has an extensive vibronic structure that "is only now beginning to look as though it has more correct assignments than doubtful ones."[207] Numerous studies have been reported of the absorption spectrum of anthracene trapped in aromatic hydrocarbon and Shpolskii matrices.[208–212] Some of the more prominent vibrational modes observed are listed in Table II together with their ground-state counterparts. Notice that a number of frequencies decrease in the upper state in accordance with the idea that the binding is weakened by promotion of an electron from a bonding to antibonding orbital. The thermally broadened solution spectrum has the appearance of a progression with 1400 cm^{-1} spacing due to the large number of fundamentals and combinations with frequencies in the range 1300–1500 cm^{-1}. This observation forms the basis of a convenient model for crystal calculations.

Several triplet states have been detected. The first occurs at 14,850 cm^{-1} in solution and a second, detected by triplet–triplet transitions, has been observed at 26,050 cm^{-1}.[213–214]

TABLE II
Vibrational Frequencies of Anthracene[a]

Symmetry	Ground state	First excited state
a_g	398	392
	527?	590
	621	652
	755	733
	1020	1028
	1175	1169
	1268	1239?
	1411	1396
	1562	1498
b_{3g}	917?	891
	1180	1166
	1480 (Raman)	1462

[a] Taken from Ref. 105.

2. *Crystal Spectra of Anthracene*

The literature on the singlet crystal spectra of anthracene is too large to attempt review of it here. Crystalline anthracene[215,216] is readily cleaved along the *ab* plane [(001) crystal face] and thin crystals prefer to grow such that this is the developed face. It is not surprising, therefore, that the greatest efforts have been expended in obtaining the absorption spectra of this face. These spectra are invariably reported for light linearly polarized parallel to the *a* and *b* crystal axes. In the region of the first singlet absorption 25,000–29,000 cm^{-1} all except the thinnest of crystals are practically opaque, and crystals have to be less than a micron in thickness before the details of the absorption bands can be discerned. It has been found from these transmission experiments that the *b* component is rather broad, even at 4°K.[217] In fact, it is suspected that near ϵ_{max} complete penetration of the crystal by the light beam has not been obtained, thus casting doubt on the accuracy of the measured Davydov splittings.[218] This is only one of the problems; the handling of thin crystals is plagued with difficulties associated with the preparation, mounting, and cooling, and their effects through strain on the spectrum.[219] Most of these difficulties can be circumvented by measuring the reflection spectra of thick crystals, with the bonus that many crystal faces are usually available.

The first work on the spectrum of the *ab* face reported low ($\simeq$30 cm^{-1}) values for the Davydov splitting of the 0–0 transition near 4000 Å, presumably due to incompletely polarized light,[220,10] In 1958 H. C. Wolf[221] reported that the 0–0 transition was split by 220 cm^{-1} at 90°K, and in the next few years similarly "high" values were reported by groups working in the USSR, England, and Australia.[222–224] Values in excess of 300 cm^{-1} have also been suggested by Lacey and Lyons (380 cm^{-1}),[225] Matsui (310 cm^{-1}),[226] and McRae (380 cm^{-1})[227]; however, most of the other reported splittings are around 200 cm^{-1}. The total splitting summed for all vibronic transitions appears to be in the range 300–450 cm^{-1}.

In a redetermination using a strainfree mounting Jetter and Wolf[219b] found the splitting to be 170 cm^{-1} with the *a*- and *b*-polarized components located at 25,360 and 25,190 cm^{-1}.

The problems in observing the first singlet absorption at 3800 Å have not encouraged examination of the higher singlets, which are much more intense. Craig and Hobbins[228] located two peaks polarized along the *b* axis at 37,300 and 38,500 cm^{-1}, but were unable to find an *ac*-polarized peak before cutoff at 43,500 cm^{-1}. Lyons and Morris[206] found strong *ac*-polarized peaks between 45,000 and 51,500 cm^{-1} and also strong *b*-polarized bands at 45,000 and 52,000 cm^{-1}. This implied a very large splitting for the second singlet in the range 8,000–15,000 cm^{-1}.

In a recent paper Rice, Morris, and Greer[218] reexamined the data available for the *b*-polarized 0–0 component of the 3800 Å transition of the *ab* face of crystalline anthracene. They suggested that because photon penetration depths were small for anthracene, surface disorder resulted in a breakdown of the crystal selection rule $\Delta\mathbf{k} = \mathbf{q}$, and that transitions to all **k** in the Brillouin zone were allowed. This being the case, the width at half height affords a measure of the total width of the exciton band. From the practical point of view it is important to know whether symmetry breaking is a perturbation that just broadens the absorption line or is strong enough to cause a significant shift in the absorption maximum. We shall return to this important problem later, after describing the results of recent reflection measurements.

For weak transitions symmetry breaking due to surface disorder is unimportant since, on the average, photons penetrate past the surface layers and are absorbed in the bulk region of the crystal.

Reflection spectra of crystalline anthracene have been measured by a number of groups over the last ten years.[229–238] The most extensive study is that of Clark,[155,235,237] who has recorded reflection spectra from 20 to 80 kK for a number of faces. The reflection spectrum of face (001) is now known at least approximately down to 440 Å (28 eV). Two striking phenomena have been observed; the first is the change in Davydov splittings with direction of the wave vector, and the second is the observation of the massive metallic reflection bands in the near-UV.[155] Fox and Yatsiv[239] first pointed out that the splittings should be **k**-dependent, and the first experimental indication that this effect exists can be seen in the reflection spectra reported for the 3800 Å system by Morisova.[231] The first full study, however, is that described by Clark and Philpott[155] and carried out at 298°K.

In the paper by Brodin et al.[234] the *b* component of the vibrationless transition [face (001)] was observed at 4°K to have a complex structure starting with a "zero-phenon" line at 25,098 cm^{-1}. No structure was observed in the *a*-polarized reflection band. Reflectivities for *b* were around 0.65, which are much higher than the room temperature values of 0.35.[155] So far there seems to have been no independent verifications of this important discovery.

The reflection spectrum of face (001) taken at room temperature is shown in Fig. 21. The corresponding absorption spectrum derived by a Kramers-Kronig analysis is shown in Fig. 22. The vibrational structure of the first singlet is clearly visible, that for the *b*-polarized second singlet at 37,000 cm^{-1} much less so. There are three strong *b*-polarized peaks at 45, 53, and 63 kK that may correspond to three *M* axis π–π^* transitions. The main transitions of the *a* spectrum are not so nicely separated because strong positive exciton interactions mix the electronic states. The strong

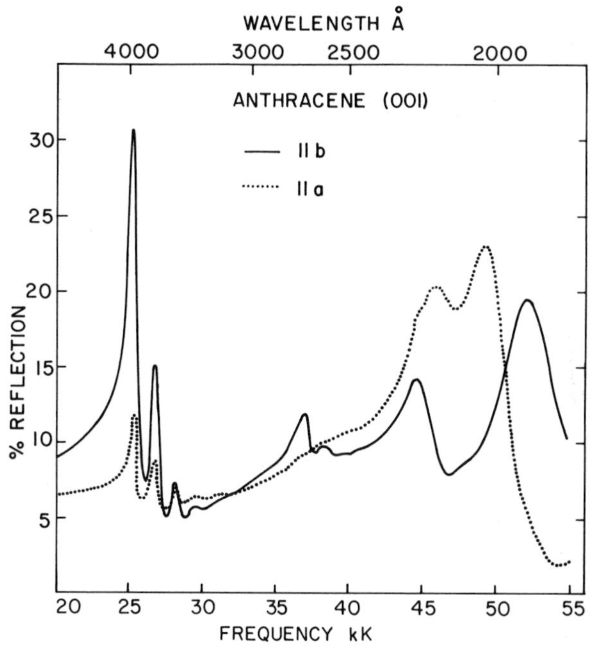

Fig. 21. Reflection spectra of the (001) face. The solid and dotted curves were obtained with the incident light polarized parallel to the *b* and *a* crystallographic axes, respectively.

absorption peaking at 49 kK probably represents the *a* component of the second singlet; however, because of strong interactions it is unlikely that any state has a unique parentage in this region. A thin crystal absorption spectrum has been reported by Lyons and Morris,[206] the main features of which are compatible with Fig. 22 up to 52 kK in the *b* spectrum and 60 kK in the *a* spectrum. The thin crystal absorption between 45–52 kK, though qualitatively similar, is broader and peaks at 51 kK and not at 49 kK. Presumably the differences are due to incomplete penetration of the thin crystal in the regions of high extinction coefficient. Above 70 kK the absorption spectrum rises steadily due to the onset of σ–π^*, π–σ^*, and σ–σ^* transitions, as can be seen in reflection spectrum of Koch et al.[240] obtained using synchrotron radiation.

The room-temperature reflection spectra of two other natural faces of anthracene, $(20\bar{1})$ and $(\bar{1}10)$, and the derived absorption spectra are shown in Figs. 23–26. Plotted on the scales shown, the splittings of the first singlet transitions are just perceptible. The eye-catching features of both reflection spectra are the massive blocks of high reflectivity due to the intense *L* axis transition. These are the metallic reflection bands of anthra-

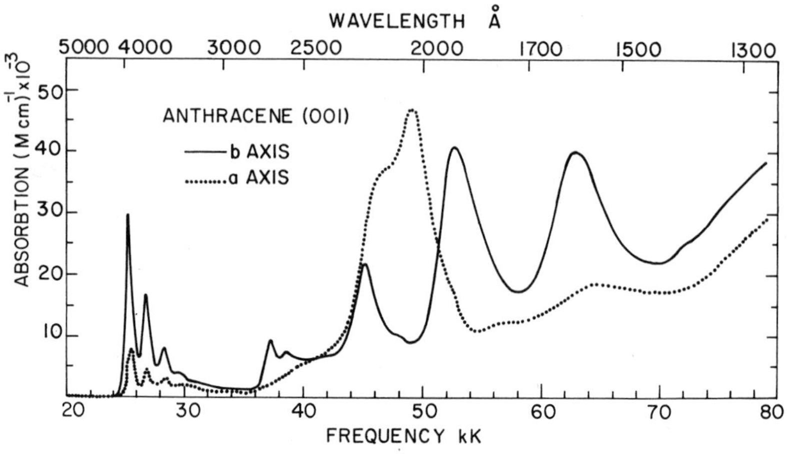

Fig. 22. The absorption spectra of the (001) face derived by Kramers-Kronig analyses of corresponding reflection curves.

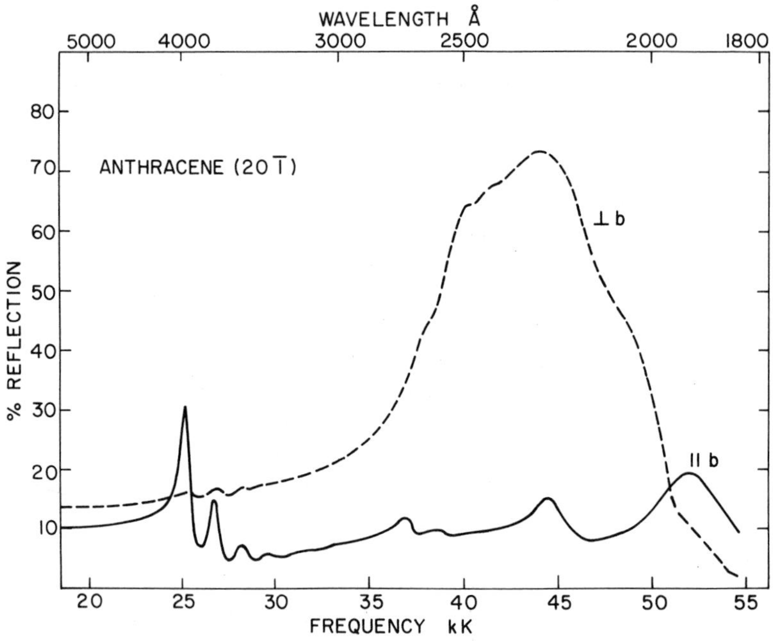

Fig. 23. Reflection spectra of the (20$\bar{1}$) face. The *b* axis spectrum (solid curve) is identical to the *b* axis spectrum of the (001) face. The dashed curve is the reflection spectrum obtained with the incident radiation polarized perpendicular to the *b* axis.

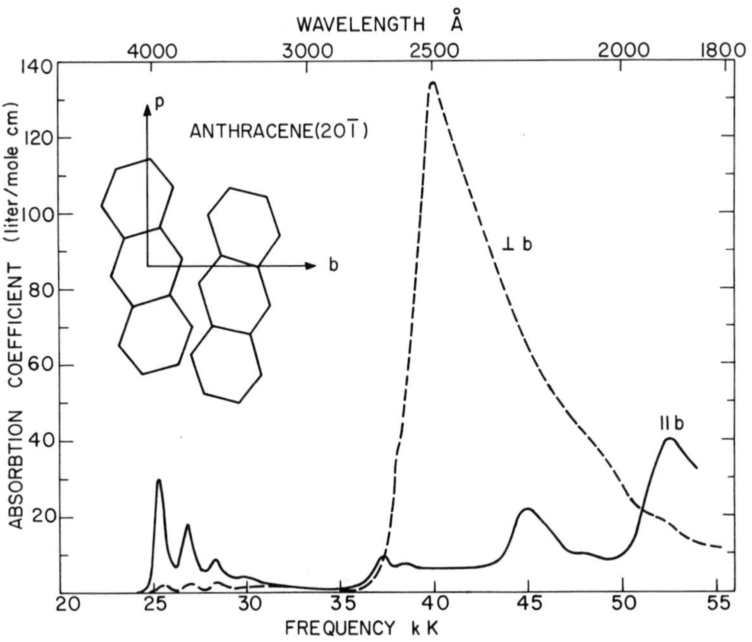

Fig. 24. Derived absorption spectra of the (20$\bar{1}$) face. The solid ($||b$) and broken ($\perp b$) curves were obtained from the corresponding reflection spectra shown in Fig. 23 via Kramers-Kronig analyses. The projection of the two molecules of the unit cell onto the (20$\bar{1}$) plane is shown.

cene mentioned previously in this article. For both the (20$\bar{1}$) and ($\bar{1}$10) faces the photon wave vector is very nearly perpendicular to the L molecular axis. This means that the macroscopic part of the dipole sum is small and the exciton interaction too weak in both a- and b-polarized spectra to invert the order of electronic states [as apparently occurs for face (001)]. The exciton–photon coupling is weak for the $||b$ and strong for the $\perp b$ since the L molecular axis is almost perpendicular to the b-crystal axis. The metallic reflection is therefore confined to the $\perp b$ spectra. In anthracene this phenomenon is more complicated than that previously observed by Simpson and co-workers for crystalline dyes.[144,145] For one thing it encompasses several less intense electronic states, whereas in the dyes the first singlet is isolated from other much weaker states by 2 eV. There is also a lower-lying singlet state and (presumably) a whole manifold of triplet levels, and all these states facilitate radiationless decay. According to (147) the maximum width of the metallic reflection band is $(4\pi/v_0)|\mu_{II}|^2$; for anthracene the first factor is 6155 cm^{-1}/Å^2 and since $\mu_{II} \simeq 1.87$ Å the

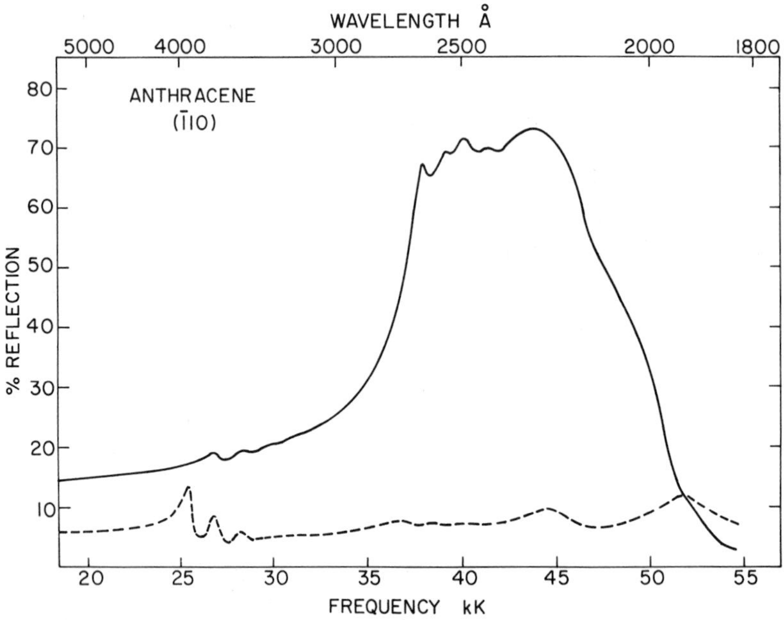

Fig. 25. Reflection spectra of the (1̄10) face. The solid and broken curves correspond to the incident radiation polarized along the R_{max} and R_{min} directions, respectively, which are indicated in the upper left-hand corner of Fig. 26.

theoretical width is about 21,500 cm^{-1}. It is evident from Figs. 23 and 25 that the experimental width is about half the theoretical estimate, suggesting a background dielectric of $\epsilon \simeq 2$.

Clark has used an ultramicrotome to cut artificial faces for which the orientation of the L molecular axis to crystal surface is intermediate between (1̄10) and (001). In this way a complete series of spectra has been obtained showing the change from normal (001) to metallic reflection (110).[241]

The development of vibrational structure on the low energy edge of the metallic reflection bands is seen in going from (201̄) to (1̄10). This is in accord with strong vibronic coupling theory, for as the photon's wave vector $\mathbf{q}$ gets closer to being perpendicular to the transition moment, the single-particle state $|\mathbf{q}\rangle$ breaks away from the multiparticle bands and the transition to $|\mathbf{q}\rangle$ becomes discrete. When $|\mathbf{q}\rangle$ lies near the bottom of the single-exciton band the nearest levels that can absorb light are one (ground) vibrational quantum away, hence the appearance of vibrational structure. This structure is not as well defined as for cationic dye

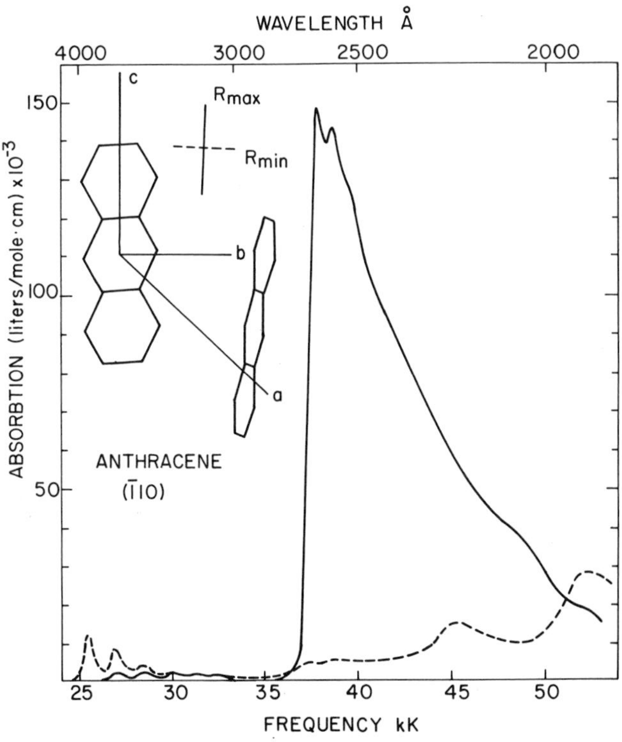

Fig. 26. Derived absorption spectra of the ($\bar{1}$10) face. The solid and broken curves were obtained from the corresponding reflection curves shown in Fig. 25 via Kramers-Kronig analyses.

crystals, presumably because of damping processes such as radiationless decay within the molecules, interbranch crossing, and symmetry breaking due to surface disorder.

The absorption band positions, factor group splittings, and polarization ratios derived from the reflection spectra of faces (001), (20$\bar{1}$), and ($\bar{1}$10) are listed in Tables III and IV.

Next we comment on the idea, described earlier, of symmetry breaking due to surface disorder.[218] For the first singlet (3800 Å) transition the theoretical penetration depth is 100 Å. This is 11 layers normal to (001) and 12 layers normal to (20$\bar{1}$). We may regard 10 layers as representative of the region sampled by the photons in a reflection experiment. If the crystal is perfect ten layers is sufficient for the reflection spectrum to yield accurate values of the energies, Davydov splittings, etc., for, owing to rapid convergence of planar dipole sums, molecules in the second layer

TABLE III

3800 Å Absorption Data for Faces (001), (20$\bar{1}$), and ($\bar{1}$10)[a]

(001)				(20$\bar{1}$)		($\bar{1}$10)			
‖ b		⊥b		⊥b		‖ R_{min}		‖ R_{max}	
ν (cm^{-1})	f	ν (cm^{-1})	f	ν (cm^{-1})	f	ν (cm^{-1})	f	ν (cm^{-1})	f
25310	0.097	25500	0.020	25580	0.006	25430	0.036	?	0.001
26820	0.071	25930	0.022	27000	0.010	26890	0.031	~27000	0.080
28300	0.038	28360	0.014	28400	0.010	28340	0.018	28430	0.010

[a] Taken from Ref. 155.

TABLE IV

Spectral Data for Absorption Regions Exclusive of the 3800 Å System[a,b]

(001)					(20$\bar{1}$)		($\bar{1}$10)				
‖ b			⊥b		⊥b		‖ R_{max}		‖ R_{min}		
	ν (cm^{-1})	f	ν (cm^{-1})	f	ν (cm^{-1})	f	ν (cm^{-1})	f		ν (cm^{-1})	f
II	37300	0.16	38800[c]	0.04	36800[c]	4.3[d]	37900	4.7[d]	II	37300	≤0.1
II	38500		40100[c]		38100[c]		38700		II	38700	
III	44900	0.6	46600[c]	0.7[d]	40100		39500[c]		III	45200	0.2
	48200	0.06	49100		48100[c]		48000[c]			47800[c]	
IV	52500	~0.8	52600[c]		52200[c]		52000[c]		IV	52300	~0.5

[a] The oscillator strengths given are rough estimates since the individual bands are extensively overlapped.

[b] Frequency uncertainty is ±200 cm^{-1} for all but the sharpest peaks.

[c] Inflection or shoulder.

[d] These oscillator strengths relate to the entire band envelopes which no doubt con-ain components of transitions III, IV, and possibly others.

see a dipole field practically indistinguishable from the value deep inside the crystal. If there is long-range disorder in a layer due to vacancies, line faults, etc., the planar sums will depend on the origin and some symmetry breaking will result. It has been observed that the reflectivity of mechanically cut surfaces is well below that of natural ones.[241] However, the spectra, though lower in reflectivity, display all the maxima and minima expected. This suggests the loss is due to diffuse scattering since microscopic examination of the artificial surfaces reveals them to be badly scored and scratched. This suggests that by using surfaces carefully selected for their freedom from observable defects, symmetry breaking due to surface disorder can be kept at a level where it acts as a perturbation on the spectrum rather than the cause of large intensity redistributions.

For the strong second singlet transition (2500 Å in molecule) the situation is more acute. The absorption coefficients obtained from Figs. 24 and 26 are around 10^7 cm^{-1}, and the corresponding penetration depths are 10 Å, or one layer. For a perfect crystal the second layer sees a dipole field close to that felt by interior layers so the reflection spectrum monitors the optical properties of the bulk crystal. For a real crystal with surface defects the photons sample more disorder than bulk crystal if the disorder extends deeper than two layers. Much obviously depends on the concentration of defects per unit area of surface. The derived absorption spectra are very asymmetric, as predicted by the symmetry breaking argument, though it should be remembered that part of the absorption is due to multiparticle states and another electronic transition. The spectra shift around in the way predicted by exciton theory, and the development of vibrational structure in going from $(20\bar{1})$ to $(\bar{1}10)$ is also in agreement with the theory. Therefore, it appears that surface disorder is acting as a perturbation which is rather stronger than that seen for the first singlet. These conclusions and the idea of symmetry breaking due to surface disorder are tentative. More conclusive arguments await the measurement of high-resolution low-temperature reflection spectra of strong transitions.

3. *Theoretical Calculations*

Three methods have been used to calculate the factor group splittings of the first singlet transition for the (001) face.[242–244,157a] We shall refer to them as the multipole, MO (molecular orbital), and dipole methods. All three methods are deficient in one way or other. The possibility of a fourth approach to the problem of strong exciton interactions will be discussed at the end of this section.

The dipole method is simply the application of dipole theory, either classical or quantum, to the calculation of energies, splittings, and polarization ratios. Its deficiency is self-evident—no electron-exchange and higher-multipole interactions are included. It is the only method to have been applied correctly; the multipole and MO methods have not been used correctly thus far, for reasons enlarged upon below. In a strict sense the dipole method is not logically distinct from either the multipole or the MO methods. However, since it has been put to wide use in the calculation of band structure and density of states, it is deemed appropriate to discuss the results of dipole theory separately.

The method of transition multipoles was originally devised to treat the factor group splittings of the first singlet states of naphthalene and benzene.[245–247] Subsequently it was used on the first singlets of anthracene and tetracene.[242,248] The transitions involved are very weak for naphthalene and benzene and in the intermediate range for anthracene and tetracene. For molecules with D_{2h} symmetry there are only two allowed transition octopoles for each of the transitions ${}^1A_{1g} \rightarrow {}^1B_{1u}$ (M axis) and ${}^1A_{1g} \rightarrow {}^1B_{2u}$ (L axis), and there are no transition quadrupoles. The lattice sums of exciton interactions contain dipole–dipole, dipole–octopole, and octopole–octopole sums. For example, the $\mathbf{k} = \mathbf{q}(|\mathbf{q}| \simeq 0)$ lattice sum of inequivalent exciton interactions for the ${}^1B_{1u}$ state of anthracene is

$$I_{12}(\mathbf{k}) = X\mu_M{}^2 + 111.85\, \mu_M \mathcal{O}_M + 2.798\, \mu_M \mathcal{O}_M{}' + 3.203\, (\mathcal{O}_M)^2 - 0.00811\, (\mathcal{O}_M{}')^2 + 0.3156\, \mathcal{O}_M \mathcal{O}_M{}' \text{ cm}^{-1}$$

where the dipole sum X is given by

$$X = 967 \text{ cm}^{-1}/\text{Å}^2 \qquad \mathbf{k} \perp (001)$$

$$X = 1117 \text{ cm}^{-1}/\text{Å}^2 \qquad \mathbf{k} \perp (20\bar{1})$$

The transition octopoles $\mathcal{O}_M$ and $\mathcal{O}_M{}'$ are given by

$$\mathcal{O}_M = Q_3{}^{1s}, \qquad \mathcal{O}_M{}' = Q_3{}^{3s}$$

in the notation of Craig and Walmsley.[8] When Craig and Thirunamachandran reported their calculation it was thought that a spherical shape best represented the boundary of the exciton wave packet formed by photon absorption and they accordingly used the value $X = 53$ cm^{-1}/Å^2. As we have discussed and argued at several places in this article the boundary conditions imposed by the radiation field dictate that the Ewald (or planewise) sum be used. Clearly the spherical sums, being so much smaller than the Ewald sums, require larger octopole moments and the values $\mathcal{O}_M = 2.4$Å^3 and $\mathcal{O}_M{}' = 20$ Å^3 were found to give the best fit to crystal spectra. Calculations based on the dipole theory with Ewald sums predict splittings larger than the range found experimentally, and the octopoles are required to *reduce* the value of the lattice sum. Silbey, Jortner, and Rice[243] found that the multipole contribution to the splitting was -180 cm^{-1}, so the multipole contribution to the inequivalent sum is approximately -100 cm^{-1}. Thus if both octopoles are of the same order of magnitude, the contribution of $\mathcal{O}_M{}'$ to $I_{12}(\mathbf{k})$ can be neglected and $\mathcal{O}_M \simeq -1.5$ Å^3. Smaller values like this one are in better agreement with molecular orbital calculations of transition octopoles.[17]

At the present time it is not known whether the introduction of transition octopoles will lead to a better theory of exciton interactions in anthracene. Calculations to remedy this situation are being contemplated.

Next we turn to the MO calculations of exciton interactions in anthracene.[243,244] In principle this is the most rigorous way of calculating the lattice sums required by the exciton theory of crystal spectra. There is one great drawback to the MO method; the wave functions used predict oscillator strengths of the free molecules that are too large by as much as a factor of five (in the case of Hückel orbitals). Therefore, the exciton interactions are scaled by a factor that brings the calculated and observed oscillator strengths into coincidence. This involves the unpleasant but unavoidable assumption that the higher multipoles, which are a more sensitive measure of transition charge density, scale the same way as the transition dipoles. The Chicago group of Rice, Jortner, and co-workers have computed the exciton states of a large number of molecular crystals using Hückel and Pariser-Parr wave functions. Their work on anthracene is particularly interesting since the exciton interactions were broken down into dipole, higher-multipole, and electron-exchange components. Tanaka and Tanaka[244] have also calculated the factor group splittings with SCF wave functions, treating the long-range part of the dipole interaction in the same way as in the calculation of Silbey, Jortner, and Rice.[243] Although we have great admiration for the effort involved in these calculations they are inadequate in two respects. First, since the (off-diagonal) crystal interactions among different electronic states are strong, the second quantized (Boson) formulation of exciton theory should have been used (the multipole calculation is also in default here). Second, via an incorrect analysis of retarded interactions, the infinite crystal contribution from long-range dipole interactions outside a large sphere of radius R_0 was given by

$$\left(\frac{4\pi\mu^2}{v_0}\right)[3(\hat{\mathbf{k}}\cdot\hat{\mathbf{u}})^2 - 1]$$

This is the Cohen-Keffer infinite-cubic-crystal result.[249] In arriving at this result the retardation correction was thrown away, which was fortunate since, as we have explained in Section II.B, it is not required. The errors introduced into the dipole sum by using the cubic-crystal approximation for the long-range part are not very severe, being about 14% and <1%, respectively, for the A_u and B_u factor group sums of the first singlet. The dipole component of the first-order splitting is too high by nearly 20%. Using the Ewald sums we estimate a first-order splitting of 545 cm^{-1} in place of the 622 cm^{-1} figure given by Silbey et al. The difference is slight (about 14%), considering all the other uncertainties in exciton theory.

Clearly these MO calculations are of great value; they have provided us with a first glimpse of the relative magnitudes of dipole, higher-multipole, and electron-exchange interactions.

The same argument may be applied to the intense long axis transition. It has been suggested that an oscillator strength of 1.6 is the best value for this transition. If this is correct then the Pariser wave function overestimates the oscillator strength by a factor of two, and the interactions should therefore be scaled accordingly. The higher multipoles are then found to contribute $-3342\ \text{cm}^{-1}$ to the inequivalent interaction. The point dipole Ewald interaction is $2661 \times (1.9)^2 = 9606\ \text{cm}^{-1}$, so that the effective resonance interaction is about $6300\ \text{cm}^{-1}$.

Next we consider the results of the dipole theory for crystalline anthracene. The calculations of Mahan[156] were the first to use both the Boson formulas and Ewald dipole sums. More extensive calculations were later performed by Philpott,[157] who used more accurate structural data and included the higher M axis transitions. The exciton band structure for the 3800 Å[250,251] and 2500 Å[250] transitions has been calculated for wave vectors pointing along the directions (100), (010), (001), $(20\bar{1})$, and $(\bar{1}10)$. Schroder and Silbey[252] have recently calculated the exciton energies for a three-dimensional net over the whole of the first zone and computed a density of states function for the 0–0 component of the 3800 Å transition. The molecular data used in most of these calculations are shown in Table V and the Ewald dipole sums are given in Tables VI and VII.

Results of dipole theory calculations for the two- and four-state models are summarized in Tables VIII and IX.[155] The second singlet dominates the spectrum and is subject to the greatest crystal interactions. According

TABLE V
Anthracene Spectral Data Used in Dipole Theory Calculations

Transition	Polarization	Energy (cm^{-1})	Transition dipole (Å)
I (3800 Å)	M	26000[a] $+n1400$	0.61
II (2500 Å)	L	39000	1.87
III (2200 Å)	M	45000	0.64
IV (1900 Å)	M	54000	0.83

[a] Vibronic components $n = 0, 1, \ldots, 4$ with Franck-Condon factors $\zeta_n{}^2 = 0.324$, 0.316, 0.218, 0.092, 0.050.

TABLE VI

Dipole Sums ($cm^{-1}/Å^2$) for Faces (001) and ($20\bar{1}$)

αX βY	$t_{\alpha X,\beta Y}(0)$	$(\hat{\mathbf{d}}_{\alpha X}\cdot\hat{\mathbf{k}})(\hat{\mathbf{k}}\cdot\hat{\mathbf{d}}_{\beta Y})$ (001)	$(\hat{\mathbf{d}}_{\alpha X}\cdot\hat{\mathbf{k}})(\hat{\mathbf{k}}\cdot\hat{\mathbf{d}}_{\beta Y})$ ($20\bar{1}$)	$T_{\alpha X,\beta Y}(\hat{\mathbf{k}})$ (001)	$T_{\alpha X,\beta Y}(\hat{\mathbf{k}})$ ($20\bar{1}$)
1*M* 1*M*	−2077	308	457	−1769	−1620
1*M* 2*M*	660	308	457	967	1117
1*M* 1*L*	329	836	−349	1165	596
1*M* 2*L*	996	836	−349	1832	1263
1*L* 1*L*	−172	2273	267	2100	−521
1*L* 2*L*	389	2273	267	2661	40

TABLE VII

Macroscopic Component of the Dipole Sums ($cm^{-1}/Å^2$) for Face ($\bar{1}10$)

αX βY	$(\hat{\mathbf{d}}_{\alpha X}\cdot\hat{\mathbf{k}})(\hat{\mathbf{k}}\cdot\hat{\mathbf{d}}_{\beta Y})$	αX βY	$(\hat{\mathbf{d}}_{\alpha X}\cdot\hat{\mathbf{k}})(\hat{\mathbf{k}}\cdot\hat{\mathbf{d}}_{\beta Y})$
1M 1*M*	461	2*M* 1*L*	440
1*M* 2*M*	−1152	2*M* 2*L*	−124
2*M* 2*M*	2880	1*L* 1*L*	67
1*M* 1*L*	−176	1*L* 2*L*	−19
1*M* 2*L*	50	2*L* 2*L*	5

to theory the splitting of the second singlet drops rapidly in going from face (001) to ($\bar{1}10$). Geometrically speaking, the direction of the photon wave vector **q** has swung around from being nearly parallel to the *L* axis to being almost perpendicular to it. Experimentally a great change is seen in the splitting which goes from 9300–15,300 to 2800 to 600 cm^{-1}. The results of dipole theory with two or four states overestimates the splitting by 1500–2000 cm^{-1} for ($20\bar{1}$) and ($\bar{1}10$). The discrepancy between theory and experiment may be due to nondipolar interactions or to a reduction in the magnitude of the dipole interaction due to dielectric shielding. A value of approximately −2000 cm^{-1} for the nondipolar interaction is in agreement with the calculations of Silbey, Jortner, and Rice.[243] From their Table III.b.1 the higher-multipole component of the inequivalent interaction is (in their notation)

$$\frac{PA}{1.4} - I_{dd'}{}^{\beta} = -1837\ \mathrm{cm}^{-1}$$

Since their value for oscillator strength is high, 2.3, we scale this last number by (1.6/2.3) and obtain −1278 cm^{-1}.[253] The higher-multipole com-

TABLE VIII

Calculated Exciton Energies (cm^{-1}) and Polarization Ratios for Faces (001), ($20\bar{1}$), and ($\bar{1}10$)

	(001) face				($20\bar{1}$) face				($\bar{1}10$) face		
Transition	a_ι	b	Splitting	PR [a]	$\perp b$	b	Splitting	PR [a]	$\perp$	$\parallel$	Splitting
I 0	25702	25498	204	3.5	25921	25499	422	181	26030	25664	366
1	27180	27092	88	2.4	27328	27092	236	113	27423	27162	261
2	28664	28618	46	2.2	28752	28618	134	141	28809	28654	155
3	30142	30124	18	2.1	30180	30124	56		30200	30138	62
4	31567	31556	11	2.0	31589	31556	33		31597	31564	33
II	53330	37040	16290		41570	37040	4530		40000	37330	2670

[a] Polarization ratio ($f_b/f_{\perp b}$).

TABLE IX

Influence of Higher Electronic Transitions on the 3800 and 2500 Å Systems

	(001) direction				(20$\bar{1}$) direction				($\bar{1}$10) direction		
Transition	a	b	Splitting	PR[b]	$\perp b$	b	Splitting	PR[b]	$\perp$	$\parallel$	Splitting
I 0	25652	25355	297	4.3	25918	25355	563	275	26030	25601	429
1	27153	27041	112	2.6	27326	27041	285	142	27423	27129	294
2	28648	28589	59	2.3	28750	28589	162	175	28809	28635	174
3	30135	30110	25	2.3	30179	30110	69		30200	30129	71
4	31562	31547	15	2.2	31588	31547	41		31596	31558	38
II	57431	36853	20578		41321	36853	4468		39050	36910	2140
III	43279	43783	−504		44942	43783	1159		45663	44579	1084
IV	50299	52408	−2109		53722	52408	1314		54905	53077	1828

[a] I, 3800 Å; II, 2500 Å; III, 2200 Å; IV, 1900 Å.
[b] Polarization ratio ($f_b/f_{\perp b}$).

ponent of the first-order splitting is twice this, namely, $-2556\ \text{cm}^{-1}$, which is close to the value estimated above from the spectra.

For the first singlet transition the observed change in Davydov splitting with crystal face is small and subject to considerable experimental error. More reliable values wait for the measurement of low-temperature reflection spectra. The dipole theory predictions are qualitatively correct but are far from being quantitative, especially when four electronic states are used. In these latter calculations the extra M-polarized states are coupling collectively and depressing the lowest b-polarized level. In the $\perp b$ spectrum the coupling is much weaker so that the lowest $\perp b$ state is depressed much less. This indicates that the crystal interactions are weaker than the dipole interactions and that the combined effects of higher multipoles and electron exchange act as a counter balance. Clearly once accurate low-temperature data is available we should be able to find the nondipolar part of the exciton interaction for the first singlet transition. The observed energy of a medium-intensity transition represents a balance reached between the dipole and nondipole interactions coupling the given state to all others. Thus at the moment it appears that the Davydov splittings of the first singlet depend not only on the magnitude of the sums $T_{1M2M}(\hat{\mathbf{k}})$ and how they change with direction $\hat{\mathbf{k}}$, but also on how the other electronic states move as $\hat{\mathbf{k}}$ changes.

Finally, we shall comment on planewise dipole sums and nondipolar exciton interactions. According to the De Wette method the sum is broken into planar components

$$\mathbf{T}_{\alpha\beta}(\mathbf{k}) = \sum_{l=-\infty}^{+\infty} \mathbf{v}_{\alpha\beta,l}(\mathbf{k})$$

where $\mathbf{v}_{\alpha\beta,l}(\mathbf{k})$ is the contribution from the lth plane of dipoles, all planes being perpendicular to $\mathbf{k}$. The two-dimensional sums $\mathbf{v}_{\alpha\beta,l}(\mathbf{k})$ fall off with an exponential-like rapidity as $|l|$ increases. As an example we show in Table X the results of some calculations for anthracene with $\mathbf{k}$ perpendicular to the ab crystal plane [(001) face], and $|\mathbf{k}| = 0$. The sums refer to unit dipoles oriented along the x ($||a$) and y ($||b$) axes. For both equivalent and inequivalent sums the $l = 0$ component is about two orders of magnitude greater than the other components. The numbers shown in Table X are dimensionless since the sums $T_{\alpha\beta}(\mathbf{k})$ are dimensionless; see (160). Now the full interaction [Eq. (131)] can also be written in planewise form and for $|\mathbf{k}| \simeq 0$ we have

$$I_{\alpha\beta}{}^{uv}(\hat{\mathbf{k}}) = \sum_{l} [I_{\alpha\beta}^{(d)}(\hat{\mathbf{k}}\,|\,l) + I_{\alpha\beta}^{(nd)}(l)]$$

where $I^{(d)}$ and $I^{(nd)}$ are the dipole and nondipole parts, respectively. Since the $I_{\alpha\beta}^{(nd)}(l)$ are sums of short-range interactions, they fall off rapidly with

TABLE X

Planewise Dipole Sums for Anthracene[a] [001]

	Equivalent		Inequivalent	
l	$\hat{\mathbf{\mu}}_x \cdot \mathbf{v}_l \cdot \hat{\mathbf{\mu}}_x$	$\hat{\mathbf{\mu}}_y \cdot \mathbf{v}_l \cdot \hat{\mathbf{\mu}}_y$	$\hat{\mathbf{\mu}}_x \cdot \mathbf{v}_l \cdot \hat{\mathbf{\mu}}_x$	$\hat{\mathbf{\mu}}_y \cdot \mathbf{v}_l \cdot \hat{\mathbf{\mu}}_y$
0	−0.1521	−0.8221	−1.2424	−0.3524
1	−0.4102(−3)	0.6709(−3)	0.3661(−3)	−0.6834(−3)
2	−0.9458(−5)	0.4701(−7)	0.9457(−5)	−0.4919(−7)
3	0.1631(−8)	0.3414(−11)	−0.1631(−8)	−0.3412(−11)

[a] Crystal structure data of D. W. J. Cruickshank, *Acta Cryst.*, **9**, 915 (1956).

$|l|$. Also they are independent of $\hat{\mathbf{k}}$ so that the summation can go over any set of crystal planes. The $I_{\alpha\beta}^{(nd)}(l)$ may be treated as empirical parameters to be found by fitting calculated spectra to the experimental ones. Once found, these parameters could be analyzed into sums of neighbor or multipole interactions and then the exciton band structure and density of states calculated.

B. Tetracene and Pentacene

In the series anthracene, tetracene (naphthacene), and pentacene the $^1A_{1g} \rightarrow {}^1B_{1u}$ (M axis, ^1L_a) transition lies lowest and shifts progressively to the red as the length of the molecule increases. The transition has approximately the same intensity ($f \simeq 0.3$) in all three molecules, and since it lies clear of the jungle of higher levels we have an important series with which to test exciton theories of moderately intense transitions. Tetracene and pentacene have electronic spectra even richer in absorption bands than anthracene, and if higher electronic states can exert any influence on the lower ones it should appear in their crystal spectra. We believe that the increase in factor group splittings along the series is due to interaction with an increasing number of higher levels along the series. Crystal spectra for the (001) face are the only ones reported in the literature.

The vapor spectrum of tetracene from 20,000 to 54,000 cm^{-1} has been reported by Morris[254] and the solution spectrum by Clar[255] and Platt.[202] The crystal spectrum of the 0–0 transition of the first singlet shows a splitting of about 700 cm^{-1} [256–259] for the high temperature triclinic crystal[260] form. Near 70°K crystalline tetracene undergoes a phase transition to what may be a monoclinic structure.[261] The splitting of the low-temperature modification is 940 cm^{-1}. There is good agreement among the reported values of the splittings. For the first three vibronic transitions (frequency spacing near 1430 cm^{-1}) the Davydov splittings are (cm^{-1}): 630, 270, 90 (Bree and Lyons); 620, 275, 125 (Prikhotko and Skorobgatko); 630, 222, 153 (Tanaka). The $||b$ component of the multiplet lies lowest as in anthracene, and it is also broad, indicating the presence of surface disorder or phonon transitions. The agreement between different workers is undoubtedly due to the large splitting and the small overlap between the $||b$ and $\perp b$ absorption bands so that none of the experimental problems due to imperfectly polarized light occur.

Dipole theory calculations[157a] have been based on the following model spectrum for tetracene: I, 22,220 + n 1430 cm^{-1}, ξ_n^2 = 0.271, 0.327, 0.209, 0.134, 0.060 (n = 0, 1, 2, 3, 4), M axis, μ = 0.6 Å; II, 35,030 cm^{-1}, M axis, μ = 0.46 Å; III, 38,850 cm^{-1}, L axis, μ = 1.88 Å; IV, 44,000 cm^{-1}, M axis, μ = 1.03 Å; V, 49,000 cm^{-1}, M axis, μ = 0.97 Å. According to dipole theory the splittings of the first singlet build up as the higher transitions are added. The final results for the splittings with all the above-listed transitions included in the calculation are 631, 265, 138, 96,, and 54 cm^{-1} for n = 0, . . ., 4. The agreement, as striking as Mahan's calculation for anthracene,[156] is fortuitous since there are uncertainties in the crystal structure, oscillator strength, and polarization of transitions IV and V that could easily destroy the coincidence. However, the fact remains that dipole interactions above all else appear to be controlling the magnitude of the splitting.

For a molecule the size of tetracene the dipole approximation is not adequate for nearest-neighbor interactions, and the nondipolar correction is expected to be large, especially for the intense transitions. Kuhn and co-workers have shown how poor the dipole approximation is for the exciton interaction between cyanine dye molecules.[262–264] The stage has been reached with the tetracene (and pentacene, see below) problem where calculations using the MO method are needed to determine how the higher-multipole and electron-exchange interactions modify the dipole coupling of higher transitions to the first singlet.

The solution spectrum of pentacene has been described by Clar and Platt. The crystal structure at room temperature is triclinic $P\bar{1}$, like tetracene and hexacene. The molecular orientations and unit cell angles α, β, α are within a few degrees the same as tetracene.[260] Likewise, the unit cell dimensions a and b agree with those of tetracene to within 0.03 Å (c axis is 3.5 Å greater).[260] In the crystal spectrum obtained by Prikhotjko and co-workers[256–267] a splitting of 1000 cm^{-1} at 77°K was measured for the 0–0 level of the first singlet. The width of the $||b$ absorption at half height exceeded that observed in anthracene and tetracene. No dipole calculations appear to have been reported for pentacene, though in view of the rapid convergence of dipole sums for face (001) and similarities in the structure of the *ab* crystal plane and similarity in electronic spectra, the dipole splitting should be at least as large as found for tetracene.

The interpretation supplied by dipole theory for the increase in Davydov splitting along the series anthracene, tetracene, pentacene is that as rings are added the number of excited π states is increased, and these depress the $\|b$ exciton branch more than the $\perp b$ branch. A splitting of 1000 cm^{-1} must mean that the exciton band arising from the 0–0 transition is not entirely composed of single particle excitations (this may be true also of tetracene) since there are several (prominent) excited state vibrational frequencies smaller than the splitting.[266] Thus along the series anthracene to pentacene a gradual change from intermediate to strong vibronic coupling occurs. The author knows of no other structurally similar crystals in which this change could be studied.

C. Naphthalene and Benzene

The properties of the exciton bands of the first singlet transitions of benzene and naphthalene have been reviewed in detail by Robinson.[26] The transitions are very weak ($f \lesssim 0.01$) and so lie outside the scope of this article, which is concerned more with the stronger transitions ($f \gtrsim 0.3$). Time and space also prevent consideration of the beautiful experiments and the detailed analyses in terms of exciton interactions of near-neighbor molecules. Some of the inferences drawn from these studies of very weak transitions have, in our opinion, wide-ranging implications for the whole of exciton theory. It is upon these implications that we wish to comment here.

The interpretation of the weak crystal spectra has been based on a one-band Frenkel exciton model; that is, there are no interactions between different vibronic exciton bands. Intensity changes away from the oriented gas values are due to such interactions. In the early analyses the restricted Frenkel approximation was used but this has now been relaxed. Experimental data from many sources (absorption, fluorescence, isotopic impurity, band–band, impurity-pair spectra) can be rather completely understood in terms of empirical near-neighbor exciton interactions. The largest of these interactions in naphthalene (-18 cm^{-1}) is between translationally inequivalent molecules in the same unit cell.[268] The observed Davydov splitting arises primarily from this interaction.

It has been shown that the theory of Craig and Walmsley, although giving a good fit for the splitting and polarization ratios in naphthalene, cannot explain the wealth of experimental data now available. In benzene where only one octopole transition moment is symmetry-allowed the wrong sequence is predicted within the Davydov multiplet.[246] Therefore some nonoctopole (dipole is negligible) interaction is an important part of the exciton coupling between benzene molecules. Electron exchange is an obvious source of the missing interaction. However, the first MO calculations of exciton interactions in the first singlet of naphthalene and benzene suggested that coupling of neutral Frenkel states to close-lying chaxge transfer (CT) (or Wannier exciton, ion-pair state) levels was the source of the observed splitting and not multipole or electron-exchange interactions. Objections raised to this idea were that the CT levels were too low in energy, that they had never been detected (not relevant because of their low f value), and that the Davydov splittings in different crystals were invariant under isotopic substitution. Later, after Geacintov and Pope[43,269] detected CT levels near their classical positions, the MO calculations were repeated[48] with the CT levels placed farther away from the Frenkel exciton band. The results of these calculations are that a third of the splitting is due to CT interactions and the rest to higher-multipole interactions. The octopole moments required are smaller (-6.8 and 9.1 Å^3), which is more in line with theoretical calculations of these quantities.

The question is now whether the nonmultipole interaction is due to electron exchange or CT interactions. Everyone agrees that the available molecular wave functions are only approximate. Robinson[26] has suggested that the exact singlet wave functions are more diffuse than the approximate ones, and that electron-exchange interactions are therefore underestimated in the calculations. To obtain agreement, diffuseness must be increased and this is accomplished by mixing a CT component into the Frenkel states. It should also be added that if the exact wave functions are more diffuse the interaction of neutral and CT exciton states may be even larger than presently estimated.

Regardless of how this question is resolved there are implications for the rest of exciton theory. We have seen that for the stronger transitions the dipole component of the exciton interaction varies with $\hat{\mathbf{k}}$, being smallest (generally) when $\hat{\mathbf{k}}$ is perpendicular to the crystal transition dipole. For these directions MO calculations, using approximate wave functions, may not give a good fit to the experimental data and it will be necessary to include CT states, unless a near-neighbor parameterization can be introduced along the lines described in Section III.A.3.

D. Dye Molecules

A summary of the electronic properties of organic dye molecules has been given by McGlynn.[270] Most dyes have intense transitions in the visible region with transtion dipoles around 2 Å. Consequently a major component of any exciton interaction in large aggregates is of the dipole–dipole type.

Dye molecules generally have low symmetry and their electronic spectra are congested through sheer numbers of totally symmetric vibrations with a wide range of frequencies. The absorption spectrum of a typcial dye consists of a short series of broad overlapping peaks spaced 1000–1500 cm^{-1} apart and diminishing rapidly in intensity; that is, the individual vibronic transitions are not resolved. It is, therefore, all the more surprising to find aggregate absorption spectra with very sharp transitions.

Experimental work has been divided between one- and two-dimensional dye aggregates and crystals of cationic and molecular dyes. Many two-dimensional dye aggregates have been assembled and studied by Kuhn and co-workers. The reflection spectra of a number of crystalline dyes have been reported by Simpson and co-workers. Over the years some extraordinary spectral phenomena have been discovered. The oldest and to some extent the most controversial one is the J-band of the presumed one-dimensional aggregate of the dye pseudoisocyanine. The absorption spectrum of monomer and aggregate (full line) are shown in Fig. 27. Since the monomer transition is intense, and the red shift comparable with the energy span of the monomer spectrum, there must be at least moderately strong vibronic coupling in the polymer. The sharpness of the transition suggests that the upper level is a single-exciton state close to the bottom of the polymer exciton band.

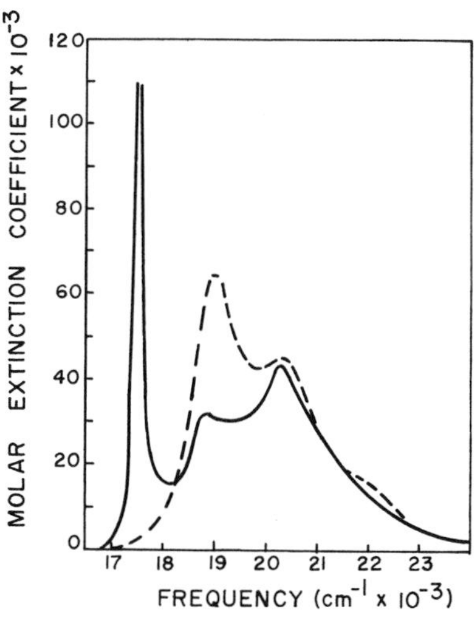

Fig. 27. Absorption spectra of monomer (broken line) and polymer (full line) form of the dye pseudoisocyanine. Concentration of the dye is approximately 10^{-5} and 10^{-3} molar for the monomer and polymer spectra, respectively.

In a series of ingenious experiments using cyanine dye chromophores attached to long hydrophobic tails, Kuhn and co-workers have made monomolecular layers of dyes in a brickwork arrangement (face-centered rectangular two-dimensional lattice). The spectral shifts observed have been calculated with good agreement using an extended dipole theory and an empirical background dielectric constant set at $\epsilon = 2.5$. In this and other work the Marburg group has put together a coherent explanation of what was previously a large collection of relatively unconnected measurements of the spectra of dye aggregates in solution and adsorbed on mica and silver halide surfaces.[271]

The polymethinium dye studied by Simpson and co-workers are structurally much simpler than those of Kuhn. In the region of the first singlet transition the crystals are opaque ($f \gtrsim 3.0$) and reflection spectroscopy was used to study the crystal spectrum. Many of these spectra exhibit broad reflection bands at least an electron volt in width. A selection of these spectra adapted from Ref. 145 are shown in Fig. 28. The top row gives the reflection spectra of the dye BDP or 1,5-bis(dimethylamino)-pentamethinium perchlorate for (*a*) face 1 at 4°K, (*b*) face 2 at 298°K, and (*c*) a second face at 298°K. The lower rows in Fig. 28 show metallic reflection spectra from crystals of the dyes ADI (or 4-acetoxy-1,7-bis(di-methylamino)heptamethinium iodide) and BDI (same as BDP, with iodide in place of perchlorate ion). Some of the spectra have large dips which start roughly 1000 cm^{-1} above the lower (see peak) reflection band edge that have been interpreted as due to absorption by two-particle excitations[272] (separate electronic and vibrational excitons). Unfortunately little can be done at present in the way of detailed calculations to prove or disprove this idea because precise structural data for the crystal are lacking. The narrow region of high reflectivity in Figs. 28*a*,*b*, and *d* is thought to due to a band of single-exciton states roughly one vibrational quantum wide. There is strong vibronic coupling so that electronic and vibrational excitons propagate separately. The lower edge of the reflection band in Fig. 28*a* where the reflectivity rises to 0.9 marks the location of the transverse exciton (transition moment $\boldsymbol{\mu}$ perpendicular to photon wave vector $\mathbf{q}$). In going from face 1 to face 2 in BDP the allowed single-particle transition moves into the two-particle continuum and is damped out by decay into two-particle states. This is in accord with the idea that the energy of allowed single-exciton transition has a large dipole component that changes with direction of $\mathbf{k}$. If the structure of these dye crystals were better known they would be excellent systems with which to study the effects of strong vibronic coupling and symmetry breaking due to phonon transitions and surface disorder.

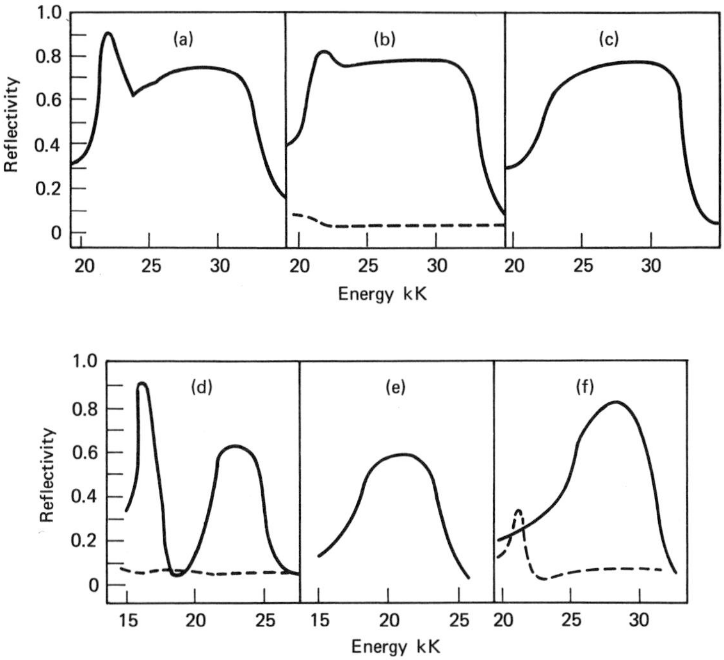

Fig. 28. Reflection spectra of some polymethinium dye crystals. Solid lines are for light linearly polarized in the direction of maximum reflectivity; broken lines, where shown, for light polarized in the direction of minimum reflectivity. (*a*) BDP, face 1, 4°K; (*b*) BDP, face 1, 298°K; (*c*) BDP, face 2, 298°K; (*d*) ADI, red crystalline form, 298°K; (*e*) ADI, blue crystalline form, 298°K; (*f*) BDI, face 1, 298°K.

IV. SUMMARY

An account has been presented of some of the recent developments in the theory of single-exciton states and related experimental work for transitions of medium and high intensity. In the last ten years the spectroscopy and theory of weak crystal transitions have moved forward rapidly, whereas the understanding of stronger transitions has remained in the doldrums. One foresees a new interest in strong transitions now that the use of reflection spectroscopy is becoming more widely appreciated among physical chemists. Whether or not we shall be able to analyze the strong transitions in the same detail achieved for weak ones remains to be seen; certainly, because of interactions between exciton bands, the problems are much more difficult. The attempt is worth the effort, for an understanding of electronic interactions between excited and ground-state molecules has widespread application outside the area of molecular aggregates.

Acknowledgment

The author's interest and comprehension of strong transitions has been stimulated by many frank and penetrating discussions with L. B. Clark and W. T. Simpson, who over the years have very graciously made their reflection results on anthracene and the polymethine dyes readily available.

References

1. D. S. McClure, "Spectra of Molecules in Crystals," *Solid State Phys.*, **8**, 1 (1958).
2. H. C. Wolf, "The Electronic Spectra of Aromatic Molecular Crystals," *Solid State Phys.*, **9**, 1 (1959).
3. J. Tanaka, "Electronic Absorption Spectra of Molecular Crystals," *Progr. Theoret. Phys. (Kyoto), Suppl.*, **12**, 183 (1959).
4. R. M. Hochstrasser, "The Luminescence of Organic Molecular Crystals," *Rev. Mod. Phys.*, **34**, 531 (1962).
5. A. S. Davydov, *Theory of Molecular Excitons*, transl. by M. Kasha and M. Oppenheimer, Jr., McGraw-Hill, New York, 1962.
6 R. S. Knox, *Theory of Excitons*, Academic Press, New York, 1963.
7. O. Schnepp, "Electronic Spectra of Molecular Crystals," *Ann. Rev. Phys. Chem.*, **14**, 35 (1963).
8. D. P. Craig and S. H. Walmsley, "Visible and Ultraviolet Absorption by Molecular Crystals," in *Physics and Chemistry of the Organic Solid State*, D. Fox, M. M. Labes, and A. Weisberger, Eds., Wiley-Interscience, New York, 1963, Vol. I, p. 585.
9. D. A. Dows, "Infrared Spectra of Molecular Crystals," in *Physics and Chemistry of the Organic Solid State*, D. Fox, M. M. Labes, and A. Weisberger, Eds., Wiley-Interscience, New York, 1963, Vol. I, p. 657.
10. A. S. Davydov, "The Theory of Molecular Excitons," *Usp. Fiz. Nauk.*, **82**, 393 (1964); *Soviet Phys.-Usp.*, **7**, 145 (1964).
11. T. N. Misra, "A Comparative Study of the Theoretical and Experimental Results on Davydov Splitting in Molecular Crystals," *Rev. Pure. Appl. Chem.*, **15**, 39 (1965).
12. D. L. Dexter and R. S. Knox, *Excitons*, Wiley-Interscience, New York, 1965.
13. M. W. Windsor, "Luminescence and Energy Transfer," in *Physics and Chemistry of the Organic Solid State*, D. Fox, M. M. Labes, and A. Weisberger, Eds., Wiley-Interscience, New York, 1965, Vol. II, p. 343.
14. V. M. Agranovich and V. L. Ginzburg, *Spatial Dispersion in Crystal Optics and the Theory of Excitons*, Wiley-Interscience, New York, 1966.
15. S. K. Lower and M. A. El-Sayed, "The Triplet State and Molecular Electronic Processes in Organic Molecules," *Chem. Rev.*, **66**, 199 (1966).
16. R. M. Hochstrasser, "Electronic Spectra of Organic Molecules," *Ann. Rev. Phys. Chem.*, **17**, 457 (1966).
17. S. A. Rice and J. Jortner, "Comments on the Theory of the Exciton States of Molecular Crystals," in *Physics and Chemistry of the Organic Solid State*, D. Fox, M. M. Labes, and A. Weissberger, Eds., Wiley-Interscience, New York, 1967, Vol. III, p. 199.
18. H. C. Wolf, "Energy Transfer in Organic Molecular Crystals: A Survey of Experiments," *Adv. At. Mol. Phys.*, **3**, 119 (1967).
19. F. Gutmann and L. E. Lyons, *Organic Semiconductors*, Wiley, New York, 1967.

20. W. H. Wright, "Ultraviolet Properties of Crystalline Anthracene," *Chem. Rev.*, **67**, 581 (1967).
21. K. Kambe, "On Some Recent Developments in the Theory of Electronic States in Crystals of Anthracene and Similar Molecules," *Progr. Theor. Phys. (Kyoto), Suppl.*, **40**, 136 (1967).
22. J. B. Birks and I. H. Munro, "The Fluorescence Lifetimes of Aromatic Molecules," *Progr. React. Kinetics*, **4**, 239 (1967).
23. D. P. Craig and S. H. Walmsley, *Excitons in Molecular Crystals*, Benjamin, New York, 1968.
24. P. Avakian and R. E. Merrifield, "Triplet Excitons in Anthracene Crystals—A Review," *Mol. Cryst.*, **5**, 37 (1968).
25. O. Schnepp, "The Spectra of Molecular Solids," *Adv. At. Mol. Phys.*, **5**, 155 (1969).
26. G. W. Robinson, "Electronic and Vibrational Excitons in Molecular Crystals," *Ann. Rev. Phys. Chem.*, **21**, 429 (1970).
27. J. B. Birks, *Photophysics of Aromatic Molecules*, Wiley-Interscience, New York, 1970.
28. A. S. Davydov, *Theory of Molecular Excitons*, Plenum Press, New York, 1971.
29. E. F. Sheka, "Electronic Vibrational Spectra of Molecules and Crystals." *Usp. Fig. Nauk.*, **104**, 593 (1971).
30. R. B. Leslie, *Biological Membranes*, D. Chapman, Ed., Academic Press, New York, 1968, p. 289.
31. W. A. Little, *Phys. Rev.*, **134**, A1416 (1964).
32. The coupling of exciton and conduction states has been considered by D. B. Chesnut, *Mol. Cryst.*, **1**, 351 (1966).
33. J. Bardeen, L. N. Cooper, and J. R. Schrieffer, *Phys. Rev.*, **108**, 1175 (1957).
34. S. D. Colson, R. Kopelman, and G. W. Robinson, *J. Chem. Phys.*, **47**, 27 (1967).
35. Oscillator strengths used in this article are defined by $f = 2m\omega|\boldsymbol{\mu}|^2/(\hbar e^2)$ where ω and $\boldsymbol{\mu}$ are the frequency and dipole moment of the transition.
37. J. Frenkel, *Phys. Rev.*, **37**, 17 (1931).
37. J. Frenkel, *Phys. Rev.*, **37**, 1276 (1931).
38. J. Frenkel, *Phys. Z. Sowjetunion*, **9**, 158 (1936).
39. J. Jortner, S. A. Rice, J. L. Katz, and S. Choi, *J. Chem. Phys.*, **42**, 309 (1965).
40. R. Silbey, J. Jortner, S. A. Rice, and M. T. Vala, Jr., *J. Chem. Phys.*, **42**, 733 (1965).
41. G. H. Wannier, *Phys. Rev.*, **52**, 191 (1937).
42. R. S. Berry, J. Jortner, J. C. Mackie, E. S. Pysh, and S. A. Rice, *J. Chem. Phys.*, **42**, 1535 (1965).
43. M. Pope, in *Optical Properties of Dielectric Films*, N. N. Axelrod, Ed., The Electrochemical Society, Inc., New York, 1968, p. 79.
44. R. Silbey, S. A. Rice, and J. Jortner, *J. Chem. Phys.*, **43**, 3336 (1965).
45. R. Silbey, J. Jortner, M. T. Vala, Jr., and S. A. Rice, *J. Chem. Phys.*, **42**, 2948 (1965).
46. P. Sarti-Fantoni, *Mol. Cryst.*, **1**, 457 (1966).
47. R. Silbey, J. Jortner, M. Vala, Jr., and S. A. Rice, *Mol. Cryst.*, **2**, 385 (1967).
48. W. L. Greer, S. A. Rice, J. Jortner, and R. Silbey, *J. Chem. Phys.*, **48**, 5667 (1968).
49. D. M. Hanson, R. Kopelman, and G. W. Robinson, *J. Chem. Phys.*, **51**, 212 (1969).
50. See also comments on p. 465 of Ref. 26.
51. Photon Umklapp processes involving nonzero reciprocal lattice vectors are ignored here.
52. W. T. Moffit, *J. Chem. Phys.*, **25**, 467 (1956).
53. W. T. Moffit and J. T. Yang, *Proc. Natl. Acad. Sci. U.S.*, **42**, 596 (1956).
54. W. T. Moffit, *Proc. Natl. Acad. Sci. U.S.*, **42**, 736 (1956).

55. W. T. Moffit, D. D. Fitts, and J. G. Kirkwood, *Proc. Natl. Acad. Sci. U.S.*, **43**, 723 (1957).
56. I. Tinoco, Jr., "Theoretical Aspects of Optical Activity: Polymers," *Adv. Chem. Phys.*, **4**, 113 (1962).
57. I. Tinoco, Jr., "The Exciton Contribution to the Optical Rotation of Polymers," *Radiation Res.*, **20**, 133 (1963).
58. J. T. Yang and T. Samejima, "Optical Rotatory Dispersion and Circular Dichroism of Nucleic Acids," *Progress in Nucleic Acid Res. and Molecular Biol.*, **9**, 223 (1969).
59. J. Brahms and S. Brahms, "Circular Dichroism of Nucleic Acids," in *Structure of Proteins and Nucleic Acids*, G. D. Fasman and S. N. Timasheff, Eds., Dekker, New York, 1970.
60. W. Rhodes, "Linear Response Theory of the Optical Properties of Biopolymers," in *Spectroscopic Approaches to Biomolecular Conformation*, D. W. Urry, Ed., AMA Press, Chicago, 1970.
61. L. G. S. Brooker, "Sensitizing and Desensitizing Dyes," in *The Theory of the Photographic Process*, C. E. K. Mees and T. H. James, Eds., Macmillan, New York, 1966, 3rd ed., Chap. 11, p. 198.
62. W. West and B. H. Carroll, "Spectral Sensitivity and the Mechanism of Spectral Sensitization," in the *Theory of the Photographic Process*, C. E. K. Mees and T. H. James Eds., Macmillan, New York, 1966, 3rd ed., Chap. 12, p. 233.
63. E. E. Jelley, *Nature*, **138**, 1009 (1936).
64. E. E. Jelley, *Nature*, **139**, 631 (1937).
65. H. Zimmermann and G. Scheibe, *Z. Elektrochem.*, **60**, 566 (1956).
66. W. Cooper, *Chem. Phys. Lett.*, **7**, 73 (1970).
67. W. West, S. P. Lovell, and W. Cooper, *Phot. Sci. Eng.*, **14**, 52 (1970).
68. E. G. McCrae, *Australian J. Chem.*, **14**, 354 (1961).
69. N. S. Bayliss, *Rev. Pure Appl. Chem.*, **1**, 64 (1954).
70. F. M. Loxsom, *J. Chem. Phys.*, **51**, 4899 (1969).
71. J. W. Raymonda and W. T. Simpson, *J. Chem. Phys.*, **47**, 430 (1967).
72. R. H. Partridge, *J. Chem. Phys.*, **49**, 3656 (1968).
73. R. H. Partridge, *J. Chem. Phys.*, **52**, 2485, 2491, 2501 (1970).
74. J. W. Raymonda, *J. Chem. Phys.* (1971); L. Edwards and J. W. Raymonda, *J. Am. Chem. Soc.*, **91**, 5937 (1969).
75. (a) S. Katagiri and C. Sandorfy, *Theoret. Chem. Acta*, **4**, 203 (1966); (b) B. A. Lambos, P. Sauvageau, and C. Sandorfy, *J. Mol. Spectry.*, **24**, 253 (1967).
76. E. F. Pearson and K. K. Innes, *J. Mol. Spectry.*, **30**, 232 (1969).
77. C. Tric and O. Parodi, *Mol. Phys.*, **13**, 1 (1967).
78. A. Witowski and W. Moffit, *J. Chem. Phys.*, **33**, 872 (1960).
79. R. L. Fulton and M. Gouterman, *J. Chem. Phys.*, **35**, 1059 (1961).
80. R. L. Fulton and M. Gouterman, *J. Chem. Phys.*, **41**, 2280 (1964).
81. M. H. Perrin and M. Gouterman, *J. Chem. Phys.*, **46**, 1019 (1967).
82. See for example, A. J. McCaffery, S. F. Mason, and B. J. Norman, *J. Chem. Soc. A*, **1969**, 142.
83. S. D. Colson, D. M. Hanson, R. Kopelman, and G. W. Robinson, *J. Chem. Phys.*, **48**, 2215 (1968).
84. E. R. Bernstein, S. D. Coloson, R. Kopelman, and G. W. Robinson, *J. Chem. Phys.*, **48**, 5596 (1968).
85. V. K. Dolganov and E. F. Sheka, *Fiz. Tverd. Tela*, **12**, 1450 (1970); *Soviet Phys.—Solid State*, **12**, 1138 (1970).
86. V. L. Broude, E. I. Rashba, and E. F. Sheka, *Phys. Status Solidi*, **19**, 395 (1967).

87. In particular for those levels lying below the threshold of two-particle excitations. For more details see M. R. Philpott, *J. Chem. Phys.*, **55**, 2039 (1971).
88. T. Azumi and S. P. McGlynn, *J. Chem. Phys.*, **41**, 3131 (1964).
89. T. Azumi, A. T. Armstrong, and S. P. McGlynn, *J. Chem. Phys.*, **41**, 3839 (1964).
90. P.-O. Löwdin, *J. Chem. Phys.*, **18**, 365 (1950).
91. See, for example, K. Kumar, *Perturbation Theory and the Nuclear Many Body Problem*, North-Holland, Amsterdam, 1962.
92. D. P. Craig, *J. Chem. Soc.*, **1955**, 2302.
93. D. M. Hanson, R. Kopelman, and G. W. Robinson, *J. Chem. Phys.*, **51**, 212 (1969).
94. R. Kopelman, *J. Chem. Phys.*, **47**, 2631 (1967).
95. Because the nonanalytic part of the exciton interaction, due to the macroscopic part of the dipole sum, is proportional to $(\boldsymbol{\mu}_\alpha{}^u \cdot \mathbf{k})(\mathbf{k} \cdot \boldsymbol{\mu}_\beta{}^v)$.
96. D. M. Hanson, *J. Chem. Phys.*, **52**, 3409 (1970).
97. H.-K. Hong and R. Kopelman, *J. Chem. Phys.*, **55**, 724 (1971).
98. H.-K. Hong and R. Kopelman, *J. Chem. Phys.*, **55**, 5380 (1971).
99. D. M. Hanson, *J. Chem. Phys.*, **51**, 653 (1969).
100. C. A. Hutchison and B. W. Mangum, *J. Chem. Phys.*, **34**, 908 (1964).
101. D. P. Craig, J. M. Hollas, M. F. Redies, and S. C. Wait, Jr., *Phil. Trans. Roy. Soc. London, Ser. A*, **253**, 543 (1961).
102. D. P. Craig and J. M. Hollas, *Phil. Trans. Roy. Soc. London, Ser. A*, **253**, 569 (1961).
103. D. P. Craig and R. Gordon, *Proc. Roy. Soc. (London) Ser. A*, **288**, 69 (1965).
104. A. V. Bree and S. Katagiri, *J. Mol. Spectry.*, **17**, 24 (1965).
105. A. V. Bree, S. Katagiri, and S. R. Suart, *J. Chem. Phys.*, **44**, 1788 (1966).
106. E. I. Rashba, *Opt. Spektroskopiya*, **2**, 568 (1957).
107. E. I. Rashba, *Fiz. Tverd. Tela*, **4**, 3301 (1962); *Soviet Phys.-Solid State*, **4**, 2417 (1963).
108. E. I. Rashba, *Fiz. Tverd. Tela*, **5**, 1040 (1963); *Soviet Phys.—Solid State*, **5**, 757 (1963).
109. D. P. Craig and M. R. Philpott, *Proc. Roy. Soc. (London) A*, **290**, 583, 602 (1966).
110. R. G. Body and I. G. Ross, *Australian J. Chem.*, **19**, 1 (1966).
111. M. R. Philpott, *J. Chem. Phys.*, **49**, 4537 (1968).
112. B. Sommer and J. Jortner, *J. Chem. Phys.*, **50**, 187 (1969).
113. B. Sommer and J. Jortner, *J. Chem. Phys.*, **50**, 822 (1969).
114. G. F. Koster and J. C. Slater, *Phys. Rev.*, **95**, 1167 (1954).
115. G. F. Koster and J. C. Slater, *Phys. Rev.*, **96**, 1208 (1954).
116. G. F. Koster, *Phys. Rev.*, **95**, 1436 (1954).
117. D. P. Craig and M. R. Philpott, *Proc. Roy. Soc. (London) Ser. A*, **293**, 213 (1966).
118. O. A. Dubovskii and Yu. V. Konobeev, *Fiz. Tverd. Tela*, **6**, 2599 (1964); *Soviet Phys.—Solid State*, **6**, 2071 (1965).
119. R. E. Merrifield, *J. Chem. Phys.*, **38**, 920 (1963).
120. I. S. Osad'ko, *Fiz. Tverd. Tela*, **11**, 441 (1969); *Soviet Phys.—Solid State*, **11**, 347 (1969).
121. V. I. Sugakov, *Opt. i Spektroskopiya*, **21**, 574 (1966).
122. M. R. Philpott, *J. Chem. Phys.*, **53**, 136 (1970).
123 M. Born and R. Oppenheimer, *Ann. Phys.*, **84**, 457 (1927).
124. M. Born and K. Huang, *Dynamical Theory of Crystal Lattices*, Oxford University Press, London, 1968, pp. 166, 406.
125. $\xi_u{}^2$ is the fraction of intensity for the given electronic state observed in the 0–u vibronic transition.
126. D. P. Craig and T. Thirunamachandran, *Proc. Roy. Soc. (London) Ser. A*, **271**, 207 (1963).

127. P. Soven, *Phys. Rev.*, **156**, 809 (1967).
128 D. W. Taylor, *Phys. Rev.*, **156**, 1017 (1967).
129. T. Matsubara and F. Yonezawa, *Progr. Theoret. Phys. (Kyoto)*, **35**, 759 (1966); **37**, 1346 (1967); **39**, 1076 (1968).
130. Y. Onodera and Y. Toyozawa, *Proc. Phys. Soc. Japan*, **24**, 134 (1968).
131. P. L. Leath and B. Goodman, *Phys. Rev.*, **148**, 968 (1966).
132. B. Velicky, S. Kirkpatrick, and H. Ehrenrich, *Phys. Rev.*, **175**, 747 (1969).
133. R. N. Aiyer, R. J. Elliott, J. A. Krumhansl, and P. L. Leath, *Phys. Rev.*, **181**, 1001 (1969).
134. S. Takeno, *J. Phys. C (Solid State)*, **4**, L118 (1971).
135. I. M. Lifshitz, *Adv. Phys.*, **13**, 483 (1964).
136. H.-K. Hong and G. W. Robinson, *J. Chem. Phys.*, **54**, 1369 (1971), and ref. cited therein.
137. H.-K. Hong and R. Kopelman, *J. Chem. Phys.*, **55**, 3491 (1971).
138. J. Hoshen and J. Jortner, *Chem. Phys. Lett.*, **5**, 351 (1970).
139. J. Hoshen and J. Jortner, *J. Chem. Phys.*, **56**, 933 (1971).
140. These are not the pure crystal states but have the guest wave functions $\psi_p{}^u$ and ψ_p in place of the hosts $\varphi_p{}^u$ and φ_p, respectively.
141. M. Born and E. Wolf, *Principles of Optics*, Pergamon Press, London, 1970, 4th ed., Chap. 2, p. 76.
142. The transverse and longitudinal parts of vector fields contain'ng Dirac delta functions need careful definition; see E. A. Power, *Introductory Quantum Electrodynamics*, Longmans, Green, London, 1964, Chap. 6, p. 73.
143. $|I(\mathbf{k})|$ estimated from the width of the reflection band may exceed 1 eV in some of the cationic dye crystals. See Refs. 144 and 145.
144. B. G. Anex and W. T. Simpson, *Rev. Mod. Phys.*, **32**, 466 (1960).
145. B. M. Fanconi, G. A. Gerhold, and W. T. Simpson, *Mol. Cryst. Liq. Cryst.*, **6**, 41 (1969).
146. The wave vector **k** of the collective level in question matches **q**, that of the absorbed photon.
147. W. T. Simpson and D. L. Peterson, *J. Chem. Phys.*, **26**, 588 (1957).
148. It is here assumed that the collective level lies at the bottom of the band of the exciton states.
149. S. I. Pekar, *Zh. Eksperim. i Teor. Fiz.*, **33**, 1022 (1957); *Soviet Phys. JETP*, **6**, 785 (1958).
150. U. Fano, *Phys. Rev.*, **103**, 1202 (1956); **118**, 451 (1960).
151. J. J. Hopfield, *Phys. Rev.*, **112**, 1555 (1958).
152. V. M. Agranovich, *Zh. Eksperim. i Teor. Fiz.*, **37**, 340 (1959); *Soviet Phys. JETP*, **10**, 307 (1960).
153. BDP is short for 1,5-bis(dimethylamino)pentamethinium perchlorate: $Me_2N—CH{=}CH—CH{=}CH—CH{=}NMe_2{}^+ClO_4{}^-$.
154. M. R. Philpott, *J. Chem. Phys.*, **54**, 2120 (1971).
155. L. B. Clark and M. R. Philpott, *J. Chem. Phys.*, **53**, 3790 (1970).
156. G. D. Mahan, *J. Chem. Phys.*, **41**, 2930 (1964).
157. (a) M. R. Philpott, *J. Chem. Phys.*, **50**, 5117 (1969); (b) M. R. Philpott, unpublished calculations.
158. G. D. Mahan, *Phys. Rev.*, **43**, 1569 (1965).
159. V. M. Agranovich, N. E. Kamenogradskii, and Yu. V. Konobeev, *Fiz. Tverd. Tela*, **11**, 1445 (1969); *Soviet Phys.—Solid State*, **11**, 1177 (1969).

160. M. A. Kozhushner, *Zh. Eksperim. i Teor. Fiz.*, **56**, 1940 (1969); *Soviet Phys. JETP*, **29**, 1041 (1969).
161. O. N. Karpukhin and M. A. Kozhushner, *Dokl. Akad. Nauk SSSR*, **182**, 905 (1969); *Soviet Phys.—Dokl.*, **13**, 905 (1969).
162. N. E. Kamenogradskii and Yu. V. Konobeev, *Phys. Status Solidi*, **37**, 29 (1970).
163. C. W. Deutsche and C. A. Mead, *Phys. Rev.*, **138**, A63 (1965).
164. G. D. Mahan and G. Obermair, *Phys. Rev.*, **183**, 834 (1969).
165. M. R. Philpott, *J. Chem. Phys.*, **56**, 996 (1972).
166. J. J. Hopfield and D. G. Thomas, *Phys. Rev.*, **132**, 563 (1963).
167. J. J. Hopfield and D. G. Thomas, *J. Phys. Chem. Solids*, **12**, 276 (1960)
168. A. S. Davydov, *Zh. Eksperim. i Teor. Fiz.*, **45**, 723 (1963); *Soviet Phys. JETP*, **18**, 496 (1964).
169. A. Haug, *Phys. Rev.*, **147**, 612 (1966).
170. M. Born and K. Huang, *Dynamical Theory of Crystal Lattices*, Oxford University Press, London, 1954, p. 30.
171. F. W. De Wette and G. E. Schacher, *Phys. Rev.*, **137**, A78 (1965).
172. The planar sums fall off rapidly for planes with a high density of molecules per unit area. If the direction **k** is perpendicular to planes with low densities then the planar sums will die off more slowly and the surface will contain many layers.
173. See the appendix in G. D. Mahan, *J. Chem. Phys.*, **43**, 1569 (1965).
174. A. S. Davydov and V. A. Onishchuk, *Phys. Status Solidi*, **24**, 373 (1967).
175. M. R. Philpott, *J. Chem. Phys.*, **52**, 5842 (1970).
176. T. Holstein and H. Primakoff, *Phys. Rev.*, **58**, 1098 (1940).
177. V. M. Agranovich and B. S. Toschich, *Zh. Eksperim. i Teor. Fiz.*, **53**, 149 (1967); *Soviet Phys. JETP*, **26**, 104 (1968).
178. D. I. Lalovic, B. S. Toshich, and R. B. Zakula, *Phys. Rev.* **178**, 1472 (1969).
179. S. V. Tyablikov, *Methods in the Quantum Theory of Magnetism*, Plenum Press, New York, 1967.
180. C. Mavroyannis, *J. Math. Phys.*, **8**, 1515, 1522 (1967); **11**, 491 (1970).
181. This was checked by solving the equations for the case of one excited state for guest and hosts. The additional term results in a slightly different form for the trap-parameter. Estimated based on $\delta_v \lesssim 200$ cm^{-1} and $\Delta_u \simeq 30{,}000$ cm^{-1}.
182. $\mathcal{P}$ is the pure crystal polarization ratio provided the factor group splitting is small compared to the molecular excitation energy. Thus $\mathcal{P}$ for naphthalene first singlet is accurate to within a 1%, for second anthracene transition to within 20%.
183. If the 3800 Å transition of anthracene does have a discrete guest state it will be close to the band edge and therefore afford a good chance to see the effect, since the variations in the energy denominations will be all the greater.
184. The density function has been used in the treatment of electron traps. R. A. Faulkner, *Phys. Rev.*, **175**, 991 (1968).
185. A. M. Clogston, *Phys. Rev.*, **125**, 439 (1962).
186. T. Ando, *Progr. Theoret. Phys.* (*Kyoto*), **40**, 471 (1968).
187. F. M. Loxsom, *J. Chem. Phys.*, **51**, 4899 (1969).
188. C. W. Deutsche, *J. Chem. Phys.*, **52**, 3703 (1970).
189. F. M. Loxsom, *Intern. J. Quant. Chem.*, **3S**, 147 (1969).
190. F. M. Loxsom, *Phys. Rev. B*, **1**, 858 (1970).
191. W. Rhodes, *J. Chem. Phys.*, **53**, 3650 (1970).
192. M. R. Philpott, *J. Chem. Phys.*, **56**, 683 (1972).
193. F. M. Loxsom, L. Tterlikkis, and W. Rhodes, *Biopolymers*, **10**, 2405 (1971).
194. The quantization volume for photons is taken to be the sample volume.

195. The radiative lifetime may be increased, however, by as much as an order of magnitude since the imaginary parts of the retarded interactions are proportional to the natural width of the isolated molecule (they all add in phase).
196. M. R. Philpott, unpublished work.
197. In deriving (405) time-reversal invariance was used to identify the wavefunction of $|\mathbf{k}u\rangle$ as the complex conjugate of the of $|-\mathbf{k}u\rangle$.
198. M. R. Philpott and P. Sherman, unpublished calculations.
199. A. Carrington and A. D. McLachlan, *Introduction to Magnetic Resonance*, Harper and Row, New York, 1967, Appendix I, p. 260.
200. S. Singh, W. J. Jones, W. Siebrand, B. P. Stoicheff, and W. G. Schneider, *J. Chem. Phys.*, **42**, 330 (1965).
201. B. Sharf and R. Silbey, *J. Chem. Phys.*, **53**, 2626 (1970).
202. See, for example, J. R. Platt and co-workers, *Systematics of the Electronic Spectra of Conjugated Molecules: A Source Book*, Wiley, New York, 1964.
203. R. M. Hochstrasser, *Accts. Chem. Res.*, **1**, 266 (1968).
204. J. P. Byrne and I. G. Ross, *Can. J. Chem.*, **43**, 3253 (1965).
205. L. E. Lyons and G. C. Morris, *J. Mol. Spectry.*, **4**, 480 (1960).
206. L. E. Lyons and G. C. Morris, *J. Chem. Soc.* (*London*), **1959**, 1551.
207. I. G. Ross, "High Resolution Spectra of Large Polyatomic Molecules," *Adv. Chem. Phys.*, **20**, 341 (1971).
208. R. Ostertag and H. C. Wolf, *Phys. Status Solidi*, **31**, 139 (1969).
209. J. Sidman, *J. Chem. Phys.*, **25**, 115 (1956); *Phys. Rev.*, **102**, 96 (1956).
210. E. V. Shpolskii, *Soviet Phys. Usp.*, **6**, 411 (1963); **5**, 522 (1962); **3**, 372 (1961).
211. T. N. Bolotnikova, L. A. Klimova, G. N. Nersesova, and L. F. Utkina, *Opt. i Spectroskopiya*, **21**, 237 (1966).
212. R. N. Jones and E. Spinner, *Spectrochim. Acta*, **16**, 1060 (1960).
213. G. Porter and M. W. Windsor, *Proc. Roy. Soc.* (*London*) *Ser. A*, **245**, 238 (1958).
214. R. E. Kellog, *J. Chem. Phys.*, **44**, 411 (1966).
215. D. W. J. Cruickshank, *Acta Cryst.*, **10**, 504 (1957).
216. R. Mason, *Acta Cryst.*, **17**, 547 (1964).
217. M. S. Brodin and S. V. Morisova, *Opt. Spectry.* (*USSR*), **19**, 132 (1965).
218. S. A. Rice, G. C. Morris, and W. L. Greer, *J. Chem. Phys.*, **52**, 4279 (1970).
219. (a) See, for example, M. S. Brodin and A. F. Prikhot'ko, *Opt. Spectry.*, **7**, 82 (1959); (b) H. J. Jetter and H. C. Wolf, *Phys. Status Solidi*, **22**, K39 (1967).
220. D. P. Craig and P. C. Hobbins, *J. Chem. Soc.* (*London*), **1955**, 2309.
221. H. C. Wolf, *Z. Naturforsch.*, **13a**, 413 (1958).
222. M. S. Brodin and S. V. Morisova, *Opt. Spectry.* (*USSR*), **10**, 242 (1961).
223. T. A. Claxton, D. P. Craig, and T. Thirunamachandran, *J. Chem. Phys.*, **35**, 1525 (1961).
224. A. R. Lacey and L. E. Lyons, *Proc. Chem. Soc.* (*London*), **1960**, 414.
225. A. R. Lacey and L. E. Lyons, *J. Chem. Soc.* (*London*), **1964**, 5393.
226. A. Matsui, *J. Phys. Soc. Japan*, **21**, 2212 (1966).
227. E. G. McRae, *Australian J. Chem.*, **16**, 315 (1963).
228. D. P. Craig and P. C. Hobbins, *J. Chem. Soc.*, **1955**, 539.
229. A. Bree and L. E. Lyons, *J. Chem. Soc.*, **1956**, 2662.
230. A. Matsui, Y. Ishii, and T. Hikita, *J. Phys. Soc. Japan*, **21**, 2091 (1966).
231. S. V. Morisova, *Opt. i Spektroskopiya*, **22**, 566 (1966); *Opt. Spectry.* (*USSR*), **22**, 310 (1966).
232. A. Matsui and Y. Ishii, *J. Phys. Soc. Japan*, **23**, 581 (1967).
233. G. T. Wright, *J. Chem. Phys.*, **46**, 2951 (1967).

234. M. S. Brodin, S. V. Morisova, and S. A. Shturkhetskaya, *Ukr. Fiz. Zh.*, **13**, 353 (1968); *Ukr. Phys. J.*, **13**, 249 (1968).
235. L. B. Clark, *J. Chem. Phys.*, **51**, 5719 (1969).
236. G. C. Morris, S. A. Rice, and A. E. Martin, *J. Chem. Phys.*, **52**, 5149 (1970).
237. L. B. Clark, *J. Chem. Phys.*, **53**, 4092 (1970).
238. G. C. Morris, S. A. Rice, M. G. Sceats, and A. E. Martin, *J. Chem. Phys.*, **55**, 5610 (1971).
239. D. Fox and S. Yatsiv, *Phys. Rev.*, **108**, 938 (1957).
240. E. E. Koch, S. Kunstreich, and A. Otto, *Opt. Commun.*, **2**, 365 (1971).
241. L. Clark, private communication.
242. D. P. Craig and T. Thirunamachandran, *Proc. Phys. Soc.* (*London*), **84**, 781 (1964).
243. R. Silbey, J. Jortner, and S. A. Rice, *J. Chem. Phys.*, **42**, 1515 (1965).
244. M. Tanaka and J. Tanaka, *Mol. Phys.*, **16**, 1 (1969).
245. D. Fox and O. Schnepp, *J. Chem. Phys.*, **23**, 767 (1955).
246. T. Thirunamachandran, Ph.D. Dissertation, University of London, 1961.
247. D. P. Craig and S. H. Walmsley, *Mol. Phys.*, **4**, 113 (1961).
248. D. P. Craig and T. Thirunamachandran, *Boll. Sci. Fac. Chim. Ind. Bologna*, **21**, 15 (1963).
249. See Eq. (18) in M. H. Cohen and F. Keffer, *Phys. Rev.*, **99**, 1128 (1955).
250. M. R. Philpott, *J. Chem. Phys.*, **54**, 111 (1971).
251. A. S. Davydov and E. F. Sheka, *Phys. Status Solidi*, **11**, 877 (1965).
252. J. Schroder and R. Silbey, *J. Chem. Phys.*, **55**, 5418 (1971).
253. These are solution oscillator strengths.
254. G. C. Morris, *J. Mol. Spectry.*, **18**, 42 (1965).
255. E. Clar, *Aromatische Kohlenwasserstoffe*, Springer-Verlag, Berlin, 1952.
256. A. Bree and L. E. Lyons, *J. Chem. Soc.* (*London*), **1960**, 5206.
257. J. Tanaka, *Bull. Chem. Soc Japan*, **38**, 86 (1964).
258. A. F. Prikhotjko and A. F. Skorobogatko, *Opt. i Spektroskopiya*, **20**, 65 (1965); *Opt. Spectry.* (*USSR*), **20**, 33 (1966).
259. L. E. Lyons and G. C. Morris, *J. Chem. Soc.* (*London*), **1965**, 2764.
260. R. B. Campbell, J. M. Robertson, and J. Trotter, *Acta Cryst.*, **15**, 289 (1962).
261. A. F. Prikhotjko and A. F. Skorobogatko, *Fiz. Tverd. Tela*, **7**, 1259 (1965); *Soviet Phys.—Solid State*, **7**, 1017 (1965).
262. V. Czikklely, H. D. Försterling, and H. Kuhn, *Chem. Phys. Lett.*, **6**, 11 (1970).
263. H. Bucher and H. Kuhn, *Chem. Phys. Lett.*, **6**, 183 (1970).
264. V. Czikklely, H. D. Försterling, and H. Kuhn, *Chem. Phys. Lett.*, **6**, 207 (1970).
265. A. F. Prikhotjko and A. F. Skorobogatko, *Ukr. Fiz. Zh.*, **10**, 350 (1965).
266. A. F. Prikhotjko, A. F. Skorobogatko, and L. I. Tsikora, *Opt. Spectry.* (*USSR*), **26**, 115 (1969).
267. A. F. Prikhotjko and L. I. Tsikora, *Opt. Spectry.* (*USSR*), **25**, 242 (1968).
268. R. Kopelman, *Record Chem. Progr.* (*Kresge-Hooker Sci. Lib.*), **31**, 211 (1970).
269. N. Geacintov and M. Pope, *J. Chem. Phys.*, **45**, 3884 (1966).
270. S. P. McGlynn, T. Azumi, and M. Kinoshita, *The Triplet State*. Prentice-Hall, New Jersey, 1971.
271. G. R. Bird, K. S. Norland, A. E. Rosenhoff, and H. B. Michaud, *Phot. Sci. Eng.*, **12**, 196 (1968).
272. M. R. Philpott, *J. Chem. Phys.*, **54**, 2120 (1971).

AUTHOR INDEX

Numbers in parentheses are reference numbers and show that an author's work is referred to although his name is not mentioned in the text. Numbers in *italics* indicate the pages on which the full references appear.

Jones, R. P., 201(43), 213(62), 216(65), *224*

SUBJECT INDEX